Deactivation and Regeneration of Zeolite Catalysts

CATALYTIC SCIENCE SERIES

Series Editor: Graham J. Hutchings *(Cardiff University)*

Published

Vol. 1 Environmental Catalysis
edited by F. J. J. G. Janssen and R. A. van Santen

Vol. 2 Catalysis by Ceria and Related Materials
edited by A. Trovarelli

Vol. 3 Zeolites for Cleaner Technologies
edited by M. Guisnet and J.-P. Gilson

Vol. 4 Isotopes in Heterogeneous Catalysis
edited by Justin S. J. Hargreaves, S. D. Jackson and G. Webb

Vol. 5 Supported Metals in Catalysis
edited by J. A. Anderson and M. F. García

Vol. 6 Catalysis by Gold
edited by G. C. Bond, C. Louis and D. T. Thompson

Vol. 7 Combinatorial Development of Solid Catalytic Materials: Design of High-Throughput Experiments, Data Analysis, Data Mining
edited by M. Baerns and M. Holeňa

Vol. 8 Petrochemical Economics: Technology Selection in a Carbon Constrained World
by D. Seddon

Vol. 9 Deactivation and Regeneration of Zeolite Catalysts
edited by M. Guisnet and F. R. Ribeiro

CATALYTIC SCIENCE SERIES — VOL. 9

Series Editor: Graham J. Hutchings

Deactivation and Regeneration of Zeolite Catalysts

Michel Guisnet

Fernando Ramôa Ribeiro

Technical University of Lisbon, Portugal

ICP

Imperial College Press

Published by

Imperial College Press
57 Shelton Street
Covent Garden
London WC2H 9HE

Distributed by

World Scientific Publishing Co. Pte. Ltd.
5 Toh Tuck Link, Singapore 596224
USA office: 27 Warren Street, Suite 401-402, Hackensack, NJ 07601
UK office: 57 Shelton Street, Covent Garden, London WC2H 9HE

British Library Cataloguing-in-Publication Data
A catalogue record for this book is available from the British Library.

DEACTIVATION AND REGENERATION OF ZEOLITE CATALYSTS
Catalytic Science Series — Vol. 9

ISBN-13 978-1-84816-637-0
ISBN-10 1-84816-637-0

Typeset by Stallion Press
Email: enquiries@stallionpress.com

Printed by FuIsland Offset Printing (S) Pte Ltd. Singapore

PREFACE

In industrial practice, the progressive deactivation of solid catalysts is a major economic concern and mastering their stability has become as essential as controlling their activity and selectivity. For these reasons, there is a strong motivation to understand the mechanisms leading to any loss in activity and/or selectivity and to find out the efficient preventive measures and regenerative solutions that open the way towards cheaper and cleaner processes.

This book covers both the fundamental and applied aspects of solid catalyst deactivation and encompasses the latest state of the art in the field of reactions catalysed by zeolites. This particular choice is justified by the widespread use of molecular sieves in refining, petrochemicals and organic chemicals synthesis processes, by the large variety in the nature of their active sites (acid, base, acid-base, redox, bifunctional) and, especially, by their peculiar features, in terms of crystallinity, structural order, molecular size of their pores, etc. which make them ideal models for understanding deactivation phenomena.

The covered topics range from methods and techniques of characterizing aged catalysts and deactivating species, to the mechanisms of poisoning, coking and thermal alterations, to kinetic modelling, then to the means developed to prevent deactivation and optimize regeneration. In the last part of the book, a series of selected refining, petrochemical and fine chemical processes provide the reader with theoretical insights and practical hints on the deactivation mechanisms and draw attention to the key role played by the loss and regeneration of activity in process design and industrial practice.

Review papers are regularly published on this topic of great economic importance, in particular in the Proceedings of the International Symposiums on Catalyst Deactivation that have been periodically organized since 1980. However, only a few books have been published on this subject so far, the last one dated 1988. This means that most of the graduates, post-graduate students and researchers active in heterogeneous catalysis cannot rely on recent bibliographic data when dealing with this key topic. When preparing this book, our primary aim was to fill this gap, providing reference text not only for scholars carrying out their studies in the field of industrial chemistry and chemical engineering, but also for professionals involved in academic and industrial research, coping with everyday issues of deactivation and regeneration of solid catalysts.

We would like to thank all the authors who, by their fundamental and/or industrial expertise, have significantly served this ambitious objective. We hope that this book will contribute both to the improvement of industrial processes and to original and fruitful research on the multifaceted aspects not yet fully understood of catalyst deactivation. Special thanks are due to Professor Avelino Corma, a worldwide renowned expert in catalysis and in zeolites who has agreed to introduce this work.

Thanks are also due to all those who have kindly collaborated with us on this book, especially to Nuno Fonseca who has devoted a significant part of his time to the hard task of preparing many figures and formatting all the chapters of this book.

Michel GUISNET
Fernando RAMÔA RIBEIRO
May 2010

FOREWORD

It is possible to say that catalysis in general, and solid catalysts in particular, are key elements for developing efficient and environmentally benign chemical processes. In this sense, industry is asking for selective catalysts able to perform a high number of turnovers. Essentially, this is another way of saying that the optimal catalysts should minimize the formation of sub-products, while producing sufficient kilograms of the desired product per kilogram of catalyst to render the process economically feasible. However, it is most common that solid catalysts also catalyse some undesired reactions, leading to products that remain adsorbed on the surface, poisoning the active sites or blocking access to them, and gradually causing catalyst activity decay.

Sometimes one may accept a loss of activity if catalyst poisoning also results in an improvement of selectivity. However, when a point is reached at which the remaining activity is too low, the process becomes impracticable.

While it is obvious that the main objective of scientists and industrialists is to design and synthesize solid catalysts with well defined homogeneous single sites, active only for the desired reaction, it is also clear that this objective is difficult to fulfill and 100% selectivity with 'clean' catalytic events is rarely achieved. In other words, most solid catalysts become deactivated sooner rather than later and they should be regenerated to keep the process economical.

Catalyst regeneration can be performed continuously with moving bed processes when the lifetime of the catalyst is short, or it can be done *in situ* for fixed bed processes when the lifetime of the catalyst is long enough. There are other cases, however, wherein the catalyst is irreversibly deactivated and has to be discharged and replaced.

Regardless of the regeneration system used, the fact that the catalyst deactivates and needs to be regenerated implies an economic cost either through further investment in the unit (moving bed systems) or because of the decrease in production during the programmed shutdown of the unit for regeneration. It is therefore obvious that improving catalyst life and regeneration is a key issue and that a better understanding of the catalytic processes resulting in catalyst deactivation and regeneration is of paramount importance.

Zeolites are probably the most successful solid catalysts in oil refining, petrochemicals and chemicals industries, and more recently in the treatment and purification of exhaust from mobile sources. Zeolites, which consist of crystalline microporous molecular sieves with pore diameters in the range of many reactant molecules, have introduced unique shape selectivity effects in a variety of catalytic processes.

The presence of regular pores of molecular dimensions can be a blessing or a curse from the catalyst decay point of view. Indeed, in some cases shape selectivity effects avoid the formation of poisons or molecules that may block the pores, and catalyst life improves with respect to the amorphous counterpart catalyst over which the above molecules can be performed without space constraints. On the other hand, and depending on pore-cavity shape and dimensionality, bulky molecules could build up inside the crystals and impede the diffusion of reactants to the active sites by blocking the pores, leading to rapid deactivation of the catalyst. It appears, then, that zeolites with their crystalline structure and well defined pores are excellent catalysts to study the structure–activity–selectivity–decay relationship.

In this book, a number of authors well-known in the field of solid catalyst deactivation and regeneration have treated the deactivation and regeneration of zeolite catalysts in a global way. They have described at the molecular and macroscopic level the chemical and physical processes leading to catalyst deactivation. To do this they have characterized the chemical species responsible for catalyst deactivation and to follow they elaborate on the reactions and mechanisms of formation involved. Kinetic models for deactivation are then established, with special emphasis on deactivation by coke formation.

The understanding of the mechanisms of deactivation is the basis upon which the authors have proposed methods for preventing or diminishing the rate of deactivation and, when necessary, they show how to better achieve regeneration.

Though the fundamentals and the methodology on catalyst deactivation and regeneration presented in this book can be used for any catalytic process, a series of industrially relevant case studies are also presented. The specific processes have been very well selected since they illustrate different situations ranging from catalyst deactivation occurring in a few seconds up to several years, from reversible to irreversible deactivation, and from mono- to bifunctional catalysis. Finally, they have introduced a chapter of much interest today that involves the deactivation of molecular sieves in the synthesis of organic chemicals.

This book is very rationally planned and constructed by the editors, who have great experience and knowledge on the subject. Additionally, the different authors have done a super job in developing their chapters, making a special effort to cover such a broad discipline in a relatively limited space.

The result is a well written and informative book that is an excellent introduction to zeolite catalyst deactivation and remediation for those wishing to learn or do research on the subject, as well as a reference book for those already working in the field, regardless of whether they come from industry or academia.

Avelino CORMA
Polytechnical University of Valencia (Spain)
August 2010

AUTHORS

H.S. Cerqueira
Consultant of Petrobras Engineering Unit
GERIS Engenharia e Serviços LTDA
Av. Rio Branco, 128/18 20040-002
Rio de Janeiro, RJ, Brasil

D. Chen
Department of Chemical Engineering
Norwegian University of Science and Technology (NTNU)
N-7491 Trondheim, Norway

K. Moljord
Department of Chemical Engineering
Norwegian University of Science and Technology (NTNU)
N-7491 Trondheim, Norway
Statoil Research Center
Postuttak, N-7005 Trondheim, Norway

A. Holmen
Department of Chemical Engineering
Norwegian University of Science and Technology (NTNU)
N-7491 Trondheim, Norway

M. Guidotti
CNR – Istituto di Scienze e Tecnologie Molecolari
IDECAT-CNR Unit, Dip. Chimica IMA "L. Malatesta"
via Venezian 21, 20133 Milano, Italy

B. Lázaro
Departamento de Química Orgánica
IDECAT – ICMA Unit, Instituto de Ciencia de Materiales de Aragón
Facultad de Ciencias, Universidad de Zaragoza
c/ Pedro Cerbuna s/n, E-50009 Zaragoza, Spain

M. Guisnet
IBB-Institute for Biotechnology and Bioengineering
Centre for Biological and Chemical Engineering
Instituto Superior Técnico
Universidade Técnica de Lisboa
Av. Rovisvo Pais, 1049-001 Lisboa, Portugal
Laboratoire de Catalyse en Chimie Organique
UMR6503 CNRS – Université de Poitiers

F. Ramôa Ribeiro
IBB – Institute for Biotechnology and Bioengineering
Centre for Biological and Chemical Engineering
Instituto Superior Técnico, Universidade Técnica de Lisboa
Av. Rovisvo Pais, 1049-001 Lisboa, Portugal

N.G. Fonseca
IBB – Institute for Biotechnology and Bioengineering
Centre for Biological and Chemical Engineering
Instituto Superior Técnico
Universidade Técnica de Lisboa
Av. Rovisvo Pais, 1049-001 Lisboa, Portugal

C. Henriques
IBB – Institute for Biotechnology and Bioengineering
Centre for Biological and Chemical Engineering
Instituto Superior Técnico, Universidade Técnica de Lisboa
Av. Rovisvo Pais, 1049-001 Lisboa, Portugal

M.F. Ribeiro
IBB – Institute for Biotechnology and Bioengineering
Centre for Biological and Chemical Engineering
Instituto Superior Técnico
Universidade Técnica de Lisboa
Av. Rovisvo Pais, 1049-001 Lisboa, Portugal

J.-F. Joly
IFP – Institut Français du Pétrole – Lyon
Rond-point de l'échangeur de Solaize
BP3, 69360 Solaize, France

E. Sanchez
IFP – Institut Français du Pétrole – Lyon
Rond-point de l'échangeur de Solaize
BP3, 69360 Solaize, France

K. Surla
IFP – Institut Français du Pétrole – Lyon
Rond-point de l'échangeur de Solaize
BP3, 69360 Solaize, France

M.-F. Reyniers
Laboratorium voor Chemische Technologie, Ghent University
Krijgslaan 281 S5, B-9000 Gent, Belgium

J.W. Thybaut
Laboratorium voor Chemische Technologie, Ghent University
Krijgslaan 281 S5, B-9000 Gent, Belgium

G.B. Marin
Laboratorium voor Chemische Technologie, Ghent University
Krijgslaan 281 S5, B-9000 Gent, Belgium

ACKNOWLEDGEMENTS

Acknowledgements are due to the following publishers for their kind permission to reproduce part of previous works in this book:

Elsevier

Figs. 1.1, 4.2, 4.3, 4.5, 4.6, 4.7, 4.8, 4.9, 4.10, 4.11, 4.12, 4.13, 4.14, 4.15, 4.16, 5.1, 5.2, 5.3, 5.4, 5.5, 5.6, 6.1, 6.3, 7.1, 7.3, 7.4, 7.5, 7.6, 7.7, 7.8, 7.9, 7.10, 7.11, 8.1, 8.2, 8.3, 9.1, 9.2, 9.3, 9.4, 9.5, 9.6, 9.7, 9.8, 11.2, 11.3, 11.4, 11.7, 12.1, 12.2, 12.3, 12.4, 12.5, 12.6, 12.7, 14.3, 15.2, 15.3, 15.4, 16.2, 17.8, 17.11

Tables 2.2, 7.1, 8.1, 9.2, 9.3, 9.5, 11.1, 11.3, 14.2, 17.1, 17.2, 17.4, 17.5, 17.7

American Chemical Society

Figs. 6.4, 15.1, 15.6

Tables 4.2, 9.4, 17.3

Fundação Calouste Gulbenkian

Figs. 6.2, 7.2, 13.1

EDP Sciences

Fig. 11.6

Table 14.1

Wiley-VCH Verlag GmbH & Co. KGaA

Fig. 15.5

Table 11.2

Springer

Fig. 4.4

Imperial College Press

Fig. 14.1

Kluwer Academic Publishers

Fig. 11.1

CONTENTS

Part I: Introduction to Zeolites and to Deactivation Phenomena

Chapter 1

DEACTIVATION AND REGENERATION OF SOLID CATALYSTS

M. Guisnet and F. Ramôa Ribeiro

1.1. Introduction

Most processes in life and industry would not practically be possible without catalysis, which makes particularly essential the thorough understanding of all the factors which determine the catalytic properties.

A catalyst is defined as a substance (gas, liquid or solid) that increases the rate and selectivity of a chemical reaction without being consumed. One might therefore believe that this substance will keep eternally the same catalytic properties. It is not so: in practice, all the catalysts undergo mechanical, physical and/or chemical alterations with, as a consequence, a more or less fast loss in activity and often in selectivity. This alteration of the catalytic properties, called deactivation, occurs both in homogeneous catalysis and heterogeneous catalysis even though this theme of crucial industrial importance is the subject of more attention in heterogeneous catalysis, partly because of a larger number of factors affecting the stability of solid catalysts.

Catalyst deactivation is a problem of great technical, economic and ecologic concern in the practice of industrial chemical processes [1–4]. In petrochemicals and bulk chemicals production which mainly occurs through heterogeneous catalysis, the importance of deactivation phenomena have been acknowledged for a long time and a relevant section of literature deals with methods to minimize them by an appropriate design of catalysts and processes. On the contrary, in the fine and specialty chemicals industry, which uses less solid catalysts, the importance of their lifetime remains underestimated [5]. Actually, as the production of these high added-value chemicals requires the use of valuable and costly reactants, the solid catalyst is often considered as a consumable reagent. The spent catalyst is thus disposed of and substituted by a fresh one. Another reason for that is the relatively low amount of each solid catalyst used, which makes the regeneration process costly. However, by following this simplistic point of view, the benefits due to the substitution of a heterogeneously catalysed process for a stoichiometric one are lost. The ever-increasing attention towards the economical and environmental sustainability of organic chemical production is thus prompting the scientific community to develop solid catalysts that are less prone to deactivation.

The purpose of this general chapter which is essentially based on some review papers [1–4] is to present the different causes of solid catalyst deactivation and the means which are used to prevent, limit and cure this deactivation. To illustrate this presentation, examples will be naturally chosen in the field of zeolite-catalysed processes.

1.2. Causes of Deactivation

1.2.1. *General presentation*

The many causes of deactivation of solid catalysts and their mechanisms were reviewed in recent papers by Moulijn *et al.* [1] and Bartholomew [2]. The causes of deactivation can be grouped into five types [1]:

(i) Poisoning by strong chemisorption of species (often feed contaminants) on the active sites.
(ii) Formation of deposits on the catalyst either by simple deposition (fouling) or by catalytic or thermal transformation of feed components (coking).
(iii) Chemical and structural alterations of the catalysts, e.g. noble metal sintering initiated by sulphur poisoning, dealumination and collapse of zeolite framework at high temperature in the presence of water added or formed from dehydroxylation.
(iv) Mechanical alterations due to the planned or accidental mechanical stress suffered by the catalyst particles: crushing, attrition, abrasion or erosion.
(v) Leaching of active species which particularly occurs during the liquid-phase synthesis of fine chemicals.

It should be noted that in commercial processes, catalyst deactivation is often due to several of these causes with interdependent effects. This is particularly the case in refining processes whose feedstocks are generally constituted with many reactants and contaminants and which are carried out at high temperatures. Thus the deactivation of the zeolite catalyst used in fluid catalytic cracking (FCC) results from (i) the poisoning of the acid active sites by the polyaromatic and the basic nitrogen-containing feed molecules and by the coke molecules trapped within the micropores; (ii) the deposit of heavy feed components on the surface or at the micropore mouth or the formation of coke within the micropores with blockage of the access of reactant molecules to the active sites; (iii) chemical and structural alterations of the zeolite catalyst, e.g. dealumination and partial collapse of the zeolite framework which provokes a decrease in the concentration of active acid sites (Chapter 13). The situation can also be more complex in the case of bifunctional (e.g. hydrogenating-acid) catalysts that are largely used in the refining and petrochemical industry. Indeed, deactivating species and physical alterations can affect differently the two types of active sites whose concentrations and characteristics have moreover different effects on the catalyst activity.

Another important remark is that the changes in mechanical, physical and chemical features of the catalyst responsible for deactivation cannot only occur during the reaction but also during the regeneration. A typical example is the regeneration of FCC catalysts by coke combustion at high temperatures which induces chemical and structural zeolite alterations (Chapter 13).

By considering the causes of deactivation, it clearly appears that with most of them (except type iv), chemical steps play a major role. Thus deactivation by poisoning involves chemisorption on the active sites of poison molecules in competition with the reactant molecules. These poison molecules are generally feed contaminants but can also be solvents used in liquid-phase reactions, heavy and polar reactants (e.g. polyaromatics in catalytic cracking) and undesired products (e.g. coke). Deactivation can even result from site poisoning by desired reaction products (autoinhibition) and this often occurs when the products possess a higher polar character than that of the reactants which is very frequent in the synthesis of fine chemicals. Coke formation (ii) implies many successive chemical transformations of reactant molecules, deactivation by coke resulting from poisoning (i.e. competition for chemisorption on the active sites between reactant and coke molecules in favour of the latter) or from blockage of access to the pores, hence to the sites herewith. Factor (iii) includes reactions of feed contaminants, reactant and/or product molecules with the solid catalyst. Leaching (v) which occurs essentially during the liquid-phase synthesis of functionalized organic molecules results from the presence in the reaction medium of highly polar species, often bearing complexing and/or solvolytic moieties which favour the dissolution of the active components of the catalyst (this is typically the case with metal sites).

It should however be underscored that in addition to chemical steps, physical processes often occur in catalyst deactivation. In particular, mass transport limitations can play a major role in poisoning (i), fouling (ii), structural alterations (iii) and leaching (v). Furthermore, coke is a non-desorbed secondary reaction product. Hence the formation of coke molecules requires not only chemical steps, but also their retention within the pores or on the outer surface of the catalyst. This retention may be due to chemical factors, e.g. strong chemisorption, but more frequently to physical ones: low volatility (gas-phase reaction) or solubility (liquid-phase reaction), steric blockage (trapping) within micropores, etc. (Chapter 7).

Deactivation can be reversible or irreversible under the process conditions. A reversible deactivation caused by leaching of active components from the catalyst in a continuous process (e.g. loss of Cl from alkane isomerization or reforming chlorinated alumina catalysts) can be coped with by adding the leached compound (generally precursors) to the feed. An irreversibly deactivated catalyst has to be either recycled or disposed of or to be subjected to a reactivation treatment: rejuvenation or regeneration. Rejuvenation corresponds to a reactivation treatment which can be carried out in the reactor used for the catalytic process under conditions similar or not very different from the operating conditions. A typical example is the removal of non-polyaromatic carbonaceous compounds deposited under mild conditions (e.g. relatively low temperatures) on hydroprocessing catalysts which are still reactive to be hydrocracked under more severe conditions, e.g. slightly higher temperature and/or hydrogen flow rate [3]. When rejuvenation treatments are not efficient enough to restore the catalytic performances, the catalyst is regenerated, this reactivation treatment being carried out either *ex situ* (often by a specialized company) or *in situ* [4]. Coking and sintering of metals are typical examples of deactivation processes which are irreversible under the process conditions. Indeed, coke removal as well as metal redispersion has to be carried out in an oxidative atmosphere which is incompatible with the catalytic transformation.

Table 1.1. Deactivation of solid catalysts: origin, mechanism and typical examples.

Origin	Mechanism	Examples
Poisoning	Chemisorption (or reaction) of feed impurities or reaction products on the active sites, limiting or inhibiting reactant chemisorption Reversible or irreversible	*Metals:* S compounds, CO, polyaromatics, coke *Acid oxides:* bases, polyaromatics, coke
Fouling and coking	Deposit of heavy compounds: feed impurities or secondary products (coke) on the active surface Only reversible by oxidative regeneration	*Metals:* coke (e.g. Pt), carbon (e.g. Ni) *Acid oxides:* FCC: deposit of heavy feed components, catalytic or thermal coke
Chemical and structural alterations	Decrease of the active surface, e.g. increase of crystal size, partial framework collapse Irreversible	*Metals:* sintering *Acid oxides:* e.g. dealumination then collapse of the zeolite framework under hydrothermal conditions
Mechanical alteration	Loss of catalyst caused by attrition or erosion. Loss of surface area by crushing. Irreversible	Fracture, erosion, e.g. in fluidized beds i) from collisions of particles with each other or with reactor walls ii) or due to high fluid velocities
Leaching	Loss of active component, e.g. by dissolution in reaction medium Most common in liquid-phase fine chemicals synthesis Often reversible	e.g. dissolution of metal framework (e.g. Cr in CrS-1) component of metallosilicate molecular sieves

Table 1.1 summarizes the main characteristics of catalyst deactivation for each of the five main causes. In addition, poisoning (i), deactivation by coke formation (ii) and chemical and structural catalyst alterations (iii) are discussed in Chapters 6–9 in the case of reactions over zeolite catalysts whereas leaching (v) is naturally presented in Chapter 17 dedicated to the deactivation of molecular sieves in the synthesis of organic chemicals. Mechanical alterations (iv) which typically appear in catalysts designed for long life cycles are scarcely treated in open literature papers, and hence will not be discussed in detail in this book.

1.2.2. *Catalyst deactivation, a decrease in activity and/or in selectivity*

Catalyst deactivation is by definition a loss in activity. Depending on the commercial process, deactivation can occur from a few seconds (FCC) to more than ten years (ammonia synthesis) [2]. In many cases, this loss in activity is accompanied by a change in selectivity, often (but not always) a decrease in the selectivity to the desired product.

The loss in catalyst activity can be related to a decrease in a) the number of active sites, b) the activity of the sites (their turnover frequency or TOF) and c) their accessibility by reactant molecules. Thus, over strongly acidic catalysts at low reaction temperatures, poisoning of acid sites by NH_3 is often irreversible with therefore a decrease in the number of acid active sites (a), whereas over weakly acidic catalysts at high reaction temperatures, the site poisoning will be reversible with a decrease in the average TOF value depending on the competition between reactant and NH_3 molecules for chemisorption on the acid sites (b). Note that the situation is more complicated owing to a large distribution in strength of the catalyst acid sites. Furthermore, depending on their composition and location, coke deposits can compete with reactant molecules for chemisorption (b), irreversibly poison the active sites (a) or block the access to pores in which sites are located (c).

The loss in activity due to poisoning (and coking) is often related to the concentration of deactivating species on the catalyst. When considering the plots of the ratio between the activity of poisoned and fresh catalysts (i.e. the residual activity) versus the normalized poison concentration (Fig. 6.1), three kinds of poisoning behaviour can occur [2]. Poisoning can be non-selective: the activity loss is proportional to the poison concentration; it is selective when the most active sites are preferentially poisoned and anti-selective when the sites of lesser activity are initially poisoned. Note that poisoning selectivity can be considerably affected by diffusion limitations for either the desired reaction or poisoning process or both [6].

In addition to its negative effect on the catalyst activity, deactivation can either affect differently the reactions leading to the desired and to the secondary products (selective deactivation) or affect similarly both reaction types (non-selective deactivation). This information is important to specify the catalyst characteristics that are affected by deactivation. Because of differences in conversion on fresh and deactivated catalysts, it is often difficult to discriminate between these two possibilities from the change in product distributions with deactivation. Indeed, reactant conversion often significantly affects the product distribution. Therefore, this distinction is only possible when the distributions are established at identical values of the reactant conversion (and under identical conditions of temperature and pressure).

A simple procedure which respects this condition was proposed by Chen *et al.* [7] and applied to the deactivation by coking of consecutive transformations: $A \xrightarrow{k_1} B \xrightarrow{k_2} D$. To simplify the presentation, first-order reactions and uniform coke distribution through the catalyst bed were assumed and the rate constants k were expressed as a function of C_C, the catalyst coke content:

$$k_1 = k_1^0 \exp(-\alpha_1 C_C) \qquad k_2 = k_2^0 \exp(-\alpha_2 C_C)$$

The change in conversion of A with coke formation depends mainly on the α_1 value, and the change in product selectivity on the α_2/α_1 ratio. Selective deactivation occurs when this ratio is different from 1 and non-selective deactivation when it is equal to 1. Note that a positive effect of coke on the selectivity to the desired product (generally B) can be obtained when α_2/α_1 is higher than 1, the reverse occurring when this ratio is lower than 1. Therefore, the loss in activity (i.e. the catalyst deactivation) is not always accompanied by a decrease in the selectivity to the desired product but sometimes by a selectivity increase. Many examples of selectivity improvement by poisoning or coking have been reported in the literature sometimes with application in commercial processes, e.g. increase in selectivity to para isomers of aromatic alkylation over HMFI zeolite catalysts by bulky bases [8] or coke deposits [9], in selectivity to ethene of acetylene hydrogenation over Pd catalysts by silver [10] or by sulphur compounds [11], etc.

To discriminate between selective and non-selective deactivations, the classical selectivity plots, i.e. product yields versus reactant conversion were used for comparing fresh and deactivated catalysts. With the fresh catalyst, the conversion and yield values were drawn from experiments carried out at different contact times for very short time-on-stream (TOS) values; the corresponding plot describes the selectivity pattern in the absence of catalyst decay. For the deactivated catalysts, the selectivity plot was generated from values obtained at a fixed contact time for different time-on-stream values (hence different coke contents).

In ethylene oligomerization over a HMFI zeolite, this method shows that deactivation is non-selective; indeed the selectivity curves for the primary products (B = C_3–C_7 olefins) and for the secondary products

(D = aromatics + paraffins) are quite identical with the fresh and coked catalysts. On the contrary, large differences were found between these curves in methanol transformation over a SAPO-34 catalyst which occurs through the following scheme:

$$\text{Methanol} \xrightarrow{1} \text{Dimethyl ether} \xrightarrow{2} C_2{-}C_6 \text{ olefins} \xrightarrow{3} \text{Coke}$$

Lower values on the coked catalyst than on the fresh one were found for the primary dimethyl ether product and higher values for the secondary products, olefins and coke. Deactivation is then selective, affecting reaction 2 and especially reaction 3 (i.e. the undesired coke formation) more than reaction 1 [7].

1.2.3. *Characterization of solid catalyst deactivation*

The understanding of deactivation phenomena is often made difficult both by the diversity and complexity of the deactivation causes and their effects and by the need to study them in conditions close to those of the industrial process. Fortunately, the investigators can now rely on a large range of techniques to characterize the deactivating species, in particular coke as well as the changes in the physicochemical catalyst features (active sites, porosity) [12]. Thus, classical techniques such as elemental analysis, temperature-programmed oxidation, physisorption of inert adsorbates (nitrogen, alkanes, etc.) differing by their molecular size, etc., continue to be used to get basic information on the composition of coke as well as on its effect on the active sites and the micropores of zeolite catalysts. However, additional information is now currently provided from a great variety of spectroscopy techniques, frequently operated *in situ* and even sometimes in *operando* mode, the most popular being Fourier transform infrared spectroscopy (FTIR). It should be added that spectroscopic techniques have the important advantage of being non-destructive, which allows the successive use of several of them for characterizing deactivated samples.

Note that promising results have been recently realized in the difficult characterization of highly polyaromatic coke thanks to the use of (matrix-assisted) laser desorption/ionization time-of-flight mass spectrometry (MALDI-TOF MS) [13–14], a technique which had been developed for polymer analysis. Furthermore, as a result of recent technical developments, scanning transmission electron microscopy-electron energy-loss spectroscopy (STEM-EELS) provides the possibility to determine the location, amount and nature of carbonaceous deposits (coke) in zeolite crystals [15]. It should however be remarked that simple and somewhat laborious chemical techniques can be more informative than the elegant spectroscopic techniques, a typical example being that used to establish the composition of coke trapped within zeolite micropores: mineralization of the zeolite by HF solution, recovery of coke by extraction with methylene chloride, analysis of the organic solution through classical organic methods, namely gas chromatography-mass spectrometry coupling (GC-MS) [16].

Significant progress has also been made in the modelling of catalyst deactivation. Catalyst decay was first expressed as a function of time-on-stream, then as a function of a variable related to the cause of deactivation such as the amount of coke deposited on the catalyst [17]. However, this latter expression does not consider the chemistry of the reaction steps and more particularly the interactions of coking intermediates with other species. It is why detailed kinetic models for both the main reactions and the coke formation have been recently developed that explicitly describes the competition between coking and the formation of desorbed products. In Chapter 9, the single-event microkinetic (SEMK) methodology [18] based on the carbenium and carbonium chemistry is applied to describe the effect of catalyst deactivation by coking during fluid catalytic cracking.

A difficulty, frequently encountered in the tests carried out at laboratory scale to determine the kinetics and mechanism of catalyst deactivation, originates from the too high or inversely too low rate of deactivation when operating the reaction under conditions similar to those of the commercial plants. In the second case which is the most frequent, accelerated decay tests have to be carried out [19]. The main difficulty originates from the frequent coexistence of several causes of deactivation with different sensitivities to the operating parameter used for acceleration. For such or such deactivation cause, some accelerated tests are generally better than others. Thus, with increasing the operating temperature, the test is potentially valid if deactivation is due to sintering, solid state reaction or fouling; with decreasing the space velocity, it is the case for fouling or coking, etc. [20]. The methods used to simulate catalyst deactivation are presented in Chapter 3.

1.3. How to Prevent, Limit and Cure the Deactivation of Solid Catalysts?

The significant advances which were made in the knowledge of the causes of deactivation and of their effect on the mechanical, physical and chemical catalyst characteristics during commercial processes had opened the way to appropriate solutions to prevent, to limit and to cure the deactivation of solid catalysts. The actions to be taken range from the design of catalysts with optimal characteristics to the development of the most suitable reactor-regenerator system and to the optimization of the engineering practices. It is however obvious that while general guidelines can be advanced, the actions have to be tailored to each particular commercial process.

1.3.1. *Tailoring the catalyst features*

The primary way to prevent and limit the deactivation of solid catalysts is to optimize their chemical, physical and mechanical features. Indeed this action does not require capital investment and therefore is often cheap and economically viable. With zeolite catalysts, the main physicochemical characteristics can be tailored by a large variety of methods. These methods are presented in Chapter 2 and their application to the limitation of deactivation is shown on industrial examples in Chapters 13–17 (Part V).

A fine-tuning of the acid properties of protonic zeolites can be realized by ion exchange with various cations or by dealumination through many different ways. It is therefore possible to limit the strength and density of the protonic acid sites to the values necessary to obtain a sufficiently high rate for the desired reaction with relatively slow coke formation. Furthermore, the use of nano-sized crystals, the creation of inner mesopores by dealumination or desilication facilitates the diffusion of reactant, product and coke precursor molecules with, as a consequence, high reaction rates and a limited coke formation (Chapter 11).

Structural alterations, e.g. collapse of the zeolite framework and metal sintering, and mechanical alterations which are generally irreversible are more easily prevented than cured [1, 2]. These alterations which generally occur during catalyst regeneration by coke combustion owing to the high temperatures and the presence of water (e.g. FCC) can be limited by exchange of the FAU zeolite catalysts with rare-earth metals (La, Ce, Pr). The resulting stabilization is generally related to the pillaring of sodalite cages by polynuclear cations containing oxygen-bridged rare-earth ions. The matrix of FCC catalysts can also act as a source of aluminium species that can react with the defect precursors of the framework collapse. Contamination by V has a dramatic effect on FCC catalysts, the vanadic acid formed during the oxidative regeneration destroying the zeolite framework. One solution to this problem is to use basic traps such as La_2O_3 that react with vanadic acid to form stable vanadates [21, 22]. Note that to limit these structural alterations, promoters in small amount (e.g. Pt) are generally added to the catalysts so as to facilitate complete coke combustion at lower temperatures.

As it was shown with Pt-KLTL aromatization catalysts, sulphur compounds can significantly increase Pt sintering. One strategy that was attempted to limit this process was to introduce promoters which might stabilize crystal growth, e.g. Ni cations that act as anchoring sites for Pt particles (Chapter 16).

1.3.2. *Choice of the reactor-regenerator system*

When the reaction is carried out in continuous which is generally the case, except in fine chemical synthesis, the rate of catalyst deactivation is a key factor in the selection of process options. The time scale for solid catalyst deactivation varies considerably among the different industrial processes. Thus, for zeolite-catalysed processes, the time scale ranges from few seconds between the regeneration cycles in FCC to seven years in C_5-C_6 alkane hydroisomerization. Note that the record lifetime (more than ten years) belongs to the catalysts used in ammonia synthesis [2].

When the time scale for deactivation is one year or longer, a non-regenerative fixed-bed technology is chosen, i.e. the reaction is carried out in a fixed-bed reactor, and after the operating period, the deactivated catalyst is discharged from the reactor and generally sent to a company specializing in catalyst regeneration, recycling and safe disposal. The loss of availability of the unit related to the catalyst change is about ten days (only 3% of a year). Moreover, the cost of catalyst change can be reduced when the corresponding period can coincide with the periodic safety inspection of the unit. When catalyst life is about half a year, a semi-regenerative technology is generally adopted. After stopping the reaction, the catalyst is kept in the reactor which is then used for the catalyst regeneration. In both cases, the effect of the relatively slow deactivation is generally compensated by an increase in temperature so as to maintain the desired level of conversion and/or the quality of the product [3].

With a shorter time scale for deactivation, the required frequency of regeneration imposes either switching of the reactor between reaction and regeneration modes (fixed-bed swing reactors) or continuous circulation of the catalyst from the reactor to regenerator vessels (continuous

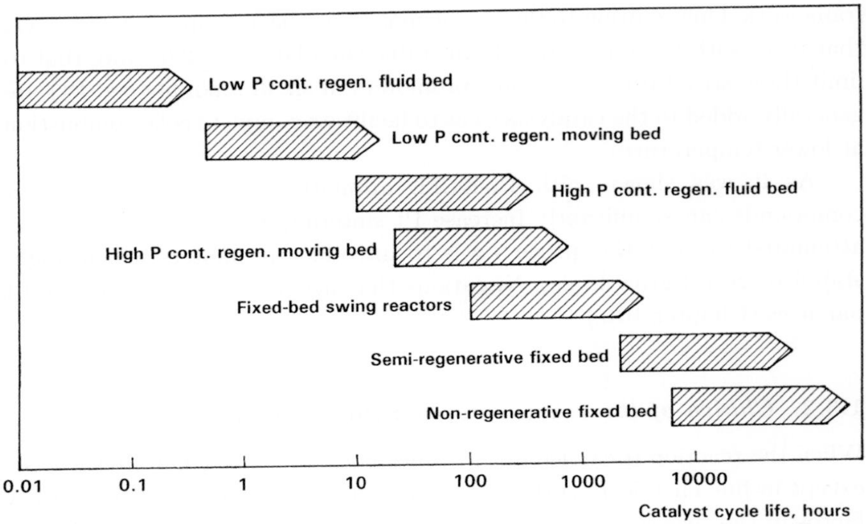

Fig. 1.1. Catalyst life, pressure (P) and selected process technology [3].

regeneration). The continuous regeneration system which is used depends on the required time of catalyst circulation from the reactor to the regenerator and vice versa. Moving beds with catalyst grains are used for periods of weeks, fluidized-bed reactors and entrained-flow reactors for minutes to days and seconds respectively, with the fluidized catalyst (in the form of small particles, e.g. $\sim 60\,\mu$m in FCC) which can be moved much faster than catalyst grains [3]. Another important factor to be considered is the operating pressure (Fig. 1.1).

1.4. Good Engineering Practice

Good engineering practice essentially implies optimization of the operating conditions both in reactor and regenerator. However, when the reactor is fed with complex mixtures like in refining processes, the first action taken is generally to decrease their concentration of contaminants. Thus, in the two-stage hydrocracking configuration, NH_3 and H_2S are eliminated from the first-stage effluents; deactivation by poisoning of the acid sites of hydrocracking catalysts used in the second stage is then avoided and noble metals can be used as hydrogenation components instead of mixed sulfides [23]. With fixed-bed reactors, the procedure which is commonly adopted is to use the first part as a guard bed to trap the feed contaminants, the rest of the bed being thus protected.

Deactivation being highly dependent on the operating conditions, their optimization determines for a large part the catalyst lifetime in commercial processes. Thus, aromatic transalkylation, e.g. toluene disproportionation, toluene-C_9^+ aromatics transalkylation, which however occurs through acid-catalysed reactions have to be carried out under high hydrogen pressure over acid zeolites doped with metals so as to limit coke formation, hence deactivation [24]. In many liquid-phase syntheses of functionalized organic compounds over acid zeolite catalysts, deactivation is essentially due to the strong chemisorption of molecules of desired products within the micropores, owing to their high polarity and their bulkiness; the resulting long contact time of these molecules with the inner acid sites leads to their transformation into carbonaceous compounds ("coke"). An appropriate selection of the operating conditions (as well as the judicious selection of the zeolite and of the type of reactor) favours desorption of the desired products, limiting significantly the catalyst deactivation. Thus in acetylation of aromatic substrates with acetic anhydride, the following operating conditions have to be chosen: high substrate/acetic anhydride molar ratio, eventual use of a suitable solvent, temperature high enough to favour the desorption but low enough to avoid extensive secondary reactions, and high space velocity values to enhance the sweeping of the products out of the zeolite catalyst [5].

Operating conditions of regeneration by coke combustion also have to be chosen so as to limit the structural alterations of zeolite catalysts related to the severe hydrothermal conditions. On the basis of the scheme of coke oxidation, a two-stage regeneration system has been developed and it is used in FCC units devoted to the treatment of residues. The first stage operates at a temperature just sufficient to allow the combustion of most of the H atoms and only part of the C atoms, the second stage at a higher temperature completing the combustion. Therefore, owing to the relatively low temperature in the first stage and to the low partial water pressure in the second, structural alterations of the catalysts are significantly limited (Chapter 12).

It should be underscored that in commercial processes in which catalyst deactivation is relatively slow, the operating conditions in fixed-bed reactors have to be adapted as a function of time-on-stream so as to maintain a pseudo stationary level of activity. For that, the operating temperature is gradually increased up to a value generally imposed by the product quality.

Note that, very often, this increase in temperature not only compensates for the decrease in activity but also causes a decrease in selectivity and accelerates deactivation [1, 3].

1.5. Conclusion

Although the deactivation of solid catalysts is a general phenomenon, it has long been neglected by most of the academic researchers, who concentrated their efforts on catalytic activity and selectivity. Fortunately, such situation has significantly changed, in particular, thanks to the periodic organization, since 1980, of international symposiums on the topic (Catalyst Deactivation 1980, 1987, 1992, etc., the last of the series being held in October 2009 in Delft) and to the publication of various review papers. Therefore, in most academic laboratories involved in heterogeneous catalysis, mastering catalyst stability has now become nearly as important as controlling activity and selectivity. However, progress in this way has still to be realized in the field of heterogeneously catalysed syntheses of fine chemicals in which the catalyst is very often simply considered, at least by academic researchers, as a consumable reagent (Chapter 17).

Very significant progress has been made in the understanding of deactivation phenomena that is essential to develop optimal ways to prevent, to limit and to cure catalyst deactivation. In particular, an ever-increasing care was placed on the choice of operating conditions to mimic the conditions of industrial applications for investigating catalyst deactivation (Chapter 3). Efficient techniques were developed to characterize the deactivating species such as coke, their mode of formation and their effects as well as the physical and chemical alterations of zeolite catalysts (Chapters 4, 5). The general tendency is to use mainly spectroscopic techniques often carried out *in situ* and even in *operando* mode. That is why it does not seem superfluous to remind that to understand the complex deactivation phenomena, a multi-technique approach, coupling them to simpler and often more laborious chemical methods, remains indispensable.

The significant advances in the understanding of the deactivation mechanisms are shown in Chapters 6–9. The three main causes of zeolite catalyst deactivation — poisoning, coking and structural alterations — are successively considered. In the modelling of deactivation by coking, kinetic models were recently developed that considered the chemical steps involved in both the formation of coke and of desorbed products, then including the interactions between coke and product precursors. These models which provide a wealth of information on the effects of operating conditions on catalyst activity, selectivity and stability are powerful tools for optimizing the behaviour of industrial reactors.

The characteristics of deactivation, in particular its rate, play an important part in the selection of the reactor-regenerator system (Chapter 10). Various measures can be taken to prevent or limit deactivation (Part IV). The tuning of the catalyst features, the optimization of operating conditions and the pretreatment of the feed to eliminate contaminants are the easier factors to be set up. Rules for selecting and optimizing zeolite catalysts and operating conditions in hydrocarbon processing in order to limit the rate of coking and the deactivating effect of coke are proposed in Chapter 11. The mode of regeneration depends naturally on the cause(s) and on the rate of deactivation. As coke formation is the main cause of deactivation during most zeolite-catalysed reactions, the mechanism of coke removal by oxidation at high temperatures is presented in Chapter 12 as well as the detrimental effect of this regeneration treatment on the zeolite characteristics.

The last chapters (13–17) are devoted to the description of deactivation and regeneration of zeolite catalysts during commercial processes of refining, petrochemical and fine chemical industries. Another criterion for the process selection deals with the type of catalysis, i.e. monofunctional catalysis over acid, noble metal and oxidation sites and bifunctional associating metal and acid sites. These examples offer the opportunity to underscore the complexity and the interdependence of the modes of catalyst deactivation.

References

[1] Moulijn J.A., van Diepen A.E., Kapteijn F., (a) in *Handbook of Heterogeneous Catalysis*, Ertl G., Knözinger H., Schüth F., Weitkamp J. (Eds.), Wiley-VCH, Weinheim (2008) Chapter 7.1, 1–18; (b) *Appl. Catal. A: General,* 212 (2001) 3–16.

[2] Bartholomew C.H., *Appl. Catal. A: General,* 212 (2001) 17–60.

[3] Sie S.T., *Appl. Catal. A: General,* 212 (2001) 129–151.

[4] Dufresne P., *Appl. Catal. A: General,* 322 (2007) 67–75.

[5] Guisnet M., Guidotti M., in *Catalysts for Fine Chemical Synthesis,* Derouane E.G. (Ed.), Wiley-VCH, Weinheim, (2006) 39–67.

[6] Wheeler A., *Adv. Catal.*, 3 (1951) 307–342.

[7] Chen D., Rebo H.P., Moljord K., Holmen A., *Ind. Eng. Chem. Res.*, 36 (1997) 3473–3479.

[8] Rollmann L.D., *Stud. Surf. Sci. Catal.*, 68 (1991) 791–797.
[9] Chen N.Y., Garwood W.O., Dwyer F.G. (Eds.), *Shape Selective Catalysis in Industrial Applications*, Chemical Industries 36, Marcel Dekker, New York (1989).
[10] Zhang Q., Li J., Liu X., Zhu Q., *Appl. Catal. A: General,* 197 (2000) 221–228.
[11] Barbier J., in *Deactivation and Poisoning of Catalysts*, Chemical Industries 20, Oudar J., Wise H. (Eds.), Marcel Dekker, New York, (1985) 109–150.
[12] Bauer F., Karge H.G., in *Molecular Sieves*, Science and Technology 5, Karge H.G., Weitkamp J. (Eds.), Springer-Verlag, Berlin, Heidelberg (2006) 249–363.
[13] Feller A., Zuazo I., Guzman A., Barth J.O., Lercher J.A., *J. Catal.*, 216 (2003) 313–321.
[14] Barth J.O., Jentys A., Lercher J.A., *Ind. Eng. Chem. Res.*, 43 (2004) 2368–2375.
[15] van Donk S., de Groot F.M.F., Stephan O., Bitter J.H., de Jong K.P., *Chem. Eur. J.*, 9 (2003) 3106–3111.
[16] Guisnet M., Magnoux P., *Appl. Catal.*, 54 (1989) 1–27.
[17] Froment G.F., *Catal. Rev.*, 50 (2008) 1–18.
[18] Froment G.F., *Catal. Rev.*, 47 (2005) 83–124.
[19] Birtill J.J., *Stud. Surf. Sci. Catal.*, 126 (1999) 43–62.
[20] Haber J., Block J.H., Delmon B., *Pure & Appl. Chem.*, 67 (1995) 1257–1306.
[21] Scherzer J., *Octane-Enhancing Zeolitic FCC Catalysts,* Chemical Industries 42, Marcel Dekker, New York and Basel, (1990).
[22] Habib E.T., Zhao X., Yaluris G., Cheng W.C., Boock L.T., Gilson J-P., in *Zeolites for Cleaner Technologies*, Guisnet M., Gilson J-P. (Eds.), Imperial College Press, London (2002) 105–130.
[23] van Veen J.A.R., Minderhoud J.K., Huve L.G., Stork W.H.J., in *Handbook of Heterogeneous Catalysis,* Ertl G., Knözinger H., Weitkamp J., Schuit G.C.A. (Eds.), Wiley-VCH (2008) Section 13.6, 2778–2808.
[24] Alario F., Guisnet M., in *Zeolites for Cleaner Technologies*, Guisnet M., Gilson J-P. (Eds.), Imperial College Press, London (2002) 189–207.

Chapter 2

ZEOLITES AS MODELS FOR UNDERSTANDING CATALYST DEACTIVATION AND REGENERATION

M. Guisnet, N.G. Fonseca and F. Ramôa Ribeiro

2.1. Introduction

The first aim of this chapter is to offer basic data about the physicochemical and catalytic properties of zeolites to readers who are not involved in the field of catalysis by these materials. For those who would be interested in a deeper insight into catalysis with zeolites, a list of reference books is provided at the end of the chapter.

The second aim is to justify the choice of zeolite catalysts for presenting and discussing in this book the state of the art in the field of catalyst deactivation and regeneration. The first reason for this choice seems obvious. Indeed, since the early works on zeolite-catalysed organic reactions, an incredibly large number of transformations was shown to be catalysed by zeolites and many important commercial processes were developed in the fields of refining, petrochemicals, specialty and fine chemicals (Table 2.1) and in pollution abatement [1–4]. Additional justifications will be proposed after the main zeolite characteristics are presented.

2.2. Pore Structure

Zeolites and zeotypes are microporous crystalline materials with a framework constituted by silica tetrahedra in which part of Si atoms can be substituted by tri- or tetravalent elements: Al, Ga, B, Fe, Ti, etc. Similar materials with a framework formed by Al and P (aluminophosphates: AlPOs), Al, P and Si (SAPOs) and sometimes by transition metal elements (MeAPOs) can also be synthesized. The pore structure, which is very

Table 2.1. Commercial processes using zeolite-based catalysts.

Category	Large pore					Medium pore					Small pore	
Structure code	BEA	FAU	LTL	MOR	MTW	AEL	EUO	FER	MFI	MWW	CHA	RHO
Refining Processes:												
FCC		A							A			
Hydrocracking		B										
Isodewaxing						B						
Dewaxing									B			
Light paraffin isomerization				B								
Hydrodearomatization		B										
Olefin (C_4, C_5) isomerization								A				
Olefin oligomerization					A				A	A		
Petrochemical Processes:												
Ethylbenzene synthesis									A	A		
Cumene synthesis	A			A								
p-Ethyltoluene synthesis									A	A		
p-*t*-Butylethylbenzene synth.					A							
C_8 aromatic isomerization				B			B		B			
Toluene disproportionation				A					A			
Toluene, C_9^+ transalkylation				A								
Naphtha aromatization			R									
LPG aromatization									B			
Methanol-to-olefin											A	
Chemical Processes:												
Amination				A					A			A
Hydration									A			
Chlorination/isomerization		A							A			
Beckmann rearrangement									A			
Acylation	A	A										
Phenol hydroxylation									R			
Cyclohexane aromatization									R			
Propene epoxidation									R			

A, B, R stand for acid, bifunctional and redox catalysis, respectively.

open, is constituted by a uniform network of channels and cages with nano-sized dimensions, i.e. with sizes similar to those of organic molecules of commercial interest. These materials can therefore be considered (and furthermore used [2]) as molecular sieves.

Most of the presently known zeolites and zeotypes (which correspond to ~180 different structures [1, 5]) can be classified into three categories according to the number of T atoms (Si and Al for zeolites) in the largest pore apertures: large-pore (12 T atoms, free diameter of 0.6–0.8 nm),

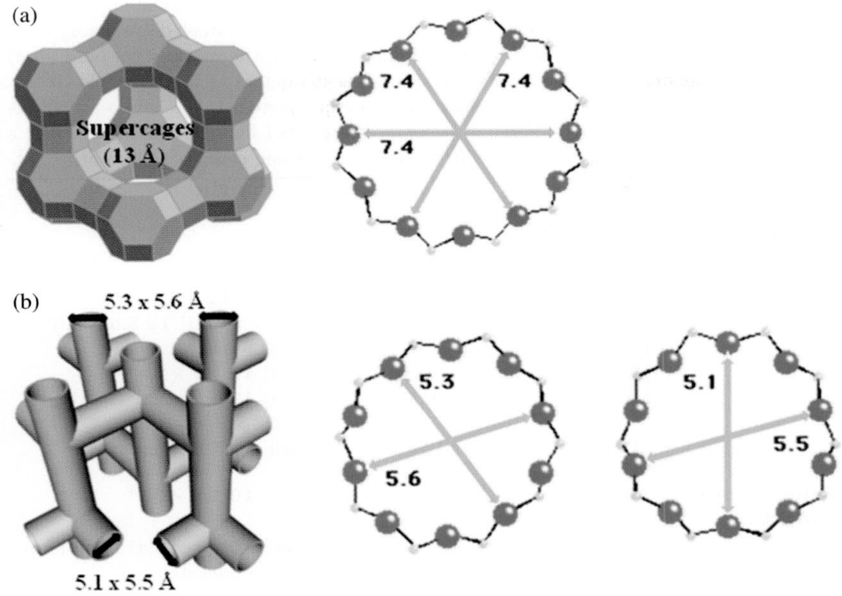

Fig. 2.1. Porous networks of the most employed zeolites: (a) FAU; (b) MFI.

medium-pore (10, 0.4–0.6 nm) and small-pore materials (8, 0.3–0.4 nm). However, ultra-large-pore materials, such as cloverite, CLO (20, 1.32 nm), have been recently synthesized [5–8]. The structure and pore openings of the most employed zeolites in catalytic processes — FAU, a large-pore zeolite with cages, and MFI, a medium-pore zeolite with channels — are shown in Fig. 2.1. Those of various other zeolites and zeotypes with commercial applications are presented in Table 2.2. Unfortunately, the short notation used in this table for the description of the pore systems does not indicate the eventual presence of cages (and their characteristics), information that is essential for the catalytic applications.

An important remark is that with most of the zeolites that are commercially used as catalysts, the crystal size is close to 1 μm. Consequently, external surface areas are much smaller than the internal surface areas. Therefore, most of the active sites are internal and the zeolite-catalysed reactions take place essentially within the micropores (channels and/or cages), hence within nano-sized reactors. Thus, zeolite catalysts can be considered as an array of nanoreactors. This has large consequences on the catalytic properties [9] owing to: the existence of a strong electric field

Table 2.2. Channel systems of the zeolites and zeotypes with commercial applications. Framework type code: three capital letters (in bold type). Channel description: each system of channels is characterized by the channel direction, the number (in bold type) of T atoms in the openings, the crystallographic free diameters of the channels (in Å). The number of asterisks indicates whether the channel system is one-, two- or three-dimensional. Interconnecting channel systems are separated by a double arrow (↔); a vertical bar (|) means that there is no direct access from one channel to the other [5].

Category	Framework type code (IUPAC)	Name	Channel systems
Large pore (12-MR)	BEA	Beta	<100> **12** 6.6 × 6.7** <−> [001] **12** 5.6 × 5.6*
	FAU	Faujasite, X, Y	<111> **12** 7.4 × 7.4***
	LTL	Linde Type L	[001] **12** 7.1 × 7.1*
	MOR	Mordenite	[001] **12** 6.5 × 7.0* <−> [001] **8** 2.6 × 5.7***
	MTW	ZSM-12	[010] **12** 5.6 × 6.0*
Medium pore (10-MR)	AEL	SAPO-11	[001] **10** 4.0 × 6.5*
	EUO	EU-1	[100] **10** 4.1 × 5.4* with large side pockets
	FER	Ferrierite	[001] **10** 4.2 × 5.4* <−> [010] **8** 3.5 × 4.8*
	MFI	ZSM-5	{[100] **10** 5.1 × 5.5 <−>[010] **10** 5.3 × 5.6}***
	MWW	MCM-22	⊥[001] **10** 4.0 × 5.5**|⊥ [001] **10** 4.1 × 5.1**
	TON	Theta 1	[001] **10** 4.6 × 5.7*
Small pore (8-MR)	CHA	SAPO-34	⊥[001] **8** 3.8 × 3.8***
	ERI	Erionite	⊥[001] **8** 3.6 × 5.1***
	LTA	Linde Type A	<100> **8** 4.1 × 4.1***
	RHO	Rho	<100> **8** 3.6 × 3.6***|<100> **8** 3.6 × 3.6***

that can pre-activate the reactant molecules; the concentration of reactant molecules within the nanoreactors with a positive effect on the reaction rates; the participation of the micropores in the orientation of the reactions to the desired product (shape selectivity), in addition to that due to the active sites.

There is however a negative consequence: the narrow size of the nanoreactors prevents the use of zeolite catalysts for the transformation or the synthesis of bulky molecules. To make the active sites of molecular sieves more easily accessible to bulky molecules, three strategies were followed.

Naturally, the first one was the research of molecular sieves with larger pores. However, up to now, the advances in this field, i.e. synthesis of mesoporous molecular sieves and of some ultra-large-pore zeotypes, have not yet led to catalytic applications. Indeed, the first structures show no short-range order and therefore have acid and catalytic properties similar to those of amorphous silica-alumina. With the second ones, the high cost of the structure-directing agents, the low thermal stability and often the monodimensionality of their pore system (e.g. VPI-5) constitute serious drawbacks [8, 9]. More promising results were obtained following two other strategies:

(i) The synthesis of hierarchical pore architectures, combining microporosity and mesoporosity, with, as a consequence, an easier diffusion of the bulky molecules to the active sites. This can be done by introducing a certain degree of mesoporosity into zeolite crystals or the opposite [8]. The many different ways which can be used were recently reviewed [8, 10, 11]. In addition to the traditional ones, i.e. dealumination by steaming and/or acid treatment, various new ways were proposed: basic desilication, incorporation of C spheres or nanotubes with mesoscopic dimensions in the zeolite synthetic mixture then elimination by combustion, doping of the synthesis gel by accelerators of nucleation (e.g. Ge), etc.

(ii) The increase of the outer/internal surface ratio, hence of the amount of outer active sites. This can be obtained by synthesizing nanocrystalline zeolites, i.e. zeolites with crystal sizes of less than 100 nm [12–15] and/or by delaminating lamellar zeolites [16, 17].

2.3. Active Sites

The active sites of zeolites can be located either in framework or in extraframework positions. In the first case, and whether their type is acidic (Brønsted or Lewis), basic, redox, the active sites can be identical and isolated one from each other. Moreover, the appropriate geometry and electronic environment for stabilizing the transition state of the reaction to be catalysed can be approached by a judicious choice of the zeolite structure. Even if the reality is more complex, this means that the characteristics of framework active sites cannot be too far from ideality. As most of the applications of zeolites in commercial catalytic processes imply acid or redox sites, alone or in cooperation (bifunctional catalysis), this paragraph is limited to the presentation of the characteristics of zeolite acid and redox sites.

2.3.1. *Acid sites*

Most of the hydrocarbon reactions as well as many transformations of functionalized compounds are catalysed by protonic sites only [18] and their rate depends evidently on the concentration and activity of these sites. This activity depends on the site characteristics: strength (a stronger acid site is more active), location (accessibility) and proximity.

An important feature of zeolites is their stronger acidity compared to amorphous aluminosilicates, which is related to an enhanced donor-acceptor interaction in zeolites [19]. This interaction was extended by Rabo and Gajda [20] into a resonance model of the Al(OH)Si bond structure with bridging hydroxyls (I) and terminal silanols (II) as extreme limits:

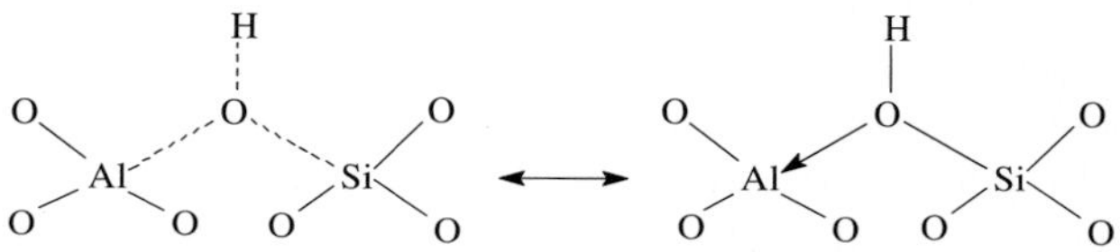

With zeolites, the OH groups are primarily bridging; with amorphous aluminosilicates, they are primarily terminal.

The preparation of protonic zeolites requires (i) the hydrothermal crystallization of zeolite structures from an aqueous solution containing silicic and aluminic compounds and either inorganic (often alkaline) or organic (quaternary ammonium) cations (M), acting as structure-directing agents; then (ii) the substitution of the M cations by protons. Alkaline cations can be directly exchanged with H^+ by acid treatment but more generally are exchanged by NH_4^+, being then eliminated by calcination at high temperatures >650 K (1); the quaternary ammonium cations can be directly decomposed upon calcination (2).

The maximum number of bridging hydroxyls, hence of strong protonic sites, is equal to the number of framework Al atoms. However, their real number is always smaller due to an incomplete exchange of cations and to dehydroxylation, dealumination and even collapse of the framework during high-temperature treatments before catalytic experiments: calcination, activation, etc.

As demonstrated by IR spectroscopy, bridging hydroxyls (hence strong protonic sites) also exist in zeolites presenting bi- or trivalent cations, which explains the high activity in catalytic cracking (FCC) of rare-earth exchanged FAU zeolite catalysts. These OH groups result from dissociation

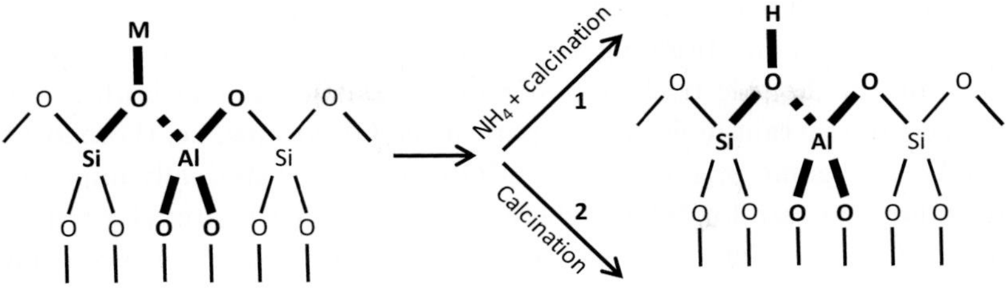

of water molecules (Eqs. 2.1 and 2.2) under the effect of the electrostatic field associated with the RE^{3+} cations, which obviously cannot be located in the close neighbourhood of three O^-.

$$La^{3+},\ 3ZO^- \overset{H_2O}{\rightleftharpoons} [La(OH)]^{2+},\ 2ZO^- + H^+,\ ZO^- \quad (2.1)$$

$$[La(OH)]^{2+},\ 2ZO^- \overset{H_2O}{\rightleftharpoons} [La(OH)_2]^+,\ ZO^- + H^+,\ ZO^- \quad (2.2)$$

In addition, hydroxyl groups with weaker acidity than the bridging ones: silanol groups, hydroxylated extraframework aluminium species, etc., generally exist in zeolites, formed in particular during the calcination treatments. The situation is still more complex due to the existence in certain zeolites of different types of bridging hydroxyls as demonstrated by IR spectroscopy. Thus, HFAU zeolites (Fig. 2.2a) exhibit two bands at 3630–3660 cm^{-1} and 3540–3560 cm^{-1} associated with the bridging OH of supercages and hexagonal prisms, respectively. HMOR zeolites exhibit

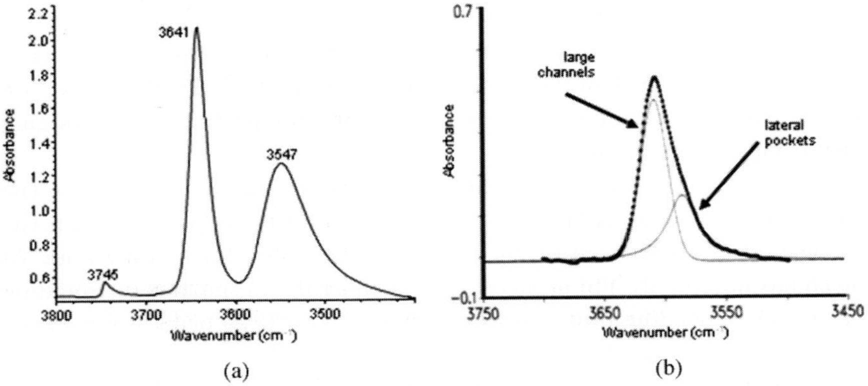

Fig. 2.2. IR spectra of HFAU (a) and HMOR (b) zeolites: (a) IR bands corresponding to the bridging OH groups of supercages (3641 cm^{-1}) and hexagonal prisms (3547 cm^{-1}); (b) deconvolution of the IR band corresponding to the bridging OH groups of the large channels and of the lateral pockets of HMOR.

one asymmetrical band that can be deconvoluted into two bands (Fig. 2.2b) corresponding to the bridging OH located in the large channels (~3610 cm^{-1}) and in the side pockets (~3585 cm^{-1}). Note that in both zeolites, the access of reactant molecules to part of the OH groups, i.e. those located in HFAU hexagonal prisms and in HMOR pockets, is sterically impossible. This is not the case for HMFI where all the bridging hydroxyl groups are located at channel intersections (IR band at ~3610 cm^{-1}), hence equally accessible by reactant molecules.

The main parameters that control the acid strength of the bridging OH groups, and so their activity, are presented below.

When different zeolites with similar framework compositions were compared in base adsorption followed by calorimetry, different initial heats of chemisorption were obtained, indicating differences in acid strength. A relation between the acid strength and the TOT bond angles (T = Al or Si) was found. The greater the angle, the stronger the sites: HMOR (bond angle of 143–180°) > HMFI (133–177°) > HFAU (138–147°). This increase in strength was related to a decrease in the deprotonation energy caused by the augmentation of the TOT angle [20].

The synthesis of metallosilicates containing trivalent elements (T^{III}) in the framework other than Al (B, Ga, In, Fe) is of interest in designing the acid strength [21]. The measured values can be directly related to the average Sanderson electronegativity of the structure: the higher the average electronegativity, the higher the acid strength:

$$\mathrm{B(OH)Si} < \mathrm{In(OH)Si} << \mathrm{Fe(OH)Si} < \mathrm{Ga(OH)Si} < \mathrm{Al(OH)Si}$$

Note however that this refers to the average acid strength, and hence becomes valid for each of the protonic sites only when the local composition is identical, i.e. for high Si/T^{III} ratio, hence at high dilution of the bridging OH groups in the framework.

Indeed, both theoretical and experimental approaches led to the conclusion that the acidity of a given bridging OH group depends on the number of the next-nearest-neighbours (NNN) [20]. Each framework Al atom has necessarily 4 Si atoms (Lowenstein's rule) in the first surrounding layer and, depending on the zeolite topology, 9 to 12 Al or Si atoms in the second layer. The acid strength is maximum at 0 Al NNN and minimum at full occupancy of the NNN sites with Al. Thus a change in the framework Si/Al ratio affects not only the total number but also the acid strength of the bridging OH groups [22]. This explains why for many reactions, the zeolite activity is maximum for a value of the Al molar fraction (x_{Al}) higher

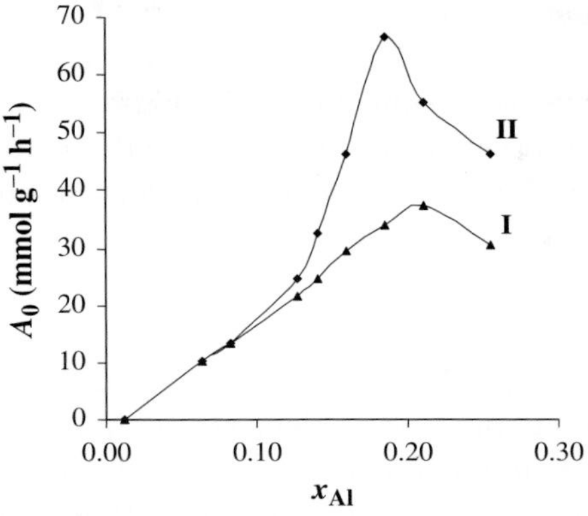

Fig. 2.3. n-Heptane cracking over HFAU zeolite samples: initial activity of dealuminated samples, A_0 vs. molar fraction of framework aluminium, x_{Al}. Samples I and II, with small and large amounts of extraframework Al species (EFAL), respectively.

than the one estimated for site isolation (0 Al as NNN), e.g. at ~0.2 instead of ~0.15 in n-heptane cracking on HFAU (Fig. 2.3) [23]. The situation is still complicated in the case of polar molecules as reactants or products, with an additional effect of the hydrophilicity-hydrophobicity properties of the zeolites on the activity, hence on the position of the maximum [23].

Figure 2.3 shows also that extraframework Al species (EFAL) generated during steam treatment by extraction of framework Al atoms have a positive effect on activity. This increase in activity was ascribed to the generation of Brønsted sites exhibiting "enhanced" acidity as demonstrated by the appearance of additional IR OH bands at lower frequency. This increase in acid strength was related to interaction between bridging hydroxyl groups and neighbouring small EFAL species with Lewis acidity [24].

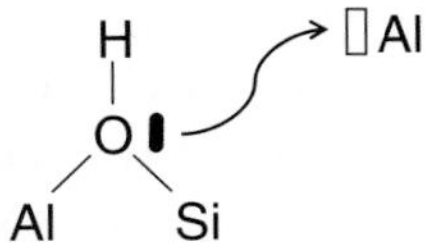

Note however that a large variety of EFAL species can be created during dealumination by steaming. Indeed, these species can be cationic or neutral, monomeric, oligomeric or polymeric, hydroxylated or not and can affect differently the catalytic activity: either positively by increasing the acid strength or negatively by exchange of the protonic sites (EFAL cationic species) or by pore blockage (polymeric species) [23]. Furthermore, it should be underscored that under mild steaming, the framework Al atoms can be partially and reversibly disconnected, acting as Lewis acid sites. The creation of these sites was claimed to be responsible for the very significant increase (by several orders of magnitude) caused by mild steaming of the activity of HBEA in Meerwein–Ponndorf–Verley reactions [25].

2.3.2. *Redox sites*

Redox sites can be introduced into zeolites, enabling them to catalyse redox or bifunctional processes. Elements with redox character can be either incorporated in the framework during hydrothermal synthesis or through post-synthesis treatments, e.g. Ti zeotypes used for oxidation of organic molecules, or introduced in a post-synthesis treatment generally for preparing bifunctional catalysts, e.g. creation of small metallic Pt clusters within the micropores of H zeolites.

In the metallosilicate family, Ti zeotypes were the most intensively studied. The reason for this is that titanium silicalite-1 (TS-1), i.e. MFI structure with Ti (IV) isomorphously substituted by Si in the framework, catalyses a broad range of selective oxidations with hydrogen peroxide (H_2O_2) under mild conditions: hydroxylation of aromatics, ammoximation of cyclohexanone, oxidation of alcohols, ethers, amines and sulphur-containing compounds, epoxidation of linear alkenes, and oxidation of alkanes to alcohols and ketones. This outstanding efficiency can be attributed to: (i) the resistance of Ti sites to extensive hydrolysis (no leaching); (ii) their isolation which prevents the undesired decomposition of H_2O_2, which is induced and catalysed by pairs of adjacent titanium sites; and (iii) the hydrophobic character of the silica framework with few defects, enabling the preferential adsorption of the hydrophobic substrates in the zeolite micropores in the presence of water [26]. Nevertheless, one of the main disadvantages of TS-1 is the relatively narrow channel systems of the MFI structure (Fig. 2.1b) which prevent its use for the synthesis of large molecules. To overcome this drawback, many research teams attempted to incorporate titanium into large-pore zeolites — BEA, MWW and MOR — or in ordered mesoporous materials, mostly of the MCM-41 type [26, 27]. However, up to now, despite some significant advances, e.g. hydrophobic TiBEA [28], there is no commercial application of these materials.

The direct hydroxylation of benzene to phenol by nitrous oxide (N_2O) over Fe-containing MFI catalysts — the AlphOx process from Solutia in collaboration with the Boreskov Institute of Catalysis of Novosibirsk — constitutes another relevant example of commercial application of catalytic oxidation over metal zeolites. The reaction is carried out at 623 K and high benzene conversions (>30%) and excellent selectivity to phenol (up to 99%) are obtained. The catalytic cycle would involve three steps: decomposition of N_2O into molecular nitrogen and an active oxygen atom that promptly reacts with benzene, forming a phenol species adsorbed on an α site, then phenol desorption with restoration of the α site [29]. The nature of α sites is still a controversial topic. Depending on the authors, they would be either Fe sites in extraframework position, framework defects or EFAL species, all these species being however Lewis acid sites.

Pt and Pd zeolite catalysts are used in many refining and petrochemical commercial processes. In most of them, the zeolite is acidic and the reaction occurs through a bifunctional scheme with hydrogenation and dehydrogenation steps catalysed by the metallic sites and rearrangement, cracking steps, etc. by the protonic sites. One exception, however, is the selective aromatization of n-alkanes C_6^+ (e.g. n-hexane into benzene) over Pt non-acidic LTL zeolite. On this catalyst, both dehydrogenation and cyclization steps are catalysed by the metallic sites [30].

The noble metal is introduced by exchange of the zeolite with a solution of complex cations, e.g. $Pt(NH_3)_4Cl_2$. This exchange is equilibrated and the addition of ammonium ions leads to a homogeneous distribution within the grain (Eq. 2.3).

$$[Pt(NH_3)_4]^{2+}_{(\mathbf{s})} + 2\,NH^+_{4_{(\mathbf{z})}} \overset{K_a}{\rightleftharpoons} [Pt(NH_3)_4]^{2+}_{(\mathbf{z})} + 2\,NH^+_{4_{(\mathbf{s})}} \qquad (2.3)$$

After decomposition of the complex by calcination under air flow in selected conditions, the reduction under hydrogen flow leads to very small platinum crystallites (<1 nm), most of them being located in the cages or channels. Another mode of introduction in bifunctional catalysts of the hydro-dehydrogenating component consists simply to associate this component generally deposited on alumina with the acidic zeolite. This method is used for preparing the catalysts employed in isomerization of the C_8 aromatic cut (e.g. Pt/Al_2O_3 — HMOR), in hydrocracking catalysts (e.g. $NiMoS/Al_2O_3$ — HFAU) and in light alkane aromatization (e.g. Ga_2O_3/HMFI), with in this latter case formation of the true dehydrogenating sites through a series of complex steps early in the reaction [31].

Complex redox species can also be synthesized in the zeolite cages (ship-in-bottle synthesis). An interesting example is related to the manganese complexes $[Mn(bpy)_2]_2^+$ synthesized in the LiY zeolite supercages that generate a high yield in the olefin epoxidation by hydrogen peroxide [32].

2.4. Shape Selectivity

This topic was and remains a subject of intense study. This short summary of the field can be completed by reading some review papers and books, such as [33–37].

2.4.1. *Introduction*

The concept of shape selectivity, first proposed by Weisz and Frilette [38], had a huge impact on the design and development of zeolite catalysed processes and it is not exaggerated to say that whatever the reactions and the type of catalysis, shape selectivity plays an essential role. This shape selectivity originates from the location of most of the zeolite active sites within cages and/or channels presenting dimensions close to molecular dimensions of the molecules, i.e. zeolites can be considered as a succession of nanoreactors. Therefore, it is easy to understand that all the reaction characteristics (selectivity but also rate and stability) will be influenced by both the characteristics of the active sites, as is the case with the classical catalysts, and those of the nanoreactors — shape but also their dimensions, their arrangement in the crystals, the shape and size of their openings — in brief by all the characteristics of the micropore system. All this means that, although universally used, this denomination of shape selectivity is somewhat simplistic [23].

2.4.2. *Main types of shape selectivity*

Shape selectivity of zeolites has three main origins: (i) molecular sieving of reactant or product molecules (with a size-exclusion limit); (ii) steric limitations on the formation of transition states; (iii) concentration or confinement of molecules within the micropores.

2.4.2.1. *Shape selectivity due to molecular sieving*

This type of shape selectivity originates from the difficulty or even impossibility for certain molecules of a reactant mixture to enter the zeolite micropores ("reactant shape selectivity") or of certain product molecules (formed over the inner active sites) to exit from the zeolite crystals

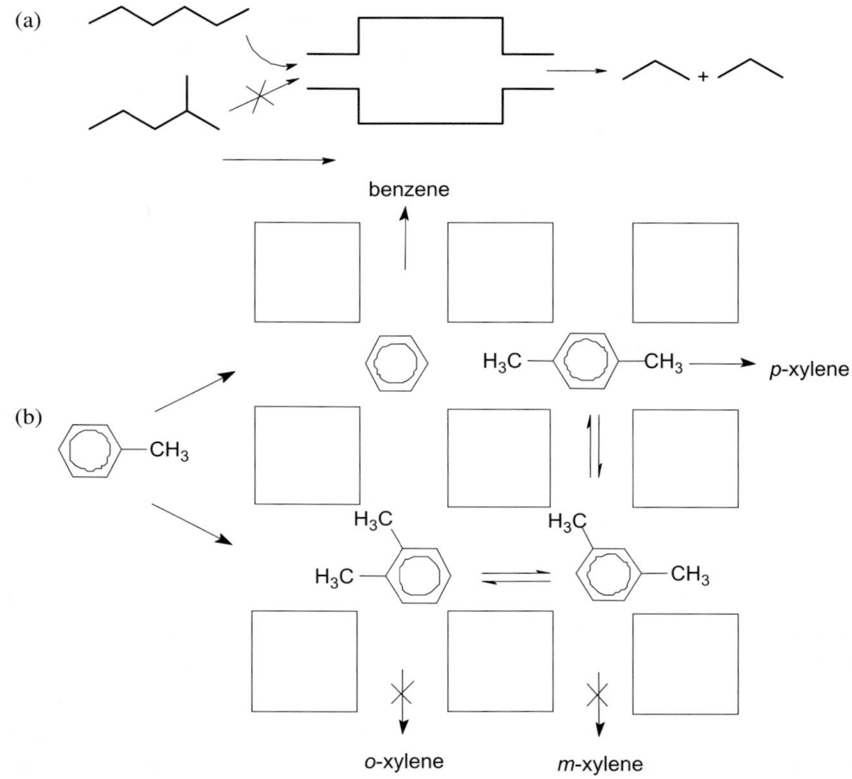

Fig. 2.4. Schematic representation of shape selectivity due to molecular sieving: (a) reactant shape selectivity (dewaxing); (b) product shape selectivity (selective toluene disproportionation).

("product shape selectivity"). Reactant and product shape selectivity are schematized in Fig. 2.4, in the example of two Mobil processes using MFI catalysts: distillate dewaxing and selective toluene disproportionation. In the dewaxing process, multiple-branched long-chain alkanes cannot enter the pores of the HMFI component of the bifunctional noble metal catalyst, while linear alkanes undergo hydrocracking (reactant shape selectivity, Fig. 2.4a). The case of product shape selectivity (Fig. 2.4b) is more complex [35]. Indeed, two conditions have to be satisfied: (i) a large difference between the rates of desorption (diffusion) of products, here the faster desorption of the smallest desired *p*-xylene (pX) isomer compared to *o*- and *m*-xylene; and (ii) the rapid transformation of these non-desorbed products into the desired one (here, their isomerization into pX). If this second condition is not satisfied, pore blockage, hence deactivation, occurs due to accumulation of non-desorbed products within zeolite micropores.

Clearly, this shape selectivity depends on the relative rates of diffusion and reaction of molecules within the micropores. Therefore, in addition to the choice of a shape-selective zeolite for the desired reaction, the parameters affecting these rates can be optimized: fine-tuning of the pore openings to the size of reactant or product molecules, e.g. by deposits of silica, magnesia, coke, etc., on the crystal outer surface, these deposits having moreover a positive effect by deactivating the non-selective outer sites; adequate choice of the operating conditions, in particular the temperature, which has a much larger effect on reaction than on diffusion rates, etc.

2.4.2.2. *Transition state selectivity*

Transition state selectivity or spatioselectivity occurs with reactions where the geometry of the pore around the active sites imposes steric constraints on the formation of transition states or intermediates. This type of selectivity was first proposed by Csicsery [39] to explain the absence in the products of dialkyl-benzenes disproportionation over H-mordenite (MOR) of 1,3,5-trialkylbenzenes although these molecules could diffuse in the large channels: the space available in these channels is too small to accommodate the bulky bimolecular intermediates and transition states of 1,3,5-trialkylbenzenes formation. The effect of spatioselectivity is particularly significant in transformations involving monomolecular (intramolecular) and bimolecular (intermolecular) reactions, being the very different sizes of the corresponding transition states and intermediates. In this case, frequently encountered in industrial processes, spatioselectivity significantly favours the monomolecular reaction, e.g. xylene isomerization without disproportionation over HMFI catalysts (Fig. 2.5). A very important example, moreover related to the topic of this book, concerns the inhibition at the channel intersections of MFI zeolites and the limitation within the micropores of many zeolites (e.g. large supercages of HFAU catalysts) of the secondary reactions leading to coke formation, with practically no (MFI) or a relatively limited (FAU) catalytic deactivation.

It should be underscored that spatioselectivity essentially depends on the spatial environment of the active sites. Unlike molecular sieving,

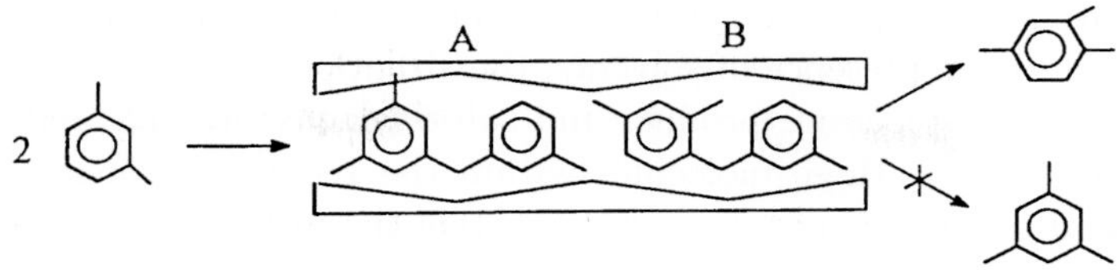

Fig. 2.5. Schematic representation of transition shape selectivity in the example of xylene isomerization over HMFI (no disproportionation).

it does not depend on the length of the diffusion path (hence on the crystal size) and on the relative rates of diffusion and reaction (hence on the characteristics of the active sites) and very little on the operating conditions. Therefore, these two types of shape selectivity can be easily distinguished. However, both can play a role in the selectivity of certain reactions.

2.4.2.3. *Selectivity due to concentration effect*

The interactions between organic molecules and the walls of pores with similar size are very strong and zeolites may be considered as solid solvents [20, 40]. Many theoretical (*ab initio* calculs) and experimental works (adsorption, reaction) show the strong interactions between zeolites and the species present in the micropores. One of the most important consequences of the solvent effect is that the reactant concentration is higher in micropores than in the gas phase [20], which has a positive effect on the reaction rates (contrary to the other types of shape selectivity), greater for bimolecular than for monomolecular reactions. This concentration effect was afterwards called the confinement effect [41], a denomination that emphasizes the role played by interactions between molecules and the micropore surface.

The concentration of reactant molecules in the zeolite micropores is largely responsible for the high activity of zeolites compared to conventional catalysts as well as for a preferential selectivity to products resulting from bimolecular reactions. A typical example of this effect on activity and selectivity can be found in catalytic cracking (FCC): REHY zeolite catalysts are 10 to 10,000 times more active than amorphous silica alumina depending on the hydrocarbon reactant and greater hydrogen transfer/cracking rate ratios (i.e. between bi- and monomolecular reactions) can be observed.

With fine chemical reactions that are often carried out in liquid phase and in the presence of solvents, the situation is complicated by the high polarity of reactant, product and solvent molecules [42]. Indeed, the "physisorption" of these molecules within the zeolite micropores is very strong and moreover very dependent on their polarity. As a consequence, within the micropores, i.e. near the active sites, the composition of the reaction mixture is generally very different from that of the liquid phase, which significantly affects the transformation process with inhibition phenomena in particular (Chapter 17).

2.4.3. *Other types of shape selectivity*

Various other types of shape selectivity were proposed: inverse shape selectivity, molecular traffic control, single-file diffusion, tunnel shape selectivity, cage or window effect, nest effect, pore-mouth catalysis, key-lock catalysis [2, 23]. The first one, inverse shape selectivity, which was proposed by Venuto *et al.* [43] in order to explain coke formation in FAU zeolites, originates from a sieving effect, and hence is simply a particular case of product shape selectivity. The others, contrary to the three types of shape selectivity described in Section 2.4.2, do not apply to all the molecular sieves but only to particular pore structures and moreover are not always clearly demonstrated. Thus, the concept of molecular traffic control is specific to zeolites, which present different interconnected pore systems with different sizes, such as MFI. According to this concept, during methanol conversion in hydrocarbons over MFI, the smallest molecules (e.g. methanol) would diffuse through the sinusoidal channels while the bulkiest (e.g. aromatics) exit through the slightly larger linear channels. The peculiarities of molecular diffusion along the channels of monodimensional molecular sieves have led to the proposal of two different types of shape selectivity. The first one (single-file diffusion) is related to the impossibility for certain molecules to diffuse side by side along the channels, and hence to undergo condensation reactions. The second one (tunnel shape selectivity) is due to the impossibility for molecules entering non-interconnected channels to desorb from them before undergoing a very large number of successive reactions. This new concept was advanced to explain the high selectivity of HFER zeolites in n-butene skeletal isomerization. Lastly, nest effect, pore-mouth catalysis and key-lock catalysis were proposed to explain the particular selectivity of certain reactions demonstrated to occur on active sites located on (or near) the external surface of medium-pore zeolites.

2.5. Conclusion

This brief presentation of zeolites shows that, contrary to what is found with conventional catalysts, general relations can be established between their main physicochemical characteristics (pore structure and active sites) and their catalytic properties. The first reason for that is the regularity of their micropore system in which the reactions generally occur. Note that this micropore system is responsible for outstanding catalytic properties (shape-selective catalysis) which are at the origin of the design and development of many commercial processes.

The second reason is the possibility of identical characteristics for the active sites, e.g. strong acid protonic sites related to the bridging OH groups, isolated framework Ti atoms responsible for oxidation activity, etc. This helps to clarify the nature of active sites in catalytic reactions (including those leading to species responsible for deactivation) as well as the effect of the site characteristics. Moreover, the possibility to finely tune the properties of both the active sites and the pore systems to the desired reaction by well-mastered methods makes this task easier.

Another characteristic which makes zeolites attractive as model catalysts to investigate deactivation phenomena is the diversity in the classes of reactions that they can catalyse: acid, base, acid-base, redox and bifunctional (redox-acid, etc.), hence the possibility to investigate, in a variety of operating conditions, the deactivation of catalysts with similar textural and structural properties.

A latter reason which will be developed in Chapter 4 is the relatively small size of the deactivating species, in particular of coke molecules, owing to steric limitations of their growth in the narrow micropores. As a consequence, these species can be generally identified, which is not possible in the case of non-microporous catalysts.

References

[1] Vermeiren W., Gilson J.-P., *Top. Catal.*, 52 (2009) 1131–1161.

[2] Guisnet M., Gilson J.-P., in *Zeolites for Cleaner Technologies*, Guisnet M., Gilson J.-P. (Eds.), Imperial College Press, London (2002).

[3] Perego C., Carati A., Ingallina P., Mantegazza M.A., Bellussi G., *Appl. Catal. A: General* 221 (2001) 63–72.

[4] Degnan T.F., *Top. Catal.*, 13 (2000) 349–356.

[5] Baerlocher Ch., McCusker L.B., Olson D.H., *Atlas of Zeolite Framework Types*, Elsevier, Amsterdam (2007).

[6] Merrouche A., Patarin J., Kessler H., Soulard M., Delmotte L., Guth J.L., Joly J.F., *Zeolites* 12 (1992) 226–232.

[7] Vogt E.T.C., Kresge C.T., Vartuli J.C., *Stud. Surf. Sci. Catal.*, 137 (2001) 1003–1027.
[8] Maschmeyer T., van de Water L., in *Catalysts for Fine Chemical Synthesis*, Microporous and Mesoporous Solid Catalysts 4, Derouane E.G. (Ed.), Wiley, Chichester, 1 (2006) 1–38.
[9] Corma A., in *Fine Chemicals through Heterogeneous Catalysis*, Sheldon R.A., van Bekkum H. (Eds.), Wiley-VCH, Weinheim (2001) 80–91.
[10] van Donk S., Janssen A.H., Bitter J.H., de Jong K.P., *Catal. Rev. Sci Eng.*, 45 (2003) 297–319.
[11] Perez-Pariente J., Diaz I., Agundez J.C.R., *Chimie*, 8 (2008) 569–578.
[12] Schoeman B.J., Sterte J., Otterstedt J.-E., *Zeolites* 14 (1994) 110–116
[13] Camblor M.A., Corma A., Martínez A., Martínez-Soria V., Valencia S., *J. Catal.*, 179 (1998) 537–547.
[14] Tosheva L., Valtchev V.P., *Chem. Mater.*, 17 (2005) 2494–2513.
[15] Larsen S.C., *J. Phys. Chem. C.*, 111 (2007) 18464–18474.
[16] Corma A., Fornès V., Pergher S.B., Maesen Th.L.M., Buglass J.G., *Nature* 396 (1998) 353–356.
[17] Corma A., Fornès V., *Stud. Surf. Sci. Catal.*, 135 (2001) 73–82.
[18] Guisnet M., *Acc. Chem. Res.*, 23 (1990) 392–398.
[19] Mortier W.J., *Proceedings 6th Int. Zeolite Conference*, Olson D., Bisio A. (Eds.), Butterworth, Guildford (1984) 734–741.
[20] Rabo J., Gajda G.J., in *Guidelines for Mastering the Properties of Molecular Sieves*, Barthomeuf D. *et al.* (Eds.), NATO ASI Series B: Physics, Plenum Press, New York, 221 (1990) 273–297.
[21] Martens J.A., Souverijns W., van Rhyn W., Jacobs P.A., in *Handbook of Heterogeneous Catalysis*, Ertl G. *et al.* (Eds.), Wiley-VCH, Weinheim, 1 (1997) 324–365.
[22] Barthomeuf D., *Materials Chemistry and Physics*, 17 (1987) 49–62.
[23] Guisnet M., Ramôa Ribeiro F., *Les zéolithes. Un nanomonde au service de la catalyse*, EDP Sciences (2006), (a) Chapter 1, 1–22; (b) Ch. 2, pp. 23–38; (c) Chapter 5, 75–88.
[24] Mirodatos C., Barthomeuf D., *J. Chem. Soc. Chem. Commun.*, (1981) 38–39.
[25] Kunkeler P.J., Zuudeeg B.J., van der Waal J.C., Koningsberger, D.C., van Bekkum H., *J. Catal.*, 180 (1998) 234–244.
[26] Bellussi G., Rigutto M.S., *Stud. Surf. Sci. Catal.*, 137 (2001) 911–955.
[27] Clerici M.G., in *Fine Chemicals through Heterogeneous Catalysis*, Sheldon R.A., van Bekkum H. (Eds.), Wiley-VCH, Weinheim (2001) 538–551.
[28] Blasco T., Camblor M.A., Corma A., Estève P., Martinez A., Prieto C., Valencia S., *Chem Commun,* (1996) 2367–2368.
[29] Panov G. I., *Cat. Tech.*, 4 (2000) 18–32.
[30] Derouane E.G., Vanderveken D.J., *Appl. Catal.*, 45 (195–88) L15–L22.
[31] Caeiro G., Carvalho R.H., Wang X., Lemos F., Guisnet M., Ramôa Ribeiro F., *J. Mol. Catal. A: Chemical*, 25 (2006) 131–158.
[32] Knops-Gerrits P.P., Toufar H., Jacobs P., *Stud. Surf. Sci. Catal.*, 105 (1997) 1109–1116.

[33] Csicsery S.M., in *Zeolite Chemistry and Catalysis*, Rabo J. (Ed.), American Chemical Society, Washington, D.C., ACS Monograph, 171 (1976) 680–713.
[34] Degnan T.F., *J. Catal.*, 216 (2003) 32–46.
[35] Chen N.Y., Garwood W.E., Dwyer F.G., *Shape Selective Catalysis in Industrial Applications*, Chemical Industries, 36 (1989).
[36] Chen N.Y., Degnan T.F., Smith C.M., *Molecular Transport and Reaction in Zeolites. Design and Application of Shape Selective Catalysts*, VCH Publishers, New York (1994).
[37] Song C., Garces J.M., Sugi Y. (Eds.), *Shape Selective Catalysis*, ACS Symposium Series 738 (1999).
[38] Weisz P.B., Frilette V.J., *J. Phys. Chem.*, 64 (1960) 382–386.
[39] Csicsery S.M., *J. Catal.*, 23 (1971) 124–130.
[40] Derouane E.G., *J. Mol. Catal. A: Chemical,* 134 (1998) 29–45.
[41] Derouane E.G., André J.M., Lucas A.A., *J. Catal.*, 110 (1988) 58–73.
[42] Guisnet M., Guidotti M., in *Catalysts for Fine Chemical Synthesis,* Microporous and Mesoporous Solid Catalysts 4, Derouane E.G. (Ed.), Wiley, Chichester, 2 (2006) 39–67.
[43] Venuto P.B., Hamilton L.A., *Ind. Eng. Chem. Prod. Res. Develop.*, 6 (1967) 190–192.

General books on zeolites:

- van Bekkum H. *et al.* (Eds.), "Introduction to Zeolite Science and Practice", *Stud. Surf. Sci. Catal.* 58 (1991); 137 (2001); 168 (2007).
- Karge H.G., Weitkamp J., *Molecular Sieves. Science and Technology,* Springer-Verlag, Berlin, Heidelberg. 7 books from 1998 to 2008.
- Barthomeuf D. *et al.* (Eds.), *Guidelines for Mastering the Properties of Molecular Sieves*, NATO ASI Series, B 221, Kluwer, Dordrecht (1990).
- Derouane E.G. *et al.* (Eds.), *Zeolite Microporous Solids: Synthesis, Structure and Reactivity*, NATO ASI Series, C, 352, Kluwer, Dordrecht (1992).
- Guisnet M., Gilson J.-P., *Zeolites for Cleaner Technologies*, Imperial College Press, London (2002).
- Guisnet M., Ramôa Ribeiro F., *Les zéolithes. Un nanomonde au service de la catalyse.* EDP Sciences, France (2006).

Books on shape-selective catalysis: refs. 35–37.

Books from the International Zeolite Association:

- Baerlocher Ch., Meier W.M. and Olson D.H., *Atlas of Zeolite Framework Types.* 6th Revised Edition, Elsevier, Amsterdam (2007).

- Treacy M.M.J. and Higgins J.B., *Collection of Simulated XRD Powder Patterns for Zeolites*, 5th Revised Edition, Elsevier, Amsterdam (2007).
- Robson H. (Ed.), Lillerund K.P., *Verified Synthesis of Zeolite Materials*, 2nd Revised Edition, Elsevier, Amsterdam (2001).
- Database of Zeolite Structures on the Web (http://www.iza-structure.org/databases).

Part II: Characterization Methods

Chapter 3

METHODS TO SIMULATE CATALYST DEACTIVATION

H.S. Cerqueira

3.1. Introduction

The study of a catalytic reaction is never complete without the description of deactivation phenomena. Indeed the catalyst lifetime plays a key role in the profitability of commercial units. Understanding how the catalyst deactivates under industrial conditions is essential in the development of new catalysts that rely on small-scale evaluation. The various heterogeneous catalytic processes are operated under widely different optimum conditions. Some reactions are favored at low temperatures, close to ambient, whereas others need high temperatures, above 773 K. The operating pressure can be close to atmospheric or above 40 bars. The reaction medium comprises two or three distinct phases in an environment that can be reducing or oxidizing. From one catalytic process to another, the feedstock is totally different, from only one or two reactants to a very complex mixture of chemicals of various families including even contaminants. Moreover, even in the same process, significant variations in the feed composition can happen over time. All of the above explains why the deactivation timescale can change from some fractions of a second to several years.

For processes with slow deactivation, as the activity (and selectivity) of the fresh and of the aged catalysts are very similar, the features of the fresh catalysts are quite representative of the industrial process. This is no more the case in processes where the deactivation occurs very rapidly, e.g. in a fraction of a second. Therefore, in this latter case, prior to laboratory evaluation, the fresh catalyst must be submitted to a preliminary step aiming at reproducing the deactivation it would undergo under commercial conditions. An alternative is to evaluate the fresh catalyst under mild conditions with regard to the commercial operation (e.g. lower temperature and/or higher space velocity). Special attention should be devoted to the feedstock composition, especially to the presence of contaminants that can impact the catalyst deactivation. The more similar the feedstock used in the laboratory test is to the commercial one, the better. In cases where the commercial deactivation is slow, a common approach is to accelerate the laboratory deactivation by selecting more severe process conditions or by doping the reactor feed with a high level of the main contaminants. In the case of slow deactivation, information about the initial catalyst activity and deactivation rate can be used to estimate the catalyst lifetime in a commercial campaign.

A detailed characterization of spent catalysts is essential to understand the complex phenomena involved in the deactivation of commercial catalysts. Unfortunately, it is not very commonly carried out because most suitable characterization techniques are costly (hence generally not available on site). Furthermore, they require an expert able to appropriately collect and interpret the data.

This chapter discusses the different types of deactivations, focusing on laboratory methods used to simulate the deactivation that occurs under commercial conditions. Aspects related to the characterization of spent catalyst samples are also briefly addressed.

3.2. Types of Deactivation

Catalyst deactivation can be divided into four distinct types [1]: parallel, in series, side-by-side or independent. Types one to three correlate to the concentration of reactants (parallel deactivation), of products (in series deactivation) or of feed impurities (side-by-side deactivation); all being influenced by gradient concentrations in the catalyst bed or inside catalyst pellets. The last case, named independent deactivation, is related to changes in the catalyst structure or sintering processes, generally influenced by exposure to high temperatures and independent of main reaction products.

In several cases, besides the presence of contaminants in the feedstock, the reaction systems can be very complex, encompassing multiple reactions. In these cases, the loss of catalytic activity does not fit into one category only. At the laboratory scale, it is common to undertake fundamental studies where a commercial feed is replaced by a few representative model compounds. Multiple runs with single compounds can be very helpful in understanding parallel deactivation, where the poisoning or pore blockage is due to products from the main reaction. On the other hand, tests with different mixtures of model compounds, including key products, are useful to study cases of the in series type of deactivation, where the poisoning or pore blockage is due to products from secondary transformation of the main reaction products.

Coke and/or poisons can have different effects on the several reactions taking place, in some cases not only reducing activity but also altering catalyst selectivity. Concentration and temperature gradients within the catalyst bed affect deactivation types. Parallel deactivation is more severe at the bed inlet and at hot spots while in series deactivation increases towards the end of the bed [2].

Deactivation can also be classified as reversible or irreversible, fast or slow. Regardless of the deactivation type, it is common for the deactivated catalyst to exhibit a selectivity different from the fresh catalyst [3]. Under constant operational conditions, selectivity changes are due to changes in the catalyst surface, such as changes in active sites' strength and/or distribution, and changes in metallic dispersion.

Particularly in refining processes, there are several causes of catalyst deactivation with different effects on activity and/or selectivity. A first example is the hydrotreating process, where the catalyst is designed to be used for several months. At the beginning of operation, coke deposition achieves relatively fast a pseudo-steady-state, whereas at the same time a slow deactivation due to the presence of metal contaminants takes place; this would eventually lead to a severe reduction of activity due to pore blockage, requiring a change of the catalyst inventory [4]. In this process, during the catalyst lifetime, the reactor temperature is increased in order to compensate for the activity loss. That is the reason why it is common to express HDT catalyst deactivation in degree of temperature.

Another example where different types of deactivation happen at the same time is the FCC process. Coke deposition (series and parallel type) is fast (happens in every reaction step) and reversible (most of the coke is burned off in the regenerator), whereas significant metal deposition (side-by-side) is relatively slow (needs several reaction/regeneration cycles) and may participate in the irreversible destruction of the catalyst component structure.

3.3. Deactivation Tests and Performance Predictions

3.3.1. *Catalyst selection for commercial processes*

A key question to be asked of every potential catalyst being developed for a commercial application is: How long can it last? To help in answering this question, process monitoring combined with a kinetic model that includes deactivation is very important to establish meaningful conditions for the small-scale tests. Laboratory deactivation protocols prior to testing are very useful to approach commercial yields.

It is important to remember that most small-scale evaluations require the catalyst to be in powder form. This fact suppresses the ability to study the effect of physical factors (e.g. attrition resistance) in the deactivation, as well to study the effects of size and shape in the case of processes where the catalyst pellets are adopted. As a matter of fact, there are specific characterization equipments developed to perform physical evaluation.

A good approach to evaluate catalysts is to identify beforehand the main poisons of the corresponding industrial process and add (if possible, at similar levels) the poisons to the catalyst before small-scale evaluation. It is also common before evaluation to submit the catalyst to the most severe conditions faced during the commercial operation. In the case of a cyclic process, where the catalyst is submitted to several cycles of reaction and regeneration during its lifetime, the same can be done in the lab in an accelerated manner.

For industrial processes where the catalyst plays a significant role in the unit profitability, the risk of selecting a sub-optimum catalyst should be reduced. A common practice is the evaluation at laboratory scale of several catalyst candidates. For a given fresh catalyst, the laboratory procedure chosen should be able to approach the steady-state industrial performance. This implies a search for a pre-treatment condition able to modify the fresh catalyst in ways that match the main physical and chemical characteristics of the commercial catalyst. It is expected that this "artificially deactivated" catalyst will present a catalytic behavior similar to the commercial catalyst when tested under similar operational conditions.

A suitable example is the cyclic deactivation of FCC catalysts. It consists of a laboratory unit where a sample of fresh catalyst is submitted to many cycles of reaction and regeneration, under conditions comparable to the target commercial unit [5]. The model feedstock used during the reaction step consists of a gasoil doped with metals at higher concentration than the commercial feedstock. The whole test is sufficiently long, so the deactivated sample presents a surface area, pore volume, micropore distribution and metal content similar to the commercial equilibrium catalyst. This is the sample that shall be submitted for evaluation.

Alternatively, a fresh catalyst can be deactivated in the laboratory under different conditions prior to small-scale evaluation. The parent commercial sample shall be evaluated under the same conditions. The obtained laboratory results of the deactivated samples are then used in an optimization algorithm, in order to determine the mixture that best mimics the overall (laboratory) performance of the commercial catalyst in terms of activity and main product yields [6]. This method allows establishing a laboratory deactivation protocol to be used for this specific unit.

3.3.2. *Accelerated deactivation tests*

Long evaluation tests can be costly and reduce laboratory throughput. A solution for this problem, intrinsically related to the cases where deactivation is slow, is to adopt an accelerated deactivation test. These tests are frequently used in order to compare the stability of different catalyst candidates or to optimize process conditions.

Accelerated deactivation tests require careful planning, since a too mild or too severe deactivation may lead to results that are useless for inferring commercial activity and/or selectivity. Parallel deactivation can be accelerated by increasing the reaction temperature and/or the space velocity. When the in series deactivation prevails, the space velocity can be reduced at constant temperature in order to increase conversion and, as a consequence, the concentration of the intermediate compound(s) responsible for the deactivation. As described before, side-by-side deactivation can be accelerated by doping the feed with a contaminant level higher than the one encountered in the commercial feedstock. Finally, in the case of independent deactivation, a high reaction temperature is the alternative [3].

One option also is to deliberately change the catalyst formulation in order to speed up the deactivation process. An example is the case of supported metal catalysts, which can be prepared with a higher amount of active metal, in order to favor sintering during small-scale accelerated deactivation tests. This approach becomes tricky when dealing with catalysts with more than one metal or more than one metal phase. Table 3.1 summarizes the conditions that allow us to study or simulate the different types of deactivation.

Table 3.1. Conditions to investigate/simulate different deactivation types.

Type	Parallel R → P R → DA	In series R → P → DA	Side-by-side R → P C → DA	Independent (structure changes, sintering)
Slow	— conditions more severe than the process	— conditions more severe than the process — feed containing P — tests using P as reactant	— tests with model feed doped with high amounts of C	— pre-treatment under conditions more severe than the process
Fast	— test with real feed collecting data at short times — tests at milder conditions than the process	— test with real feed collecting data at short times — tests at milder conditions than the process	— test with real feed collecting data at short times — tests with model feed doped with small amounts of C	— test with real feed collecting data at short times — pre-treatment under milder conditions than the process

R = reactant, P = product, C = contaminant in the feed, DA = deactivating agent.

3.3.3. *High throughput techniques*

One approach that speeds up the development of new catalysts consists in screening a large number of samples (e.g. systems with 8 or more parallel reactors) by means of high throughput techniques. After the successful application of combinatorial methods in pharmaceutical synthesis, this technique expanded to other fields, such as homogeneous catalysis, olefin polymerization catalysis, hydroconversion of alkanes [7], etc. There are a number of methodologies developed for using a large number of small samples in parallel during the last decade. With highly active catalysts and enough high reaction thermicity, a possibility is to submit a large number of different samples to a model reactant and simply follow the change in temperature by means of independent thermocouples located in each catalyst sample. To ensure a constant flow over all the parallel reactors is not trivial and is a common source of error. Spectroscopic techniques such as fluorescence, infrared, ultra-violet [8], among others, can also be used to follow specific functional groups related to the desired reaction product. A combination of different techniques is also possible.

It should be mentioned that high throughput methods are very powerful for discriminating between potential good catalyst candidates and not so promising ones, or at least to simply divide the samples into three or four distinct groups (rather than to establish a complete ranking). The more active group of catalysts identified should be submitted to further small-scale testing in order to select the best catalyst. One should have in mind that when a high throughput system is available for catalyst screening, the preparation of samples becomes the bottleneck. To overcome that, specific systems were developed, using robot arms and control systems to automate the key unit operations necessary in the preparation of catalysts, allowing the preparation of a large number of samples in a timely manner. Such techniques allowed, for instance, the development of the large-pore ITQ-33 zeolite [9] which was demonstrated to be highly selective to diesel during catalytic cracking of light vacuum gasoil.

3.4. Characterization of Industrial Catalysts

The proper characterization of spent catalysts starts with an adequate monitoring of process variables during the industrial unit campaign. Indeed, in order to understand what happens during the catalyst lifetime, it is essential to know not only the steady-state operational conditions, but especially transient regimes and variations in feed or product quality. A kinetic model that includes catalyst deactivation with parameters estimated based on laboratory experimental data can be very useful, provided there is enough industrial data available at least for its validation.

To recover a representative sample of a spent catalyst from a commercial unit is a non-trivial task. Indeed, many commercial units are simply not designed with the facilities necessary to recover catalyst samples during operation. Furthermore with the units that have this ability, the conditions of the sampling procedure have to be carefully established. The major problem is that during the recovery of the catalyst samples, significant changes can affect their characteristics. Coked catalysts are particularly sensitive to changes in coke composition, these changes often depending on the process operating temperature. With processes that operate at high temperatures, when a spent catalyst sample is collected and quickly approaches room temperature, a fraction of the vapors surrounding the catalyst may condense on its outer surface. With low-temperature processes, the risk is losing the light fractions of the coke. A sound sampling procedure, validated under known conditions, is then a key step for a valuable characterization of spent catalysts. Note that even after sampling, spent catalyst samples require careful handling; cold storage in particular is indispensable to avoid partial volatilization and/or oxidation of coke.

In many cases, it is advisable to collect spent catalyst samples at different locations in the reactor. For units that are properly monitored, for instance a fixed-bed reactor with several thermocouples inside the catalyst bed, a comparison of the characterization results of samples collected at different locations may help to establish if there is a specific region of the bed where catalyst deactivation is more severe. In order to correlate the characterization results with operational conditions, it is important to remember that the dynamics of the different variables are not always the same.

A multi-technique approach is obviously indispensable to determining all the mechanical, physical and chemical changes undergone by the catalysts during the commercial process and to understanding their deactivating effect. Here again all the steps of the characterization procedures have to be critically examined. Thus it is common to have pre-treatment steps that have no influence on a fresh catalyst sample, but can permanently change a spent catalyst sample.

The interest of a multi-technique approach is illustrated hereafter in the characterization of catalysts sampled after passing the stripper of a commercial resid FCC unit (RFCC), i.e. processing atmospheric residue [10]. A combination of several characterization techniques indicated that a large part of the coke (95%) was highly polyaromatic (insoluble in CH_2Cl_2) and essentially located in the mesopores of the catalyst matrix. Transmission electron microscopy, matrix-assisted laser desorption/ionization time-of-flight mass spectroscopy and oxidative degradation with $K_2Cr_2O_7$ have shown domains where polyaromatic and heterogeneously distributed coke was trapping saturated hydrocarbons. Infrared spectroscopy after pyridine ad/desorption detected that most zeolite acid sites of the spent catalyst samples still interact with pyridine adsorbate, suggesting that in RFCC, blocking of domains of the catalyst is the most important mode of deactivation.

3.5. Conclusion

Studies of catalyst deactivation need experimental data from small-scale tests under conditions that mimic (or at least approach) the corresponding industrial process. In some cases the deactivation happens fast and data at short time-on-stream is essential, whereas in other cases it can be a slow phenomenon. In this latter case, must be applied.

Validation of small-scale deactivation tests can be obtained by comparing the results of different characterization techniques applied to both commercial spent and lab-deactivated catalysts. Special care is needed during the sampling and handling of spent catalysts. Whenever possible, the selection of operational conditions that can help to prevent deactivation is a good approach for more stable catalytic processes.

References

[1] Levenspiel O., *J. Catal.*, 24 (1972) 265–272.

[2] Cerqueira H.S., Magnoux P., Martin D., Guisnet M., *Appl. Catal. A*, 208 (2001) 359–367.

[3] Birtill J.J., *Stud. Surf. Sci. Catal.*, 126 (1999) 43–62.

[4] Furimsky E., Massoth F.E., *Catal. Today*, 52 (1999) 381–495.

[5] O'Connor P., Verlaan J.P.J., Yanik S.J., *Catal. Today*, 43 (1998) 305–313.

[6] Casali L.A.S., Rocha S.D.F., Passos M.L.A., Bastiani R., Pimenta R.D.M., Cerqueira H.S., *Stud. Surf. Sci. Catal.*, 166 (2007) 147–162.

[7] Huybrechts W., Mijoin J., Jacobs P.A., Martens J.A., *Appl. Catal. A: Gen.*, 243 (2003) 1–13.

[8] Gao K., Yuan L., Wang L., *J. Comb. Chem.*, 8 (2006) 247–251.

[9] Corma A., Diaz-Cabañas M.J., Jordá J.L., Martínez C., Moliner M., *Nature Let.*, 443 (2006) 842–845.

[10] Cerqueira H.S., Sievers C., Joly G., Magnoux P., Lercher J.A., *Ind. Eng. Chem. Res.*, 44 (2005) 2069–2077.

Chapter 4

CHARACTERIZATION OF DEACTIVATING SPECIES

M. Guisnet

4.1. Introduction

The species responsible for the deactivation of zeolite catalysts may be very diverse: organic or inorganic, simple or very complex, present as impurities in the feed of the reactor or formed during reaction, etc. As a consequence of this diversity, these species may deactivate the catalytically active sites in different ways: by competing with reactant molecules for chemisorption on these sites, by reacting with them, by sterically blocking their access, etc. Obviously, the first step to understand and prevent the negative effect of these species is to determine their main characteristics, i.e. amount, elemental composition, distribution in size and in chemical nature, location, etc.

An important remark is that in most zeolite-catalysed processes, carbonaceous deposits, generally denominated as coke, are the main species responsible for deactivation [1–6]. Typically, coke consists of polyaromatic molecules. However, during reactions carried out at low temperature and/or on zeolites with very narrow pores, deactivation is often caused by non-polyaromatic products. This has led many authors to extend this denomination to all secondary organic products retained by the zeolite catalysts and responsible for their deactivation. To avoid any ambiguity or semantic dispute, we suggest the use of the term coke in inverted commas ("coke") to designate non-polyaromatic species having a deactivating effect.

Since coking is the major cause of deactivation of zeolite catalysts, this chapter will be naturally focused on the description of the methods employed in the characterization of these carbonaceous species. However, the application of some of these methods to characterize other deactivating species (in particular poisons) will also be emphasized.

4.2. Methods for Coke Characterization

A large variety of analytical techniques were developed for the characterization of catalytic coke. Their application to zeolite catalysts was described in a recent review paper [3]. A multi-technique approach is a prerequisite to understand the modes of coking and deactivation [2–7]. However, even by coupling judiciously chosen techniques, the key information to determine the mechanisms of coking and deactivation (Fig. 4.1), i.e. the quantitative distribution in nature and size of the coke molecules (coke composition), can be rarely established.

Indeed, most of the techniques generally employed lead only to one or two of the following data:

- The amount of coke deposited on the catalyst.
- The elemental composition, e.g. H/C ratio of coke formed during hydrocarbon transformations.
- The chemical nature of the main coke components. Spectroscopic techniques such as IR, Raman, UV-Vis, ^{1}H and ^{13}C NMR, ESR, etc. are appropriate to specify this important characteristic. Moreover, they present some additional advantages which explain their general use in coke characterization: (a) they are non-destructive, hence several of them can be successively used; (b) *operando* systems, i.e. systems in which the catalytic reaction and coke formation can be simultaneously investigated, can be developed; (c) some of them, such as IR and Raman, can also

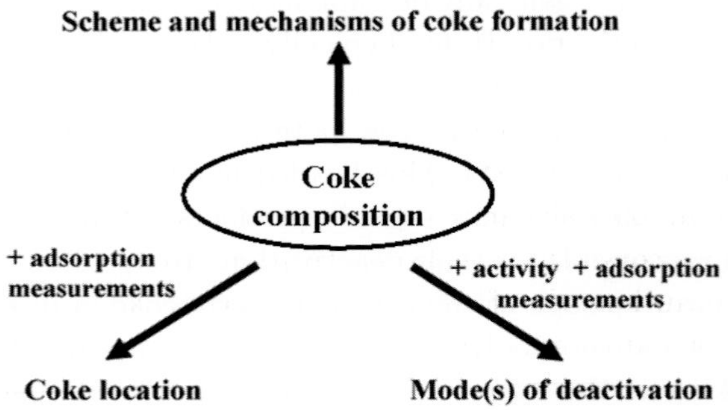

Fig. 4.1. Key role of coke composition: determination of the modes of coke formation, of coke location and of the modes of zeolite deactivation.

specify the interaction between coke molecules and active sites (e.g. acidic hydroxyl groups).

The methods used for coke characterization are detailed hereafter, in order of increasing informative significance of the obtained data.

4.3. Amount and Elemental Composition

4.3.1. *Elemental analysis*

The elemental analysis of deactivated catalysts can give simultaneously the amount of deactivating species, whether mineral or organic, and their elemental composition. The coexistence of several types of deactivating species often complicates quantification. However, the interpretation of experimental data is immediate for mineral species, which may be generally distinguished from each other. This is not the case for the organic species. For instance, basic poisons of the feed of a catalytic cracking unit can be very strongly chemisorbed on the catalyst and/or participate in coke formation; obviously, elemental analysis is quite unable to distinguish these two types of species (poison and coke) [8].

The carbonaceous compounds responsible for the deactivation of zeolite catalysts ("coke") always contain carbon and hydrogen; they may also contain other elements such as N, O, S, metals, etc. when these elements are present in the catalyst, in the reactant molecules or as feed impurities. The contents in C, H, N, S of coked catalysts can be generally obtained through the conventional elemental analysis technique which is based on the combustion of the catalyst sample at high temperature ($\sim$1300 K), with quantitative analysis of the products (e.g. CO_2, H_2O). It is by this way that the atomic H/C ratio, which for a long time was the only measured characteristic of coke, can be determined. From this simple feature, we can estimate the average degree of coke aromaticity, which is particularly useful to understand the mode of coke formation and to optimize the catalyst regeneration by oxidation at high temperatures. Unfortunately, this information loses all usefulness in the case of simultaneous formation of carbonaceous compounds with very different compositions.

Some precautions should be taken in the determination of the H/C ratio, when working with zeolite catalysts [2, 3]. Indeed, the value obtained by conventional elemental analysis can be largely overestimated because of the presence of hydroxyl groups (zeolite dehydroxylation occurs at high temperature with production of structural water: H_2O_{st}) and

Table 4.1. Determination of the atomic H/C ratio of zeolite coke.

Step 1	$M_1 = m_{\text{zeol}} + (m_{H_2O})_{\text{zeOl}} + (m_H + m_C)_{\text{coke}}$ with $(m_{H_2O})_{\text{zeol}} = (m_{H_2O})_{\text{st}} + (m_{H_2O})_{\text{ads}}$
Step 2	$(m_{H_2O})_t = (m_{H_2O})_{\text{zeol}} + (m_H)_{\text{coke}} \cdot \frac{18}{2} \quad m_{CO_2} = (m_C)_{\text{coke}} \cdot \frac{44}{12}$
Step 3	$M_2 = m_{\text{zeol}}$
Step 4	$(m_C)_{\text{coke}} = \frac{12}{44} \cdot m_{CO_2}$ $\Delta M = M_1 - M_2 = (m_{H_2O})_{\text{zeol}} + (m_H + m_C)_{\text{coke}}$ $(m_{H_2O})_t - \Delta M = 8 \cdot (m_H)_{\text{coke}} - (m_C)_{\text{coke}}$ $(m_H)_{\text{coke}} = \frac{1}{8}\left[(m_{H_2O})_t - \Delta M + \frac{12}{44} m_{CO_2}\right]$
Step 5	$\frac{\text{H}}{\text{C}} = \left(\frac{12 \cdot m_{\text{H}}}{m_{\text{C}}}\right)_{\text{coke}}$

especially of the large amount of water quickly adsorbed within the micropores ($(H_2O)_{\text{ads}}$) from the ambient atmosphere [2, 3]. To avoid this overestimation, additional data is necessary: either the amount of oxygen consumed during combustion or the mass of the sample after combustion, i.e. without coke and without water, such as in the method developed by Weitkamp [9]. This method involves five steps (Table 4.1): determination of M_1, the mass of the coked zeolite (1), of the amounts of CO_2 and H_2O formed during combustion (2), of M_2, the mass of the coked zeolite after combustion (3), of the masses of carbon and hydrogen in coke (4) and of the atomic H/C ratio (5).

It should be noted that the H/C values can be altered by desorption of low molecular or partially oxidized carbonaceous deposits during the combustion process or by an incomplete coke combustion with formation of CO [2, 3]. None of these problems, which would alter the results, generally exists with the commercial equipments that operate at high temperature and with an excess of oxygen.

Another technique used to determine the H/C ratio of coke has as its first steps those of the method developed in our laboratory to specify the composition of zeolite coke, i.e. the dissolution of the mineral part of the zeolite catalyst, for example in a solution of hydrofluoric acid, then recovery of the organic part (coke) [4]. In certain cases, coke is completely soluble in organic solvents, typically in methylene chloride (CH_2Cl_2). After drying of the solution and evaporation of the solvent, the elemental composition is established by complete combustion at high temperature. In order to avoid errors due to an incomplete elimination of the solvent, the percentage of chloride is also determined and the corresponding amount of H and C deduced from the raw values.

4.3.2. *In situ methods*

4.3.2.1. *Multi-bed reactors*

In fixed-bed reactors, the coke content of the catalyst often changes with the distance from the bed entrance. One possible reason is the presence as feed impurities of carbonaceous compounds which deposit preferentially near the entrance of the reactor. Additionally, the "coke" distribution depends on the mode of formation: primary, i.e. directly from the reactant(s), or secondary, i.e. from secondary transformation of products. In the first case, coke is preferentially formed near the entrance of the reactor; in the second, the reverse will be observed, i.e. the coke content increases with the distance from the entrance of the reactor. Reactors consisting of successive beds can be used to establish simultaneously the change in reactant conversion, product distribution, coke content and elemental composition (H/C ratio) with the distance from the entrance of the reactor. With this multi-bed reactor, the origin of coke and the mode of coke formation (primary or secondary) can be specified [10–12].

4.3.2.2. *Conventional and inertial microbalance systems*

Nevertheless, a kinetic study of coke formation and of the related deactivation requires *in situ* measurements of coke content and of the composition of the reaction mixture. This was first carried out by thermogravimetric analysis (TGA) in dynamic microbalances [12, 13]. However, these conventional balances present various drawbacks. The main one is related to the impossibility to operate in the real conditions of the catalytic experiments. Indeed, in these systems, reactants are not forced to flow through the catalyst bed like in a downflow fixed-bed reactor. In addition, the operating conditions are poorly defined and it is very difficult to estimate the value of contact time, and thus to establish the kinetic equations of reaction and coke formation. Moreover, external limitations to mass and heat transfer are frequently encountered, even if a thin bed (1 or 2 particles) and high flows can minimize them, but with the risk of catalyst elimination during the experiments. Finally, the determination of small mass changes lacks precision.

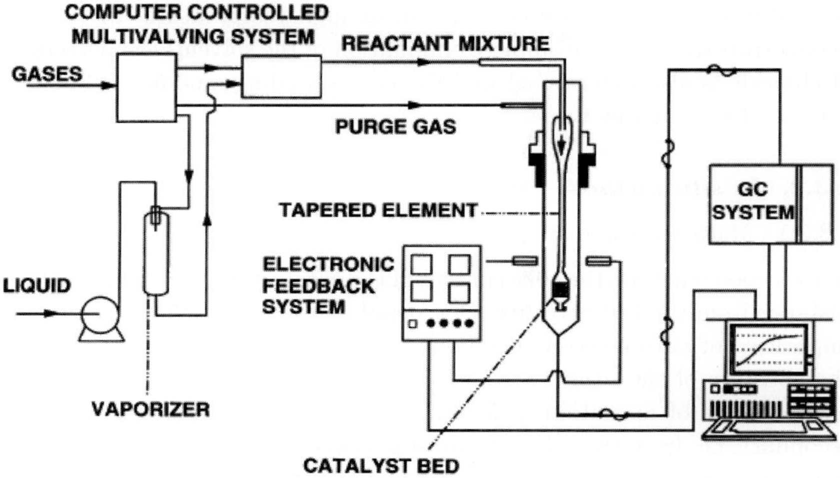

Fig. 4.2. Diagram of the flow-through vibrational microbalance (TEOM) [14].

To minimize all these drawbacks, an original microbalance reactor equipment was developed some 15 years ago [14]. In the new system, an oscillating tapered quartz element (TEOM: tapered element oscillating microbalance) replaces the conventional flow microbalance. This inertial microbalance is used to weigh a catalyst bed through which the reaction mixture is forced to move on (Fig. 4.2), the change in mass of the catalyst being monitored via change in vibrational frequency. Since all properties of a downflow fixed-bed reactor are maintained, the TEOM offers the possibility to monitor quantitatively, *in situ*, the formation rate of carbonaceous deposits, providing simultaneously valuable information concerning the catalytic reaction [14–20]. A comparative analysis of the results obtained with the TEOM and a fixed-bed reactor was done for *n*-heptane reforming over a commercial Pt-Re/Al_2O_3 catalyst: toluene and coke yields were found to be strictly identical [14].

This equipment is able to measure very little mass changes (~5 μg) with a resolution of 0.1 sec, allowing the observation of transient adsorption/desorption and of coking/decoking kinetics under realistic reaction conditions. The advantages of this system, compared to the conventional microbalance dynamics, were clearly demonstrated with the example of coke formation during ethylene oligomerization [15].

Figure 4.3 shows the remarkable efficiency of the method using the example of butene skeletal isomerization over a HFER zeolite. Indeed,

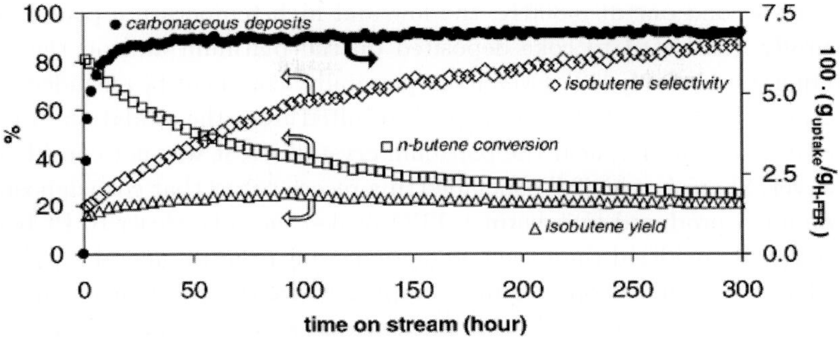

Fig. 4.3. Skeletal isomerization of n-butene over a HFER zeolite at 623 K monitored in a TEOM reactor: reactant conversion, yield and selectivity in isobutene and carbonaceous deposits ("coke") vs. time-on-stream [19].

curves giving reactant conversion, product yield and selectivity and coke formation vs. time-on-stream can be obtained with only one TEOM-GC experiment [19].

4.3.2.3. *Temperature-programmed oxidation (TPO)*

Temperature-programmed oxidation (TPO) of coked catalysts is often used to determine the total content and the elemental composition (generally only the H/C ratio) of coke. In order to identify the oxidation products, infrared, mass spectroscopy and gas chromatography are often combined with thermogravimetric measurements: TG-FTIR, TG-MS, TG-GC. However, in less sophisticated TPO equipments, the partially oxidized products (CO, desorbed hydrocarbons) exiting the oxidation reactor are transformed on a platinum alumina into CO_2 and H_2O, which are recovered in traps. The products generated during coke combustion can also be completely converted into CH_4, on a Ni catalyst with subsequent quantitative GC analysis [22].

Obviously, the oxidation rate depends on the composition of coke and on its location, in particular on the proximity to oxidizing species present on the catalysts, e.g. Pt sites of bifunctional hydroisomerization catalysts, or deposited during the catalytic process, e.g. Ni and V in catalytic cracking. Of course, it could be expected and furthermore currently used in the practice of catalyst regeneration that the presence of oxidizing species over the catalysts promotes coke oxidation. Therefore, when two peaks of production of CO_2 were observed during the TPO of coke deposited over reforming catalysts (chlorinated Pt/Al_2O_3), the first one at ~600 K and the second one at ~800 K, the low and high temperature peaks were naturally attributed to coke deposited on the platinum and on the acid alumina, respectively [22]. Even if the possibility could not be excluded that part or the totality of the low T peak resulted from the oxidation of coke deposited on alumina near the platinum crystallites, it was not considered. However, Shamsi *et al.* [23] concluded in a recent paper that coke deposited on alumina produced two distinct TPO peaks, the coke closer to Pt being more readily oxidized than the one more distant from the metal.

The situation becomes more complex when the mechanism of coke oxidation is considered. This mechanism was established on purely acidic FAU zeolites, based on the evolution of gaseous products: H_2O, CO and CO_2 as a function of temperature [24]. The oxidation was carried out with pure oxygen on a series of HFAU samples with coke deposited during propene transformation at 823 K. A stepwise programmed temperature in the range 300–925 K was chosen. The oxidation was shown to begin at relatively low temperatures (530 K), producing essentially water and oxygenated intermediates (with carbonyl, anhydride and phenol groups) retained in the zeolite. At higher temperatures (Fig. 4.4), these oxidized compounds were converted into CO and CO_2 products. From these data, it was concluded that coke combustion follows a successive scheme with rapid oxidation of the H atoms and then of the C atoms. This scheme was confirmed by oxidation of pyrene, a model component of coke, trapped in the zeolite supercages, and the main reactions involved were identified [25]. It should be noted that in the FCC units, benefit was obtained from

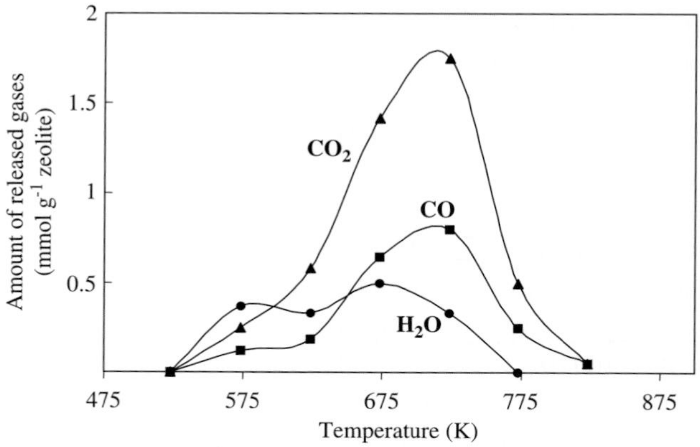

Fig. 4.4. Oxidation of coke (8.5 wt% C) deposited over a HFAU zeolite: evolution of water, carbon monoxide and carbon dioxide as function of oxidation temperature [24].

this behaviour, with the development of double regeneration systems (see Chapter 14).

This behaviour which was observed for a large series of FAU zeolites and coke contents suggests that the presence of two or more oxidation peaks in the TPO profile is not necessarily related to different locations of coke, near or far oxidizing sites. Inversely, it does not necessarily mean that the peaks correspond to cokes with different hydrogen contents; indeed, hydrogen-poor intermediates are formed by oxidation of hydrogen-rich coke components. Therefore, these conclusions have to be supported by additional arguments as it was done by Bayraktar and Kugler [26] in the characterization of coke on equilibrium FCC catalysts. Another remark related to this work is that in most of the TPO profiles, the different peaks are not well resolved, making a deconvolution treatment necessary (Fig. 4.5).

Another complication could emerge with coke formed on zeolite catalysts, during reactions carried out at relatively low temperatures (<700 K). Part of the coke, generally constituted by hydrogen-rich components, can desorb from the catalyst before oxidation, and hence

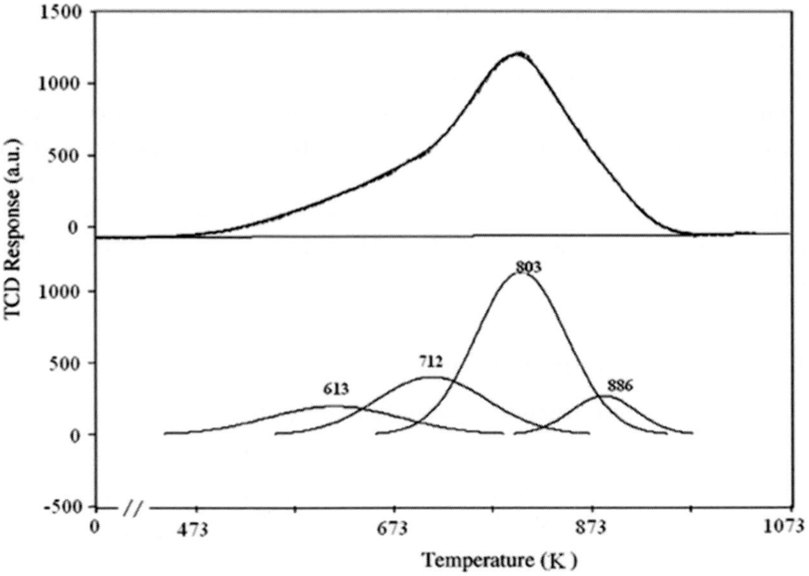

Fig. 4.5. Analysis of TPO spectrum carried out for a FCC catalyst: overlay of experimental curve and composite curve from analysis (top). Individual peaks from analysis (bottom) [26].

cannot be observed with the TPO equipments where there is total oxidation or methanation of the exit stream before analysis.

In conclusion, while the amount of coke and its atomic H/C ratio can be estimated with good accuracy by using TPO, the distribution of coke in different types has to be critically discussed with the support of other data. However, TPO is of great utility to specify the optimal conditions of the classical mode of catalyst regeneration by coke combustion. The reactivity of other gases (in particular H_2) towards coke was also tested, but temperature-programmed hydrogenation (TPH) is much less informative than TPO.

4.4. Chemical Nature of Coke Components

The chemical nature of coke components and precursors can be specified by spectroscopic characterization of coked zeolites with various techniques: infrared (IR), Raman, UV-Visible (UV-Vis), nuclear magnetic resonance (NMR), electron spin resonance (ESR) and electron energy-loss spectroscopy (EELS). Besides, this information, which remains generally imprecise (because of the large diversity of coke components), some of these techniques, in particular IR, which is the most popular, can specify the effect of the deactivating species (coke, poisons) on the active sites (Chapter 5). Moreover, some techniques can be used in *operando* mode, i.e. on the catalysts working in the conditions of classical flow reactors, with the possibility to monitor, simultaneously, the decrease in conversion and/or selectivity, the building up of coke and the modifications undergone by the active sites. Finally, these techniques do not modify the deactivated catalysts, i.e. they are non-destructive, hence successive characterization of deactivated samples by several of them is often carried out.

4.4.1. *Vibrational spectroscopy*

4.4.1.1. *Infrared spectroscopy (generally fourier transform, FTIR)*

The examples reported below refer to acid zeolite catalysts and only the data concerning coke will be presented. The effect of coke on the zeolite acid sites will be detailed in Chapter 5 for the same examples.

Bands of carbonaceous compounds deposited on catalysts can be observed in three IR spectral domains [2, 3]: 1) 3000–3200 cm^{-1}: aromatic CH stretching modes, 2) 2800–3000 cm^{-1}: CH stretching modes of paraffinic groups, 3) 1300–1700 cm^{-1}: CC stretching modes of unsaturated (olefinic, polyenyl, aromatic, polyaromatic) groups and CH bending of paraffinic groups. In this last domain, the coke bands are more intense than in the other domains and valuable information related to the amount and chemical nature of carbonaceous compounds deposited over zeolite catalysts can be easily obtained. The 3000–3800 cm^{-1} domain, which corresponds to stretching modes of hydroxyl groups, will not be considered here.

The first example concerns the effect of temperature on coke formation, during ethylene transformation over an acidic mordenite (HMOR), followed by *in situ* infrared spectroscopy [27]. A significant change in the IR spectra (domains 2 [2800–3000 cm^{-1}] and 3 [1300–1700 cm^{-1}]) of the carbonaceous compounds deposited on the zeolite ("coke") can be observed (Fig. 4.6).

- At low temperatures (300–450 K): the bands correspond mainly to paraffinic species; a band around 1600 cm^{-1} of weak intensity indicates however the presence of olefinic species. This suggests that "coke" results from ethylene oligomerization followed by secondary reactions.
- At high temperatures (>550 K): other bands, especially a very intense one at 1570–1590 cm^{-1}, is observed, which is characteristic of alkyl aromatic and polyaromatic compounds. The most intense band is shifted from around 1600 cm^{-1} to 1574 cm^{-1} with the increase in temperature and time-on-stream, indicating a change in composition.

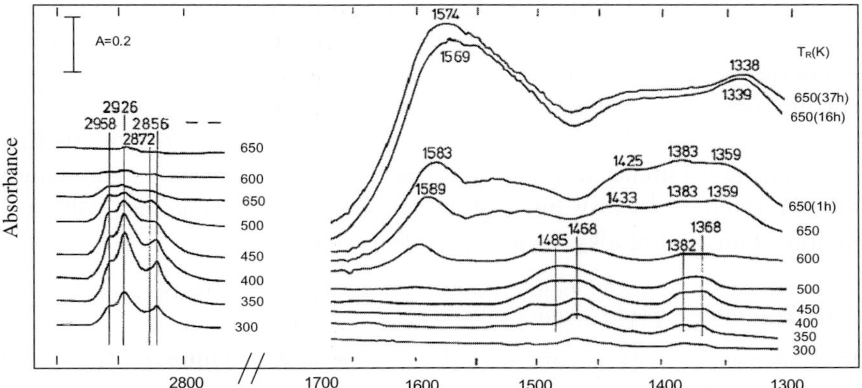

Fig. 4.6. IR spectra of the carbonaceous deposits produced during ethylene transformation on a HMOR zeolite at temperatures between 300 and 650 K [27].

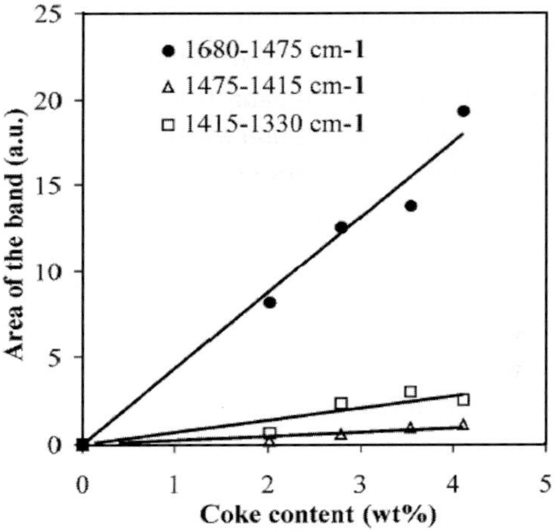

Fig. 4.7. Areas of the IR spectra bands of coke (1300–1700 cm^{-1}) vs. coke content of used MCM-22 samples [31].

Curiously, a good correlation (e.g. Fig. 4.7) between the intensity of this so-called coke band and the coke content is reported by various authors [28–31]. Therefore, after calibration, the intensity of this IR band can be used as an appropriate measure of the percentage of coke deposited on zeolite catalysts.

The influence of reaction temperature on the nature of coke was confirmed in many other studies. Thus, coke deposited over large-pore zeolites (BEA, FAU) during the liquid-phase isobutane/n-butene alkylation at temperatures <400 K was shown to be constituted by very branched aliphatic compounds, having one or several double bonds. No IR band corresponding to aromatic and polyaromatic hydrocarbons could be observed [32]. On the other hand, the IR spectrum of ferrierite (HFER) used during n-butene skeletal isomerization at 623 K presents many bands around 1500 cm^{-1} (very intense) and 3000 cm^{-1}, showing that coke was mainly composed of alkyl aromatic hydrocarbons [33].

4.4.1.2. *Raman spectroscopy*

In contrast to what occurs with IR spectroscopy, symmetric modes of weakly polar bonds such as C=C can be observed as intense bands in Raman spectroscopy, making this technique well appropriate to characterize polyolefinic "coke". However, the coked catalyst and especially coked zeolite samples exhibit strong fluorescence when the Raman scattering is excited by a visible wavelength laser. The fluorescence intensity can be 10^6 times higher than the Raman intensity, making the Raman spectrum undetectable. Fluorescence can be avoided by using the Fourier transform Raman technique but the signal intensity is lower and moreover the characterization at high temperatures impossible because of black-body emission from the samples. A more effective method to avoid fluorescence is to excite the Raman scattering by an UV laser [34]. After various improvements, the UV Raman technique was shown to be successful for the identification of carbonaceous deposits formed *in situ* during methanol conversion [35, 36].

Figure 4.8 shows the UV Raman spectra of a HMFI zeolite before reaction (a) and after methanol conversion at three temperatures (b–d).

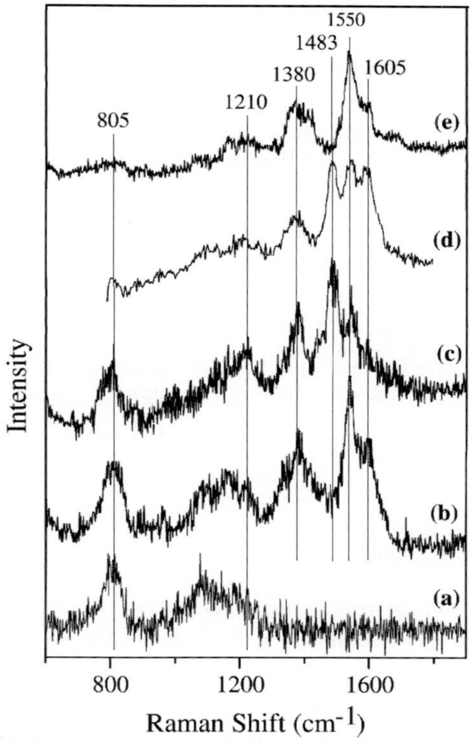

Fig. 4.8. UV Raman spectra of the reaction of methanol on ZSM-5. (a) Calcined H-ZSM-5 before reaction. (b) After dosing 720 molecules of CH_3OH per unit cell at 473 K. (c) After dosing 30 molecules of CH_3OH per unit cell at 543 K. (d) After dosing 30 molecules of CD_3OH per unit cell at 543 K. (e) After dosing 1.5 molecules of CH_3OH per unit cell at 633 K (adapted from [35]).

Two structure bands at 805 and from ~1000 to ~1200 cm^{-1} can be distinguished in the spectrum of fresh HMFI (a). In spectrum b (T = 473 K), four bands corresponding to carbonaceous deposits can be observed at 1360, 1410, 1550 and 1605 cm^{-1}. At 543 K (spectrum c), the 1550 cm^{-1} band intensity decreases and an additional band can be observed at 1483 cm^{-1}. At 633 K (spectrum d), this band is no more visible. Comparison of the spectra with those obtained with CD_3OH as reactant shows no deuterium shifts, indicating that the observed bands are associated to CC or ring stretching vibrations.

UV Raman spectra of model polyaromatic compounds indicate that the bands between 1360 and 1410 cm^{-1} and at 1605 cm^{-1} can be associated to CC or ring stretches of polyaromatic species. From the relative intensity of the bands in these two regions, a chain-like structure was proposed for these polyaromatic species. The two other bands at 1483 and 1545–1550 cm^{-1} could not be related to polyaromatic species. The only possible assignment for the last band is the C=C stretching vibration in conjugated polyenes. Finally, the 1483 cm^{-1} band, whose behaviour with temperature cannot be related to this type of structure, was attributed to conjugated C=C bonds in alkylated cyclopentadienyl species.

Coke formation was also investigated during the same reaction over HMFI, HFAU (USHY) and SAPO-34 samples at temperatures between 298 and 773 K [36]. At room temperature, Raman spectra of adsorbed methanol were almost the same for the three zeolites. In contrast, at 773 K, the Raman spectra of the surface species were quite different and the position of the bands suggests the presence of polyolefinic species on SAPO-34, of polyolefinic and monoaromatic species on MFI and of olefin and polyolefin species on SAPO-34.

4.4.2. *Ultraviolet-visible spectroscopy*

Due to the difficulty of obtaining transparent films of solids, hence the low sensitivity of this technique when operating in transmission mode, UV-Vis characterization of catalysts is generally carried out in reflectance mode (diffuse reflectance spectroscopy or DRS). Like IR spectroscopy, UV-Vis spectroscopy provides valuable data about surface intermediates formed during heterogeneously catalysed reactions. However, it is much less popular than IR, probably because spectra are complex and usually present broad and overlapping bands. Nevertheless, UV-Vis spectroscopy presents large advantages in the analysis of carbonaceous olefinic compounds ("coke"): (i) the types of double bonds can be easily distinguished due to absorption of the corresponding π–π^* transitions in quite different ranges of wavelengths; (ii) the absorption coefficients of electronic transitions of unsaturated organic compounds are more intense (at least one order of magnitude) than those associated with vibrational transitions; moreover, this technique is very sensitive to detection of carbocations [37].

The first examples of coke characterization concern acid zeolites used in methanol conversion [38]. At low temperatures (~400 K), the spectrum of a HMFI zeolite presents three bands (Fig. 4.9): the largest one at ~365 nm is attributed to polyunsaturated cations, the band at 320 nm corresponds to cyclohexenyl cations and a last band at 420 nm corresponding to bulkier species — diphenyl carbenium ions, polyalkylaromatics. At higher temperatures, the last band becomes predominant, while the one corresponding to the polyunsaturated cations disappears (Fig. 4.10). The bands observed at ~420, 320, 500 and 575 nm correspond probably to the bulky species cited above and even to very condensed polyaromatics, while the other bands at 220, 265 and 465 nm could be due to dienes and to cations of substituted benzenic compounds. Note that since the experiments were performed in static mode, some of the detected species correspond to reaction products. Another important remark is the very high sensitivity of UV-Vis to polyethylenic carbocations (360–385 nm region).

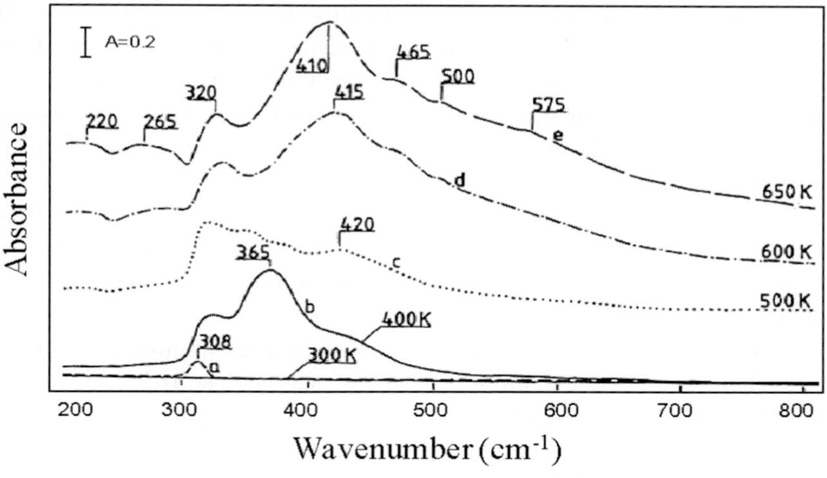

Fig. 4.9. UV-Vis spectra of the carbonaceous compounds (difference spectra), formed during methanol transformation on a HMFI zeolite at temperatures between 300 and 650 K [38].

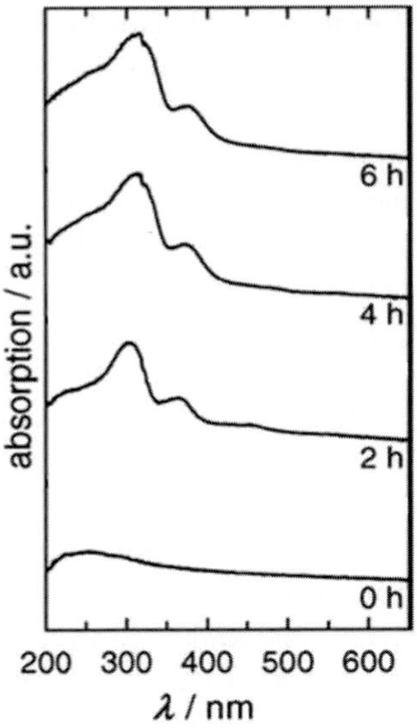

Fig. 4.10. UV-Vis spectra of the carbonaceous compounds (difference spectra) during the isobutane/butene alkylation in liquid phase on the 0.4 Pt/LaY zeolite catalyst at 350 K. Influence of time-on-stream [39].

Very recently, this technique has been used to characterize *in situ* the "coke" deposited on HFAU zeolites used in a continuous-flow stirred tank reactor to catalyse the liquid-phase alkylation of isobutane with butene [39]. Over a 0.4 Pt catalyst, at 348 K, three bands can be detected (Fig. 4.10). The most intense at 305 nm, associated to monoenyl carbocations, increases during the first 3 h on-stream and then remains constant. The second one at 370 nm, corresponding to dienyl carbocations, increases only during the first hour. Finally, the third one, which is very weak, at 450 nm, was attributed to trienyl carbocations and could only be observed at short time-on-stream. No band typical of aromatic hydrocarbons could be detected. Therefore, considering the simultaneous increase with time-on-stream in the formation of carbonaceous compounds and decrease in butene conversion and in selectivity of the desired products of trimethylpentane alkylation, the authors proposed a poisoning effect of the acid sites by irreversible adsorption of enyl cations [39].

The interaction of 1-butene with a H-ferrierite zeolite (HFER), that is an excellent catalyst for the isomerization of n-butene into isobutene, was investigated *in situ* between 300 and 700 K, using both FTIR and UV-Vis spectroscopies [40]. At 300 K, double-bond isomers were formed.

Table 4.2. Interaction of 1-butene with a HFER zeolite at increasing temperatures: attribution of IR and UV-Vis bands to organic species [40].

Temperature (K)	IR (cm^{-1})	UV-Vis (cm^{-1})	Species
300–393	2860, 2921, 2939, 2967 3016, 1629, 1643, 1657 1580, 4084, 4154	50000 32300	H-bond butenes Allyl carbocations
423	2863, 2937, 2957, 1370, 1382		Butene dimers
473–523	1450–1550	27000 23000 15000–20000 35000–39000	Dienyl carbenium ions Trienyl carbenium ions Tetraenyle carbenium ions Di- and trienes
623–673	2865, 2921, 3000–3070 1450–1650	42000–50000 27000–42000 5000–27000	Neutral and charged Methyl aromatics

Between 300 and 393 K, the presence of monoenic allylic carbocations could be observed, between 473 and 573 K, neutral and carbocationic polyenes were present, and above 623 K, mono- and polycyclic aromatic cations were found (Table 4.2).

A parallel reaction scheme was proposed: protonated butenes react with butene molecules to form either octyl carbocations, which are then transformed into isobutene (bimolecular isomerization mechanism), or polyenyl allyl carbocations (with hydride abstraction) acting as precursors of the carbonaceous compounds responsible for HFER deactivation [40].

4.4.3. *Nuclear magnetic resonance (NMR) spectroscopy*

The development in the early 1970s of magic-angle spinning (MAS) and cross-polarization (CP) techniques has opened the possibility to obtain NMR spectra of solids with a resolution degree comparable to that obtained with liquids and with intense signals, even in the case of low natural abundance of the nuclei [41]. Due to the variety of nuclei that can be involved, NMR spectroscopy is appropriate to establish the chemical nature of the deactivating organic species ("coke" or poisons), through analysis of ^{1}H, ^{13}C, ^{15}N spectra, as well as their effect on the zeolite acid (active) sites, either directly through ^{1}H, ^{27}Al, ^{29}Si spectra or via probe molecules, e.g. P-containing bases such as trimethylphosphine oxide (through analysis of ^{31}P spectra). Coke location inside the micropores can also be specified by NMR of Xenon adsorbed in zeolites (^{129}Xe spectra). Therefore, through a multinuclear solid-state NMR approach, it is possible to characterize the nature of the deactivating organic species, their location and their interaction with the acid sites [42]. Only the data concerning the nature of coke will be reported here, being related to the effect of coke on the acid properties discussed in Chapter 5.

While ^{1}H NMR is generally used for the analysis of liquid organic compounds, including the carbonaceous compounds extracted from coked zeolites (Section 5.2), ^{13}C NMR is preferred for the *in situ* characterization of zeolite coke. In fact, for several reasons, ^{1}H NMR is not the most appropriate technique because of problems related to water retention within the micropores and to overlapping and broadening of the signals corresponding to the protons of coke and of the zeolite hydroxyl groups. Regarding ^{13}C NMR, the only limitation, i.e. the low natural abundance of the ^{13}C isotope (only 1.1%), can be avoided by the use of ^{13}C enriched reactants and/or by the cross polarization (CP) technique [41].

Figure 4.11 shows the large effect of temperature on the spectra obtained after contact of 1.2-^{13}C-ethylene with a protonic mordenite (HMOR) [27]. Note that with the chosen procedure, the presence of residual reactants or products, which could perturb "coke" characterization, can be avoided. At low temperatures ($<$500 K), signals at 13, 25 and between 30 and 40 ppm, characteristic of paraffinic carbons, can be observed. Furthermore, the value, lower than 1, of the ratio between the intensities of the signals at 13 and 25 ppm and the signal observed at $\sim$40 ppm indicates the presence of many branchings. At higher temperatures, the spectra show that coke is composed by alkyl benzenes and by polyaromatic compounds with a limited number of nuclei (signals at 185 and 130 ppm). Other signals, at 310 and 245 ppm, are attributed to tertiary and allylic carbocations.

The observations obtained from the simultaneous analysis of the reaction products and of the used catalyst samples by ^{13}C NMR are very useful for clarifying the mechanisms of reactions and their change with reaction time. A particularly illustrative example is the isomerization of [1-^{13}C]-*n*-but-1-ene to isobutene, with the formation of dimers [43] and cyclopentenyl carbocation intermediates [44]. Figure 4.12 shows the ^{13}C CP/MAS NMR spectrum obtained at 523 K, with resonance signals characteristic of cyclopentenyl carbocations (152 and 252 ppm) and condensed aromatic molecules (130 and 144 ppm). The cyclopentenyl carbocations, which disappear at higher temperatures, can be considered as coke precursors; in addition, they could participate in the skeletal isomerization of *n*-butene (pseudo-monomolecular mechanism).

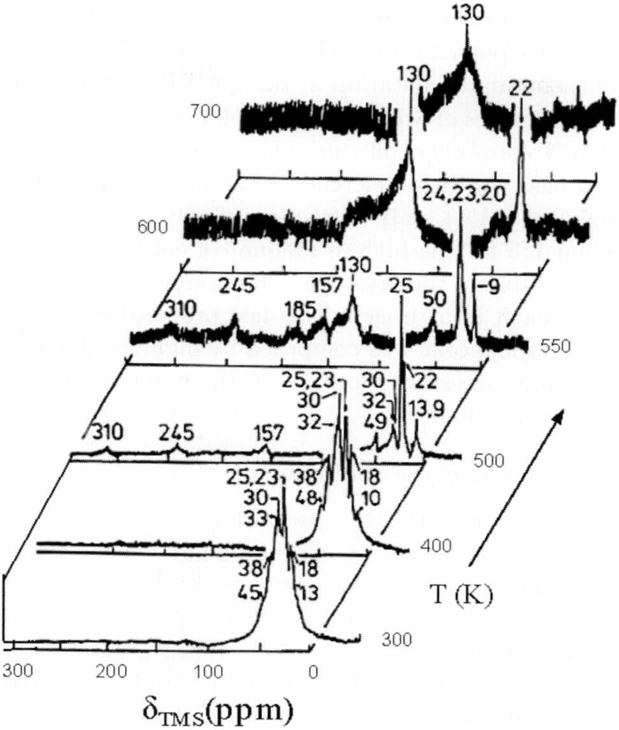

Fig. 4.11. MAS NMR ^{13}C spectra of the carbonaceous deposits produced during the ethylene transformation over a HMOR zeolite at temperatures between 300 and 700 K [27].

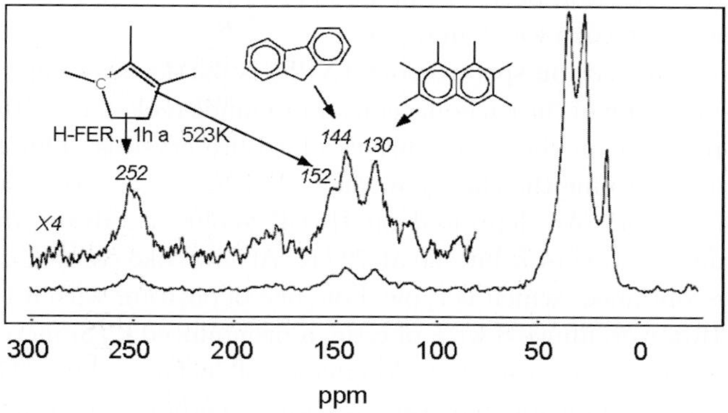

Fig. 4.12. CP-MAS NMR ^{13}C spectra of the carbonaceous deposits produced over a HFER zeolite during the transformation of 1-^{13}C-but-1-ene after heating for 1 h to 473 K and then to 523 K [44].

As referred to above, the sensitivity of ^{13}C MAS NMR can be enhanced by using the cross polarization technique. Unfortunately, this technique makes difficult the implementation of the quantitative analysis. This was shown [45] using a series of MFI zeolite samples coked (from 0.4 to 2.4 wt%) during methanol conversion into gasoline at 643 K. The comparison of the surface intensity with a reference compound shows that, except for low levels of coke (<1%), only a fraction of the "coke" was detected by NMR: less than 1/3 for the highest amount of coke. This loss of C signal intensity could be due to the presence of free radicals or to the existence of coke regions deficient in hydrogen. This last proposal leads the authors to conclude that invisible coke was composed by highly condensed polycyclic aromatic molecules, necessarily located on the outer surface of the zeolite crystals due to their bulkiness.

4.4.4. *Other spectroscopic methods*

Electron spin resonance (ESR) spectroscopy was essentially used to characterize *in situ* the carbonaceous compounds ("coke") formed during methanol or olefin conversion over different acid zeolites [3]. The great effect of the reaction temperature on coke composition was confirmed by the large differences in the shape of the signals, hence by the nature of radicals generated during coke formation. It should be noted that only a small fraction of the C atoms of coke generates radicals, e.g. only one radical per 500–10,000 C atoms of high-temperature coke is formed during ethylene transformation over HMOR [46], which could limit the application of this indirect technique of characterization. Nevertheless, a good linear correlation between the number of radicals and the amount of high-temperature coke was found [46].

X-ray photoelectron spectroscopy (XPS) is particularly appropriate to specify the location of the carbonaceous compounds (coke), i.e. either within the micropores or on the outer surface of zeolite crystals. This technique was used to determine the change in the C/Si signal intensity as a function of the amount of coke deposited on HFER zeolite catalysts, during the skeletal isomerization of *n*-butene at 723 K. At low coke contents, low C/Si values were obtained, which is typical of coke deposition within the zeolite channels. However, above 8 wt% of coke, a pronounced C/Si increase could be observed, which suggests a predominance of surface coking [47].

Some more sophisticated spectroscopic methods such as electron energy-loss spectroscopy, performed in a scanning transmission electron microscope (STEM-EELS) and laser desorption/ionization time-of-flight mass spectroscopy (LDI-TOF MS) were also used. The first application of STEM-EELS aimed to characterize the highly polyaromatic coke deposited on the outer surface of HFAU, HMFI and HOFF zeolite catalysts, during *n*-heptane cracking at 723 K [48]. Comparing the EELS spectra with the signals obtained for reference compounds, it was found that the structure of coke deposited on HOFF and HMFI was similar to that of coronene (i.e. pregraphitic), while the structure of coke formed on HFAU was comparable to that of pentacene (i.e. a linear polyaromatic). As a result of technological improvements, detailed spatial information can now be obtained concerning the amount and nature of coke. STEM-EELS measurements were performed on a HFER zeolite coked during *n*-butene skeletal isomerization [49]. At high coke content, a complete blockage of the 8-MR pores was observed, while the 10-MR channels remained partially accessible to the reactant molecules, even with alkyl aromatic species deposited near their entrances [49].

(MA)LDI-TOF MS (matrix-assisted) and LDI-TOF MS (laser desorption/ionization time-of-flight mass spectroscopy) were developed to characterize bulky bioorganic molecules. These techniques, which differ only by the adding (or not) of a matrix (typically an aromatic hydrocarbon) to the sample to be characterized, were recently applied in the characterization of heavy carbonaceous compounds ("coke"), either directly on used zeolite catalysts or on coke recovered from the catalyst. Their main advantage is related to the low energy input during the laser desorption process. As a consequence, it could be expected that the obtained mass spectrum reflects directly the molecular weight distribution of the coke components. However, the results obtained by applying LDI-TOF MS and (MA)LDI-TOF MS (with dihydroxybenzoic acid as matrix) on LaHX zeolite samples deactivated during isobutane/butene alkylation at low temperature were more complicated than expected; in particular MALDI detected selectively oxidized products. Nevertheless, the molecular weight distribution estimated by these techniques was in good agreement with the distribution measured by GC/MS of coke recovered after demineralization of deactivated catalysts [50].

4.5. Chemical Composition of "Coke"

4.5.1. *Method of determination of coke composition*

The only way to determine the chemical composition of carbonaceous deposits, i.e. the complete distribution of their components, is to remove "coke" from the catalyst and to analyse it with appropriate techniques [4, 51]. Such a method was developed 20 years ago in our laboratory [52]. This method has been used to specify the composition of coke formed on various zeolite catalysts during the transformation of different organic compounds (mainly hydrocarbons and oxygenated species). In most cases, the mode of coke formation could be deduced from the change of coke composition with reaction time.

This method includes four main steps (Fig. 4.13):

- In the first step, the molecules of reactant and product, which are retained on the zeolite, are desorbed from the catalyst, usually by treatment at the reaction temperature under a stream of inert gas (generally nitrogen). After that, the catalyst sample is removed from the reactor and kept at low temperature in closed containers, to avoid coke oxidation.
- The second step consists in the soxhlet treatment of the catalyst sample with methylene chloride (CH_2Cl_2) as solvent. Through this treatment,

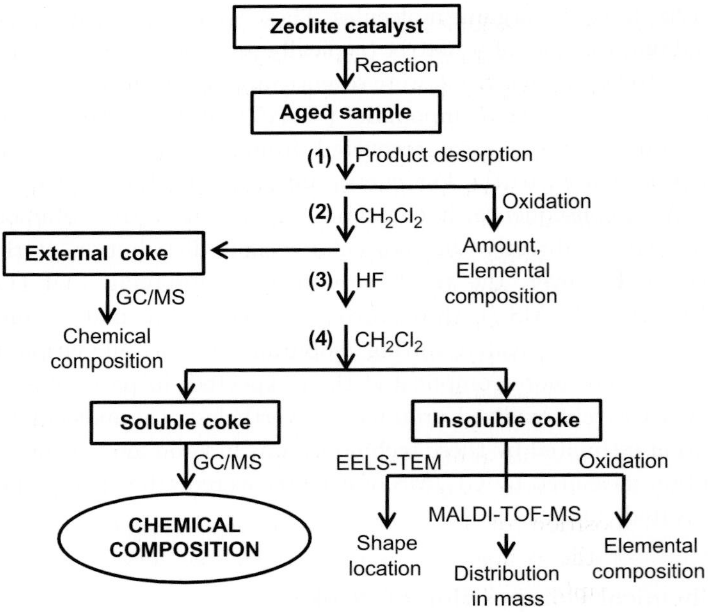

Fig. 4.13. Method used to determine the composition of zeolite coke. EELS: electron energy loss spectroscopy; TEM: transmission electron microscopy; (MA)LDI-TOF MS: (matrix-assisted) laser desorption/ionization time-of-flight mass spectrometry (adapted from [53]).

the carbonaceous compounds retained on the outer surface of the zeolite crystals and which are soluble in CH_2Cl_2 can be recovered and their amount and composition determined using classical organic analysis techniques. It should be noted that only a small part of the total coke, often less than 5%, can be recovered in this second step.

- In the third step, the remaining carbonaceous compounds are liberated from the catalyst by dissolution of the zeolite, at room temperature, with a hydrofluoric acid solution (40%).
- In the last step, the bottom phase which contains the soluble coke fraction is recovered, then passed through a bed of sodium bicarbonate in order to neutralize the residual acidity. The soluble coke fraction obtained after CH_2Cl_2 evaporation is quantified and analysed. In certain cases, a fraction of coke remains insoluble (black particles), and is often difficult to be totally recovered for analysis.

In step 1, particular attention should be paid in order to avoid any change of coke composition. In fact, during the treatment under nitrogen flow, condensation or cracking of coke molecules can occur in addition to the desired desorption of reactant and product molecules [52]. However, these secondary reactions, which are generally slower than desorption, can be limited to a small extent by choosing short treatment times. Moreover, when there are oxidizing elements in the catalysts (e.g. Pt in hydroisomerization catalysts) any contact with air must be avoided. Possible changes in coke composition could also occur in step 3, due to the use of a strong acid solution. However, blank tests with samples of an inert solid (SiO_2) impregnated with compounds that are very reactive in acid catalysis: 1-tetradecene, 1-methyl naphthalene, etc. show that it is not the case. The GC of the compounds recovered in CH_2Cl_2 was identical to the GC of the starting materials [51]. This lack of reactivity may seem unexpected, but it should be noted that the temperature of treatment by HF is low, the contact time of the acid solution with the coke components very short and the contact area between mineral and organic phases very small.

The composition of the soluble part of coke, i.e. the complete distribution of the components, can be easily determined by coupling gas chromatography to mass spectrometry (GC/MS). Other techniques, such as elemental analysis, H and ^{13}C NMR, IR, chromatography (GC or HPLC), taking reference samples for comparison, etc., are often used for confirmation. In more complicated cases, chemical transformations of coke are sometimes tested. Thus, the degree of insaturation of coke molecules formed at low temperatures on a zeolite HFAU during isobutane/butene alkylation could be estimated by hydrogenation [32].

Unfortunately, the characterization of insoluble coke is much more restricted. Only the elemental composition and some information about the chemical nature, the size and the shape can be obtained. However, significant advances have been recently achieved with the use of LDI and (MA)LDI-TOF MS techniques, that were developed to characterize bulky molecules of polymers or bioorganic compounds.

In the case of zeolite catalysts, this limitation in the characterization of insoluble coke does not generally constitute an important handicap since most of the active sites are located within their inner pores, called micropores, but which are in reality nanopores. Indeed these micropores are constituted of nano-sized channels and cages with apertures of molecular size (generally lower than 1 nm). For example, FAU, the most used zeolite, has spherical cages of 1.3 nm diameter accessible through apertures of 0.74 nm. As a consequence, the growth of coke molecules is sterically limited by the walls of the nanopores, being the largest part of coke soluble in organic solvents. At low reaction temperatures (<600 K), all coke molecules are generally soluble; it is also the case at higher temperatures but only for low values of coke content. Moreover, it is important to note that generally the insoluble coke molecules result from the transformation of soluble coke molecules, which provides valuable information on their nature.

4.5.2. *Composition of soluble coke*

The application of the method is shown here in the example of n-butene skeletal isomerization carried out at 723 K, over a HFER zeolite disposed in a fixed-bed reactor [54]. The coke composition was determined for three values of coke content: 1.8, 6.5 and 7.5 wt%, corresponding to three values of time-on-stream (TOS): 0.5, 7.5 and 20 h. Only soluble coke was found in the two first samples, whereas in the third one, traces of insoluble coke were observed as small black particles. The chromatogram of the coke recovered for a TOS value of 7.5 h (Fig. 4.14) shows that coke is a very complex mixture: H NMR and IRTF analysis indicate the presence of a large proportion of aromatics.

All the coke components have been identified by GC/MS and could be classified in seven families of methyl aromatics (A to G). The weight distribution of these families was estimated from the area of the corresponding chromatographic peaks. Figures 4.15a and b show that the components of families A, B, C and E are present at low coke contents

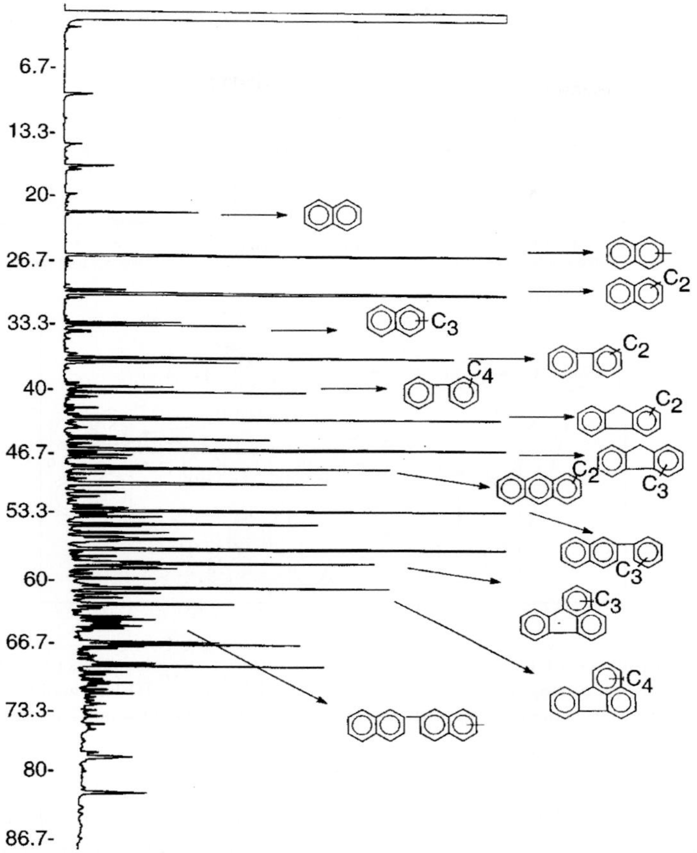

Fig. 4.14. Chromatogram of the carbonaceous compounds recovered from a HFER zeolite after *n*-butene transformation at 723 K for 7.5 h [54].

(primary products), while families D, F, G are found only at high coke contents. In agreement with a secondary mode of formation of these last coke components, a maximum can be observed in the number of molecules of A, B and C families. A scheme of coke formation was deduced from both the nature of coke molecules and the change in their number as a function of the coke content (Fig. 4.15c).

4.5.3. *Composition of insoluble coke*

As indicated above, the characterization of insoluble coke is often limited to its elemental composition. However, some techniques, such as solid-state

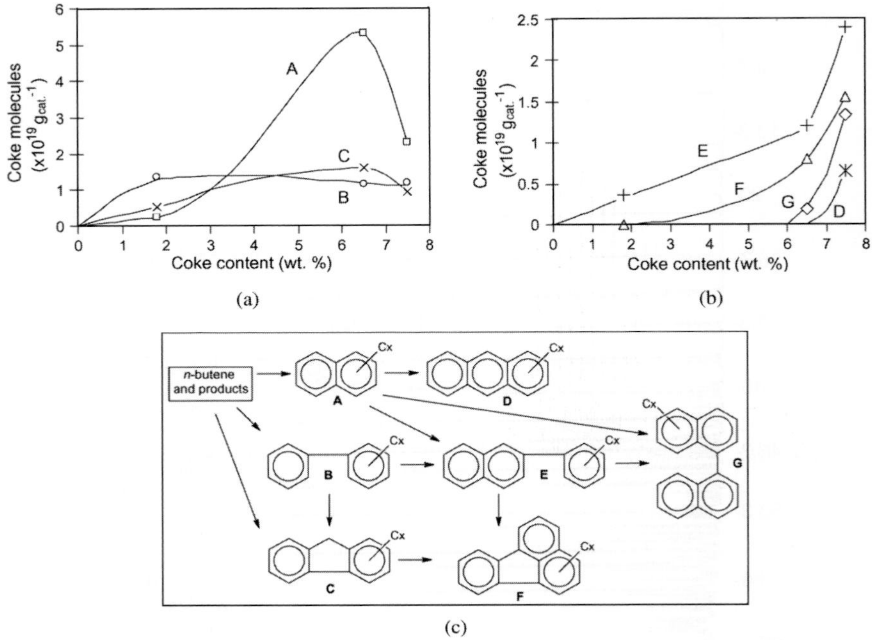

Fig. 4.15. Influence of the coke content on the number of coke molecules of various families produced during the transformation of n-butene on HFER zeolite at 623 K (a and b). Reaction scheme proposed for the formation of the different coke families (c) [54].

NMR ^{13}C, STEM-EELS and (MA)LDI-TOF MS, allow us to obtain more informative data. Thus, the use of a low-field mass spectrometer combined with the single-pulse excitation (SPE) technique has enabled us to obtain quantitative carbon skeletal parameters for FCC insoluble coke. Differences in feedstock composition were reflected in the coke structure: the coke produced by the cracking of atmospheric residue (15–20 cycles) presents more highly condensed polyaromatics than the coke formed from vacuum gas oil (8–12 cycles) [55].

The (MA)LDI-TOF MS technique was used to characterize the insoluble coke extracted by HF treatment from zeolite catalysts used in catalytic cracking [7, 56] or pre-coked during methanol conversion at 723 K [57]. The (MA)LDI-TOF MS of the insoluble coke recovered from a pre-coked HMFI (∼0.5 wt%) sample showed a wide molecular weight distribution from 202 Da to 1300 Da, with a sharp maximum at 363 Da and a broad peak at about 750 Da (Fig. 4.16). The mass increments of 24, 37 and

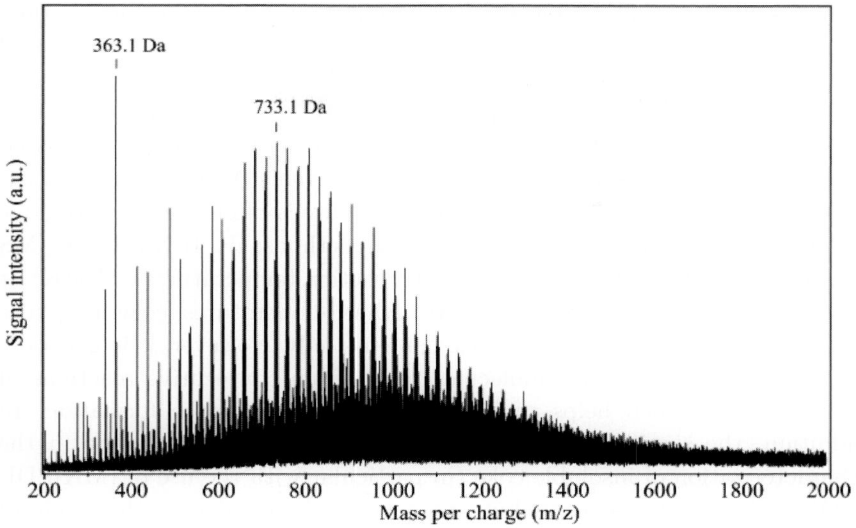

Fig. 4.16. (MA)LDI-TOF of insoluble coke recovered from a pre-coked HMFI sample [57].

50 Da which are repeatedly observed in the MS spectrum correspond to C_2, C_3H_1 and C_4H_2 entities, respectively, in agreement with the enlargement of the condensed rings. A simplified scheme was proposed to explain the growth of the polyaromatic coke components [57].

4.6. Conclusion

Many techniques were used to characterize the organic species (especially coke) responsible for the deactivation of zeolite catalysts. Generally, the characterization begins by the determination of the amount and elemental composition of coke through an oxidation treatment at very high temperature of the aged zeolite samples. Temperature-programmed oxidation (TPO), which is sometimes used for this purpose, is also very useful for specifying the optimal conditions of catalyst regeneration by coke combustion. Particular attention has to be paid in the interpretation of the data provided by these methods. On the other hand, spectroscopic techniques, which have the advantage of being non-destructive, give information about the chemical nature of coke. Unfortunately, this important information is not enough to establish the mode of coke formation. Only a destructive and relatively laborious method, requiring as a first step the acid dissolution of the zeolite, gives the coke composition, which is the key element to determining the scheme and mechanisms of formation of the complex mixture of coke molecules.

This method is well appropriate to zeolite catalysts (and adsorbents [58]), because most of the coke molecules are generally formed and located within the nanopores, and hence are relatively small. As a consequence, the main part of the coke separated from the zeolite is often soluble in organic solvents, hence easily analysed by classical methods: IR, H NMR, GC, MS and especially GC/MS coupling. At low reaction temperatures (<500 K), it is the case for the totality of coke. At higher temperatures, coke molecules are soluble for low or very low coke contents only. However, it is generally found that insoluble coke molecules result from consecutive reactions of soluble ones, which helps to deduce the scheme of coke formation. In addition, the significant advances that have been recently made in the characterization of insoluble coke (in particular with the use of (MA)LDI-TOF techniques) also contribute to a better knowledge of the mode of formation of high-temperature coke.

Therefore, the academic and industrial researchers working with zeolite-catalysed processes have now at their disposal a set of methods to characterize the species responsible for catalyst deactivation, hence to find ways to prevent or to limit their formation. When deactivation is mainly due to carbonaceous compounds ("coke"), which is generally the case in zeolite-catalysed processes (but also in separation processes [58]), the use of the only method which may give the coke composition, i.e. separation of coke from the zeolite and then characterization by organic analysis techniques, is essential. Nevertheless, a multi-technique approach associating this method with other ones, in particular with non-destructive spectroscopic methods, that are very efficient for specifying the interactions of coke molecules with the active sites, is strongly recommended. It should be underscored that the methods developed for the characterization of zeolite coke can be applied to non-microporous catalysts. However, severe limitations appear at high reaction temperatures because of the predominance from very low coke contents of bulky coke molecules, which are hence insoluble in organic solvents.

References

[1] Bartholomew C.H., *Appl. Catal. A: General*, 212 (2001) 17–60.

[2] Karge H.G., *Stud. Surf. Sci. Catal.*, 137 (2001) 707–746.

[3] Bauer F., Karge H.G., in *Molecular Sieves,* Karge H.G., Weitkamp J. (Eds.), Springer-Verlag, Berlin, Heidelberg, 5 (2006) 249–363.
[4] Guisnet M., Magnoux P., *Appl. Catal.*, 54 (1989) 1–27.
[5] Guisnet M., Magnoux P., *Stud. Surf. Sci. Catal.*, 88 (1994) 53–68.
[6] Guisnet M., Magnoux P., *Appl. Catal A: General,* 212 (2001) 83–96.
[7] Cerqueira H.S., Sievers C., Joly G., Magnoux P., Lercher J.A., *Ind. Eng. Chem. Res.*, 44 (2005) 2069–2077.
[8] Caeiro G., Magnoux P., Lopes J.M., Ramôa Ribeiro F., *Appl. Catal. A: General,* 292 (2005) 189–199.
[9] Weitkamp J., *Stud. Surf. Sci. Catal.*, 37 (1988) 515–523.
[10] Retcofsky H.L., Thompson G.P., Raymond R., Friedel R.A., *Fuel,* 54 (1975) 126–136.
[11] Marin G.B., Froment G.F., *Chem. Eng. Sci.*, 37 (1982) 759–766.
[12] Cerqueira H.S., Magnoux P., Martin D., Guisnet M., *Appl. Catal. A: General,* 208 (2001) 359–367.
[13] Dimon B., Cartraud P., Magnoux P., Guisnet M., *Appl. Catal. A: General,* 101 (1993) 351–359.
[14] Liu K., Fung S.C., Ho T.C., Rumschitzki D.S., *J. Catal.,* 169 (1997) 455–468.
[15] Chen D., Gronvold A., Rebo H.P., Moljord K., Holmen A., *Appl. Catal. A: General,* 137 (1996) L1– L8.
[16] Chen D., Rebo H.P., Moljord K., Holmen A., *Stud. Surf. Sci. Catal.*, 111 (1997) 159–166.
[17] Sánchez-Galofré O., Segura Y., Pérez-Ramírez J., *J. Catal.*, 249 (2007) 123–133.
[18] Pérez-Ramírez J., Gallardo-Llamas A., *Appl. Catal. A: General,* 279 (2005) 117–123.
[19] van Donk S., Bus E., Broersma A., Bitter J.H., de Jong K.P., *Appl. Catal. A: General,* 237 (2002) 149–159.
[20] Zhu W., Zhang J., Kapteijn F., Makkee M., Moulijn J.A., *Stud. Surf. Sci. Catal.,* 139 (2001) 21–28.
[21] Finelli Z.R, Querini C.A., Comelli R.A., *Appl. Catal. A: General,* 247 (2003) 143–156.
[22] Barbier J., Marecot P., Martin N., Elassal L., Maurel R., *Stud. Surf. Sci. Catal.*, 6 (1980) 53–65.
[23] Shamsi A., Baltrus J.P., Spivey J.J., *Appl. Catal. A: General,* 293 (2005) 145–152.
[24] Moljord K., Magnoux P., Guisnet M., *Cat. Letters,* 25 (1994) 141–147.
[25] Moljord K., Magnoux P., Guisnet M., *Cat. Letters,* 28 (1994) 53–59.
[26] Bayraktar O., Kugler E.L., *Appl. Catal. A: General,* 233 (2002) 197–213.
[27] Lange J.P., Gutsze A., Allgeier J., Karge H.G., *Appl. Catal.,* 45 (1988) 345–356.
[28] Karge H.G., Bolding E.P., *Catal. Today,* 3 (1998) 53–63.
[29] Fetting F., Gallei E., Kredel P., *Chem. Ing. Tech.*, 54 (1982) 606–607.
[30] Cerqueira H.S., Ayrault P., Datka J., Magnoux P., Guisnet M., *J. Catal.*, 196 (2000) 149–157.

[31] Laforge S., Martin D., Paillaud J.L., Guisnet M., *J. Catal.*, 220 (2003) 92–103.

[32] Pater J., Cardona F., Canaff C., Gnep N.S., Szabo G., Guisnet M., *Ind. Eng. Chem. Res.*, 38 (1999) 3822–3829.

[33] Guisnet M., Andy P., Gnep N.S., Benazzi E., Travers C., *J. Catal.*, 173 (1998) 322–332.

[34] Stair P.C., *Current Opinion in Solid State and Materials Science*, 5 (2001) 365–369.

[35] Chua Y.T., Stair P.C., *J. Catal.*, 213 (2003) 39–46.

[36] Li J., Xiong G., Feng Z., Liu Z., Xin Q., Li C., *Micropor. Mesop. Mater.*, 39 (2000) 275–280.

[37] Vedrine J.C., *Les techniques physiques d'études des catalyseurs*, Imelik B., Vedrine J.C. (Eds.), Technip, Paris (2000).

[38] Karge H.G., Laniecki M., Ziolek M., Onyestyak G., Kiss A., Kleinschmit P., Siray M., *Stud. Surf. Sci. Catal.*, 49 (1989) 1327–1334.

[39] Klingmann R., Josl R., Traa Y., Gläser R., Weitkamp J., *Appl. Catal. A: General*, 281 (2005) 215–223.

[40] Pazé C., Sazak B., Zecchina A., Dwyer J., *J. Phys. Chem. B*, 103 (1999) 9978–9986.

[41] Engelhardt G., *Stud. Surf. Sci. Catal.*, 137 (2001) 387–418.

[42] Behera B., Ray, S.S., *Catal. Today,* 141 (2009) 195–204.

[43] Philippou A., Dwyer J., Ghanbari-Siakhali A., Pazé C., Anderson M.N., *J. Mol. Catal. A*, 174 (2001) 223–230.

[44] Stepanov A.G., Luzgin M.V., Arzumanov S., Ernst H., Freude D., *J. Catal.,* 211 (2002) 165–172.

[45] Meinhold R.H., Bibby D.M., *Zeolites,* 10 (1990) 121–147.

[46] Karge H.G., Lange J.P., Gutsze A., Laniecki M., *J. Catal.,* 114 (1988) 144–152.

[47] de Jong K.P., Mooiweer H.H., Buglass J.G., Maarsen P.K., *Stud. Surf. Sci. Catal.,* 111 (1997) 127–138.

[48] Gallezot P., Leclercq C., Guisnet M., Magnoux P., *J. Catal.*, 114 (1988) 100–112.

[49] van Donk S., de Groot F.M.F., Stephan O., Bitter J.H., de Jong K.P., *Chem. Eur. J.*, 9 (2003) 3106–3111.

[50] Feller A., Barth J-O., Guzman A., Zuazo I., Lercher J., *J. Catal.,* 220 (2003) 192–206.

[51] Guisnet M., in *Handbook of Heterogeneous Catalysis,* Ertl G., Knözinger H., Weitkamp J. (Eds.), Wiley-VCH, Weinheim, 2 (1997) 626–632.

[52] Magnoux P., Roger P., Canaff C., Fouché V., Gnep N.S., Guisnet M., *Stud. Surf. Sci. Catal.,* 34 (1987) 317–334.

[53] Guisnet M., Costa L., Ramôa Ribeiro F., *J. Mol. Catal. A: Chemical*, 305 (2009) 69–83.

[54] Andy P., Gnep N.S., Guisnet M., Benazzi E., Travers C., *J. Catal.,* 173 (1998) 322–332.

[55] Snape C.E., Mcghee B.J., Andresen J.M., Hughes R., Koon C.L., Hutchings G., *Appl. Catal. A: General*, 129 (1995) 125–132.

[56] Barth J-O., Jentys A., Lercher J.A., *Ind. Eng. Chem. Res.*, 43 (2004) 2368–2375.
[57] Bauer F., Chen W.H., Bilz E., Freyer A., Sauerland V., Liu S.B., *J. Catal.*, 251 (2007) 258–270.
[58] Boucheffa Y., Joly G., Magnoux P., Guisnet M., Jullian S., *Stud. Surf. Sci. Catal.*, 139 (2001) 367–374.

Chapter 5

CHARACTERIZATION OF AGED ZEOLITE CATALYSTS

M. Guisnet

5.1. Introduction

Whatever the catalyst, two types of data are needed for understanding the deactivation process. The first concerns the cause(s) of deactivation, the second, the modification(s) that were undergone by the catalyst. The methods used to characterize deactivating species — poisons and especially coke — were presented in Chapter 4 on typical examples of reactions catalysed by zeolites. Those used to characterize the modifications that zeolite catalysts can undergo during the catalytic process are developed hereafter. These modifications can be chemical (e.g. neutralization of active sites by deactivating species), physical (e.g. blockage of the access to the zeolite micropores, degradation or even collapse of the pore structure) or mechanical (e.g. loss of active material by attrition or erosion of the particles).

The characterization of these modifications requires the comparison of the fresh and aged samples [1, 2]. An important preliminary remark is that great care has to be taken to avoid any transformation of the aged samples during the recovery (when necessary) from the reactor as well as during storage and pretreatment before characterization.

A large range of characterization techniques can be used: (i) simple weighing, e.g. for determining the loss in catalyst; (ii) physical adsorption (physisorption) of inert molecules, i.e. unable to react with the active sites and with the deactivating species, in order to estimate the limitation or blockage of the access to the zeolite micropores; (iii) chemical adsorption (chemisorption) of probe molecules on the active sites in order to quantify their residual concentration and/or the change of their characteristics (e.g. strength of the acid sites); (iv) physical methods, e.g. DRX, electron microscopy, IRTF, NMR, etc., which can provide information on the active sites or on the micropore system, etc. Note that the first three types of techniques have the advantages of being relatively simple to implement and do not require very expensive equipments.

In this chapter, examples of acid and bifunctional (hydrogenating/acid) transformations have been chosen to present the methods currently used to characterize the modifications of the zeolite catalysts caused by their deactivation.

5.2. Physisorption of Probe Molecules

The physical adsorption of inert molecules, which chemically interact neither with the catalyst nor with the deactivating species present on the aged catalysts, is the usual method employed to characterize the modifications of the zeolite micropore volume caused by deactivation. Experiments carried out at thermodynamic equilibrium allow the estimation of the total surface area and of the accessible volume of the different types of pores, especially that of the micropores. In addition, the kinetics of diffusion can be established, the obtained diffusivity values being of great use to specify limitations of the access to the micropore system. The physisorption of probe molecules is generally followed by microgravimetry. Pulsed gradient 1 H NMR technique (PFG NMR) was also largely used, e.g. by Karger *et al.* [3, 4], to determine the diffusion coefficient of methane on fresh and coked zeolites. A detailed discussion of this work was reported in Ref. [1].

As is the case with the other solid catalysts, nitrogen is classically used as an adsorbate, generally at 77 K, i.e. at its boiling point under atmospheric pressure. Alkane and aromatic hydrocarbons, often chosen with molecular sizes close to those of the reactants, are also sometimes employed. With these adsorbates, the aim is to get information on pore limitation or blockage that could be extrapolated to the reactant molecules under the reaction conditions. Note, however, that this is difficult to satisfy, physical adsorption being generally implemented at a temperature much lower than that of the reaction.

5.2.1. *Determination of coke location*

Many authors have carried out adsorption experiments over zeolite catalysts deactivated by coke deposition with as the principal aim the determination of coke location. Some typical examples are reported hereafter.

The first one deals with the deactivation of a HFAU (USHY) during m-xylene transformation carried out in a fixed-bed reactor at 523 and 723 K [5]. Catalyst samples were recovered for different values of time-on-stream. These samples were divided into several parts, serving respectively for the determination of the coke content and composition, of their acidity (pyridine adsorption, see Section 5.3.1) and of their adsorption properties (nitrogen adsorption at 77 K).

Figures 5.1a and b show that whatever the reaction temperature, coke affects the micropore volume much more than the mesopore volume. However, while the accessibility of mesopores is not affected by coke deposition at 723 K, a small decrease in the mesopore volume can be observed at 523 K. These observations suggest that coke formed is essentially deposited within the micropores. Another difference deals with

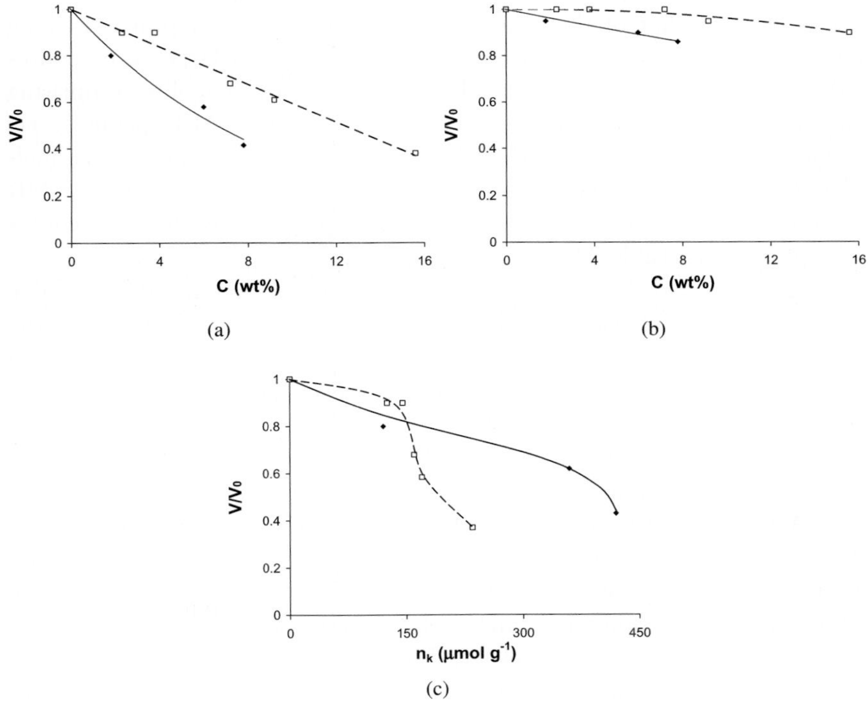

Fig. 5.1. m-Xylene transformation over a HFAU zeolite at 523 (♦) and 723 K (□). Effect of the percentage of coke (wt% C) on the zeolite on the relative micropore (a) and mesopore (b) volumes accessible to nitrogen (V/V_0); effect of the number of coke molecules n_k (μmol g^{-1} of zeolite) on the micropore V/V_0 value (c). Adapted from Ref. [5].

the decrease in the access of nitrogen to micropores that seems more pronounced at 523 K than at 723 K. The origin of these differences can be clarified from the coke compositions [6] and from derived data such as the number of coke molecules. Thus, by substituting in abscissa the amount of coke by the number of coke molecules, practically no difference was found in the effect on the mesopore volume; moreover, at both temperatures, the effect per coke molecule was very small [5]. In what concerns the micropore volume (Fig. 5.1c), the situation is more complicated with a similar effect at coke contents lower than 3 wt%, and above this value, a more pronounced effect for the coke formed at 723 K. This more pronounced effect is typical of a pore blockage, as it was furthermore confirmed by the appearance in coke of very bulky molecules.

In the second example, the effect of coke deposited during n-heptane cracking at 723 K was compared over protonic zeolites with different pore structures, by adsorption of nitrogen and of n-hexane whose molecular size is close to that of the n-heptane reactant [7, 8]. n-Heptane cracking was first carried out over the following zeolites: two large-pore zeolites (HFAU [USHY] with large cages [Fig. 2.1a], HMOR, a zeolite comprising interconnected channels with respectively 12-MR and 8-MR openings; one medium-pore zeolite (HMFI [HZSM5] with 10-MR interconnected channels [Fig. 2.1b]); and one small-pore zeolite (HERI with large cages with 8-MR openings). The characteristics of the pore systems of MOR and ERI zeolites are presented in Table 2.2.

Figures 5.2a–d show the change in the micropore volume (V) accessible to nitrogen and n-hexane vs. the amount of coke (wt% C). With all the zeolites, coke deposition causes a decrease in both the access of nitrogen and n-hexane to the zeolite micropores. However, large differences can be observed from one zeolite to another.

With HMFI and HERI zeolites (Figs. 5.2b and d), the effect of coke on the micropore volume accessible to n-hexane ($V_{n\mathrm{C6}}$) is the same as on V_{N2}. With HFAU, the effect is slightly greater with n-hexane whereas with HMOR (Fig. 5.2c), large differences can be observed with a continuous and pronounced decrease of $V_{n\mathrm{C6}}$ instead of the plateau followed by a decrease found for V_{N2}. This difference can be related to the possibility for nitrogen to diffuse in both the large channels (12-MR, 6.5×7.0 Å) and narrow channels (8-MR, 2.6×5.7 Å) while n-hexane (and n-heptane) molecules cannot enter the narrow channels as furthermore shown by the smaller value of $V_{n\mathrm{C6}}$ on the fresh zeolite. From the significance of the effect of coke on the V values, it can be concluded that there is pore blockage with HERI and HMOR

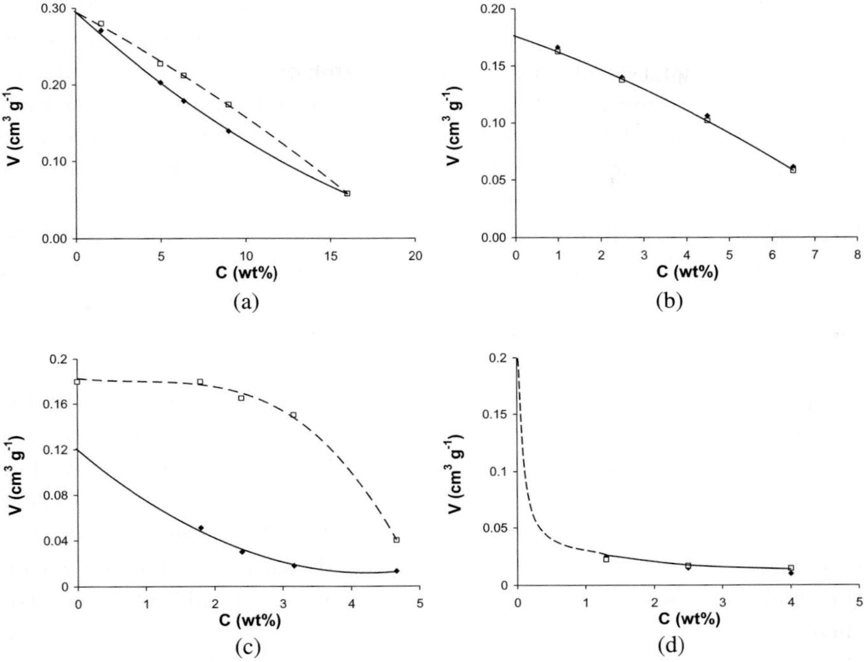

Fig. 5.2. Effect of coke (wt% C) deposited during n-heptane cracking at 723 K over four protonic zeolites on the micropore volume V (cm^3 g^{-1}) accessible to nitrogen (□) and to n-hexane (♦): HFAU (a), HMFI (b), HMOR (c) and HERI (d). Adapted from Ref. [7].

(limited to the large channels) and regular distribution of coke within the micropores of HFAU and HMFI [7, 8].

As shown in the two examples reported above, it is now well-established that in most cases, "coke" is located within the zeolite micropores rather than on the surface of the crystals and of the mesopores. This originates from the location of most of the active sites within the micropores as well as from the easier retention of coke molecules. Obviously, this conclusion is only valid when the reactant molecules can enter the micropores as demonstrated by Uguina *et al.* [9] by n-butane adsorption measurements over a MFI zeolite coked with hydrocarbons of different sizes: isobutene, toluene, which can enter the zeolite channels, and mesitylene, which is too bulky to enter them. As could be expected, coke formed from isobutene was deposited within the micropores, that originating from mesitylene on the external zeolite surface whereas coke from toluene was deposited both on the external and the internal surface.

With the above zeolites, a sole adsorbate with a molecular size close to that of the reactant was enough to get valuable information on the location of coke (within the micropores and/or on the outer surface) and on its distribution within the zeolite crystals. However, the situation becomes more complex when the zeolites have several micropores accessible by reactant molecules (e.g. OFF [10]). In this case, several adsorbates must be used. The large-pore OFF zeolite has two types of micropores: large channels (12-MR, 6.3 Å) interconnected through gmelinite cages with small openings (8-MR, 3.6×4.8 Å). The molecules of linear alkanes, such as n-hexane or n-heptane, can enter both channels whereas the bulkiest molecules, such as branched alkanes (e.g. 3-methyl pentane), can diffuse through the large channels only.

Figure 5.3 shows the effect of the amount of coke deposited on a HOFF sample during n-heptane transformation at 723 K on the pore volume accessible to n-hexane ($V_{n\text{C}6}$) and to 3-methyl pentane ($V_{3\text{mp}}$). At low coke contents (<2 wt%, Fig. 5.3, part A), coke has practically no effect on $V_{3\text{mp}}$ (large channels) but causes a decrease in $V_{n\text{C}6}$ (large channels + gmelinite cages); therefore, it can be concluded that coke is located within the gmelinite cages only. With coke contents between 2 and 5 wt% (part B), there is a decrease in both V values, $V_{3\text{mp}}$ being more affected than $V_{n\text{C}6}$ and becoming very small; this observation associated with the effect of coke

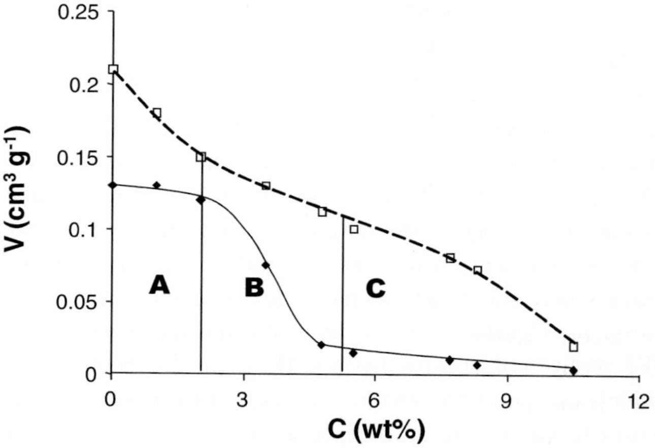

Fig. 5.3. Effect of coke (wt% C) deposited during n-heptane cracking at 723 K over a HOFF zeolite on the micropore volume V ($\text{cm}^3\ \text{g}^{-1}$) accessible to n-hexane (□) and to 3-methylpentane (♦). Adapted from Ref. [10].

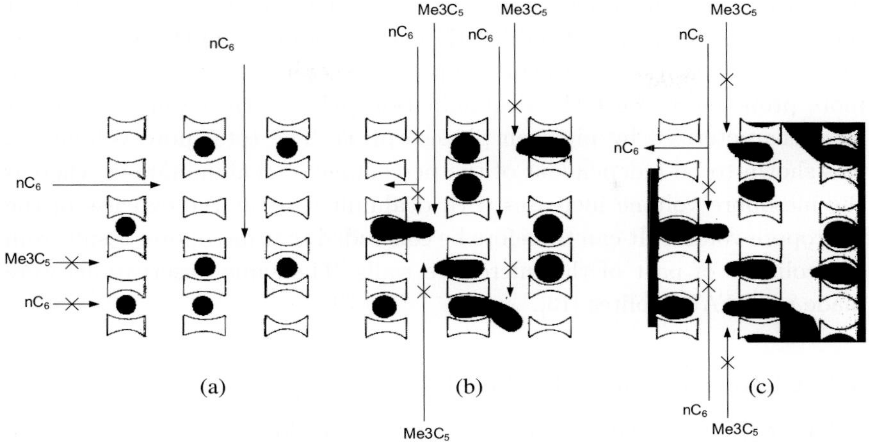

Fig. 5.4. Schematic representation of the location of coke molecules in the large channels and in the gmelinite cages of the HOFF zeolite. A, B and C correspond to the domains of coke content indicated in Fig. 5.3. Adapted from Ref. [11].

on the values of the diffusion rates of n-hexane and 3-methyl pentane along the large channels suggests that this coke limits (nC_6) or blocks (3 mp) the access to the large channels. With higher coke contents (part C), the V values decrease to practically zero, which can be related to a complete blockage of the access to the micropores owing to the coverage of the outer surface by highly polyaromatic coke molecules [11].

As shown in Fig. 5.4, this successive decrease in the access to gmelinite cages then to the large channels is in perfect agreement with the change in coke composition with the zeolite coke content [10, 11].

5.2.2. *Characterization of framework alterations*

Most of the alterations undergone by zeolite catalysts do not occur during the reaction but during the regeneration process. Indeed, regeneration of aged zeolite catalysts generally requires coke combustion which is carried out under severe conditions, namely high temperature in the presence of steam. In the following example, the effect of steaming conditions (e.g. of conditions similar to those observed during coke combustion) was shown on the porosity of a chosen zeolite: EMT, the hexagonal analogue of the FAU zeolite. The hydrothermal treatment of four NH_4NaEMT samples (Si/Al of 4) differing by the percentages of sodium exchange with ammonium cations (from 70% to ~100%) was carried out in a flow reactor in the presence of water (partial pressure of 93.5 kPa) at temperatures between 623 and 923 K. The more significant the Na exchange rate, the higher the treatment temperature, or the longer the time, the more pronounced the EMT dealumination [12]. All the resulting samples were characterized by nitrogen physisorption. The total pore volume V_t, was shown to be independent of the percentage of dealumination, whereas the mesopore volume increases with dealumination at the expense of the micropore volume. It can therefore be concluded that mesopores result from the collapse of part of the micropore walls. The same observations were made with FAU zeolites [13].

5.3. Chemisorption of Probe Molecules

5.3.1. *Effect of deactivation on zeolite acidity*

Various techniques can be used to determine the effect of deactivation on the acidity of zeolite catalysts, most of them being based on chemisorption of basic probe molecules over fresh and aged samples. NH_3 TPD and pyridine chemisorption followed by IR spectroscopy are most used to characterize the acidity of fresh or deactivated zeolites, and the origin of deactivation: poisoning or pore blockage.

Characterization through IR spectroscopy presents many advantages, leading to valuable quantitative data on both the Brønsted and Lewis acid sites, which is not the case with NH_3 TPD. In addition, the effect of deactivation on the Brønsted acid sites can be established by direct (i.e. without adsorption of probe molecules) analysis of the samples in the region of hydroxyl groups (3300–3800 cm^{-1}). However, no information on the Lewis acidity can be obtained this way, because of the absence of a specific band for the Lewis acid sites. It should be noted that with this method, only IR spectroscopy (i.e. a physical technique) is used to compare the intensity of the OH groups of the fresh and deactivated catalyst samples, which is relevant to Section 5.4. Despite that, the corresponding results will be presented here, for reasons of convenience, with those obtained from probe molecule chemisorption. A last advantage of IR spectroscopy is that catalyst deactivation, coke formation and change in acidity can be simultaneously analysed by using an IR cell reactor coupled to a chromatograph (or to a mass spectrometer) [14].

The HFAU samples deactivated during m-xylene transformation at 523 and 723 K [5], on which coke location was previously established by N_2 physisorption (Section 2.1), were characterized by pyridine adsorption at

423 K followed by IRTF. In the OH region, the spectrum was complex comprising six bands, two at 3525 and 3550 cm^{-1} which correspond to the bridging hydroxyl groups located in hexagonal prisms, two others at 3600 and 3625 cm^{-1} which correspond to bridging hydroxyl groups of the supercages, one at 3665 cm^{-1} ascribed to hydroxylated EFAL species, the last one being the silanol band at 3735–3740 cm^{-1}. Whatever the reaction temperature, coke deposits preferentially affected the bands at 3600 and 3525 cm^{-1} ascribed to bridging hydroxyls whose acid strength was exalted by interaction with EFAL species presenting Lewis acidity. This suggests that coke molecules were preferentially located near the strongest acid sites which are likely responsible for their formation [5].

The concentrations of Brønsted and Lewis acid sites able to adsorb pyridine were estimated from the intensities of respective bands at 1545 (pyridinium ions) and 1450 cm^{-1} (pyridine coordinated to Lewis sites). At both reaction temperatures, coke causes a decrease of the band at 1545 cm^{-1}, hence in the number of protonic sites (Fig. 5.5a) but, up to 8–10 wt% coke, has no effect on the band at 1450 cm^{-1}, hence on the Lewis acid sites (Fig. 5.5b). This demonstrates that Lewis sites do not intervene directly in coke formation, which is generally accepted at least during the hydrocarbon transformations. The apparent decrease of Lewis acidity observed at high coke content (Fig. 5.5b) is probably due to a partial blockage of micropores making impossible the diffusion of pyridine molecules to inner Lewis sites [5].

An important remark is that pyridine adsorption (and/or the pyridine evacuation before IRTF analysis) can cause desorption of certain

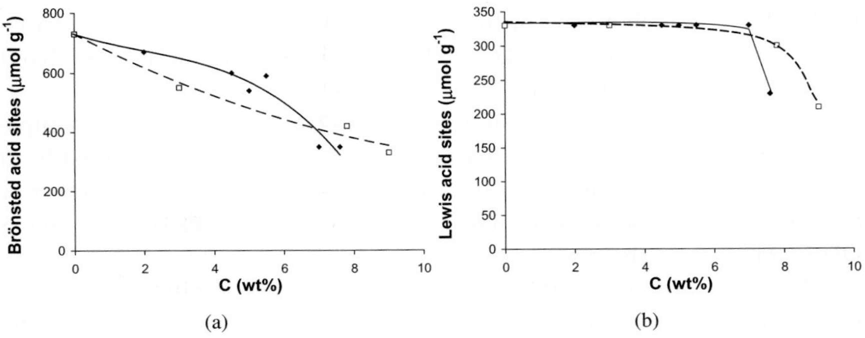

Fig. 5.5. Effect of coke (wt% C) deposited during *m*-xylene transformation over a HFAU zeolite at 523 (♦) and 723 K (□) on the concentration of Brønsted (a) and Lewis (b) acid sites (μmol g^{-1}). Adapted from Ref. [5].

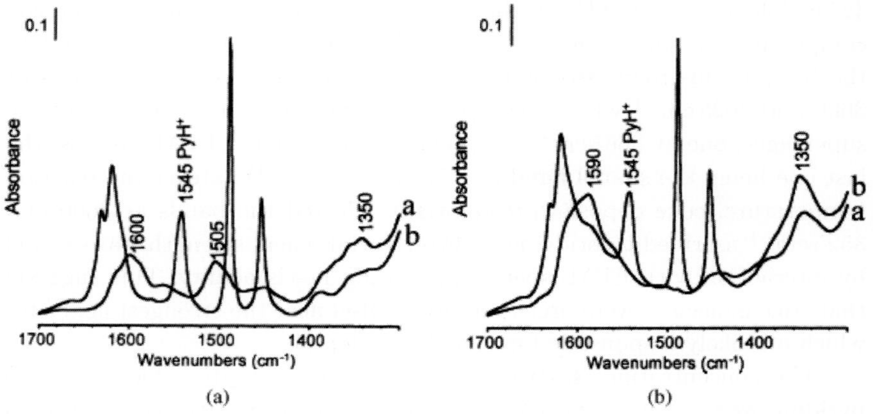

Fig. 5.6. Pyridine sorption in the zeolite with coke formed at (A) 520 K and (B) 720 K. (a) spectrum of the zeolite with coke, (b) spectrum of the zeolite with coke and sorbed pyridine [6].

coke molecules from the acidic sites, which distorts the quantitative determination of the effect of coke on acidity. This is clearly shown by the decrease of the low T (523 K) coke bands at 1350, 1505 and 1590–1600 cm^{-1} which do not overlap with the bands of pyridine (Fig. 5.6) and furthermore confirmed by elemental analysis of the coked samples before and after pyridine adsorption [6]. As could be predicted from the weaker basicity of coke molecules formed at low temperatures, this desorption is more pronounced for coke formed at 523 K than at 723 K. This coke desorption explains why the initial effect of coke on the Brønsted acidity is smaller at 523 K [5]. Owing to the large diversity of bridging OH groups in the HFAU sample tested in Ref. [5], no quantitative comparison can be made between the effect of coke on the intensities of the pyridinium band and of the bridging OH bands.

This comparison was made in the case of a HMWW (HMCM22) sample coked during m-xylene transformation at 623 K; as could be expected from the desorption of coke molecules caused by pyridine adsorption, the effect of coke was found to be more significant on the intensity of the only bridging OH band than on the pyridinium band [15].

The mode of zeolite deactivation — poisoning or pore blockage — can be drawn from the comparison between the effect of coke on the catalyst activity and on the intensity of the bridging OH band (or of the pyridinium band in the absence of coke desorption by pyridine, e.g. for high-temperature coke). Thus, as the activity of a HMOR zeolite in ethylbenzene transformation was shown to decrease very rapidly with time-on-stream while the intensity of the bridging OH band remained unchanged, deactivation was concluded to occur through pore blockage [16]. Note that as shown in Section 5.2.1, this mode of deactivation could also be proven by physisorption experiments.

Other basic probe molecules were used to characterize coked zeolite samples, their choice being often imposed by a size of micropore openings too small to allow the internal diffusion of pyridine molecules. Thus, ammonia was first used instead of pyridine for determining the effect of coke deposited during the skeletal isomerization of n-butenes at 623 K over HFER zeolites whose openings of the large channels are of only 4.2×5.4 Å [17]. However, with this probe molecule, the effect of coke on acidity is difficult to be quantified because of overlapping of the δ_{as} NH_4^+ band with the bands of coke and because of its low sensitivity. These problems do not appear with deuteroacetonitrile, which in addition, allows the characterization of Lewis sites, including secondary and tertiary carbenium ions, and has a molecular size closer to those of butene molecules and a basicity too low to displace the adsorbed coke molecules [18, 19].

5.3.2. *Effect of deactivation on the dispersion of supported noble metals*

Selective chemisorption (and desorption) of gas molecules, generally H_2, CO and O_2, is currently used to estimate the dispersion of supported metals, i.e. N_s/N_t, the ratio between the number of surface metal atoms and the total number (surface + bulk). The main problem with this method originates from the value of the adsorption stoichiometry to be used for a given gas-metal system [20]. Thus, CO can be chemisorbed dissociatively (CO: M ≥ 2) or associatively in linear (1), bridged (2) or capped forms (3). Furthermore, in H_2 chemisorption, values different from the expected stoichiometry of 1 can be obtained related to particularities of the metal/support system such as strong metal support interactions, spillover from the metal to the support, formation of bulk hydride such as with Pd, or the presence of contaminants [20]. Moreover, a significant effect of the ionicity of the support was recently shown with a series of Pt FAU zeolites [21]: the H:Pt ratio significantly decreased with the electron richness of the support oxygen atom -1.96 (Pt/NaY) >1.74 (Pt/MgY) >1.51 (Pt/LaY) >1.11 (Pt/H-USY).

It is therefore recommended to use the H:M ratio only to compare catalysts with the same support, which can be the case when comparing deactivated and fresh samples. Thus, the sulphur tolerance of Pt and Pd supported on a high silica FAU zeolite was estimated by the ratio of the metal dispersions (drawn from CO chemisorption by supposing a CO:M stoichiometry of 1) on the catalyst sample after and before the sulfuration treatment. From the obtained dispersion values, a negative effect of the increase in size of the metal particles on the S tolerance was demonstrated, this effect being more marked with Pd than with Pt [22].

H_2 chemisorption is often substituted by hydrogen-oxygen titration which has the advantage of being three times more sensitive. However, this method presents similar drawbacks, even if a procedure, which was recommended 20 years ago [23] and is generally used, eliminates for a large part the artefacts due to the strong dependence of the results on the operating conditions.

5.4. Physical Methods

Most of the physical techniques used to characterize the solid catalysts (including zeolites) may be classified according to their ability to absorb, emit or scatter photons (e.g. NMR, FTIR, EXAFS,[1] XRD, etc.), electrons (e.g. XPS,[2] EELS,[3] TEM, etc.) neutrons or ions (SIMS,[4] etc.). Many of them can be used to characterize the deactivating species and/or the alterations of zeolites caused by deactivation.

This is the case with IR spectroscopy largely employed to characterize coke molecules formed on zeolite catalysts (Chapter 4) as well as the related decrease in acidity, either directly from the bridging OH groups in the spectra of coked zeolites or indirectly from the bands corresponding to probe base molecules chemisorbed on the acid sites (Section 5.3.1). This spectroscopic technique is currently used *in situ* and even sometimes in *operando* mode.

NMR techniques are also largely employed in deactivation studies, because of the many nuclei that can be involved and the possibility to get information on various characteristics of aged zeolite catalysts: (i) the composition of the zeolite framework (^{27}Al and ^{29}Si nuclei), (ii) the

[1] Extended X-ray absorption fine structure
[2] X-ray photoelectron spectroscopy
[3] Electron energy-loss spectroscopy
[4] Secondary ion mass spectroscopy

molecular diffusion within the micropores (pulsed field gradient ^{1}H NMR), iii) the free intracrystalline volume of coked zeolites (^{129}Xe NMR, iv) the acid sites (^{31}P NMR of adsorbed alkyl phosphine oxide molecules), v) the nature of coke molecules (^{13}C NMR) [1].

During deactivation, the composition and electronic surface of the catalysts may significantly change, which can be quantitatively estimated by EXAFS analysis. In this technique, a high-energy X-ray source is transmitted through the sample yielding structural information on the surface on an atomic scale. EXAFS was used to compare the Pt clusters incorporated in a KLTL zeolite by either vapour-phase or incipient-wetness impregnation (VPI and IWI samples). For identical Pt contents, the first shell Pt-Pt and Pt-O coordination numbers (N) were respectively lower and higher on the VPI samples than on the IWI samples (e.g. for 1 wt% Pt, $N_{Pt\text{-}Pt}$ and $N_{Pt\text{-}O}$ were equal to 3.0 and 0.6 with VPI, and to 3.9 and 0.4 with IWI). This indicates that the Pt particles were smaller on VPI than on IWI samples [24]. As a high initial dispersion is beneficial to the activity and stability of the aromatization Pt/KLTL catalysts, vapour-phase impregnation will be preferred for their preparation (Chapter 16). EXAFS can also be used to estimate the sulphur tolerance of noble metals supported on zeolites [22].

Metal dispersion can be estimated from the first shell coordination numbers provided the particle size distribution is not too broad and/or not bimodal [22]. Therefore the only reliable way to determine metal dispersion is through the direct observation of the catalytic surface by electron microscopy. It should however be underscored that this technique samples only a small fraction of the catalyst, and hence that getting data representative of the entire sample requires the statistic analysis of many different areas.

Modern electron microscopy is capable of much more than producing images of particle morphology. An example shortly presented in Chapter 4 deals with the characterization of zeolite coke by electron energy-loss spectroscopy performed in a scanning transmission electron microscope (STEM-EELS). The first use of this technique [25] was limited to the characterization of highly polyaromatic coke, a large part of which was located on the outer surface of zeolite crystals. However, as a result of technical developments, information on spatial coke distribution can now be obtained within zeolite crystals as shown by STEM-EELS experiments carried out over two FER samples aged during skeletal *n*-butene isomerization [26]. Carbonaceous compounds were shown to be deposited throughout the zeolite. However, their concentration was not uniform with a preferential accumulation of coke at the entrances of the 8-MR channels. At higher coke content, the 8-MR pores were completely blocked while the 10-MR channels remained partially accessible with alkylaromatic species deposited near their entrances. This work clearly confirms the potential of this technique to determine simultaneously the location, amount and nature of carbonaceous deposits in zeolite crystals.

5.5. Conclusion

Significant progress has been made in the understanding of deactivation phenomena, thanks to the large range of chemical, physicochemical and physical techniques which can be used to characterize the features of fresh and aged catalysts.

Except when these changes are characterized in *operando* mode (which is still uncommon), the primary difficulty is to avoid any catalyst modification when the aged sample passes from the conditions of the catalytic reaction to those, often very different, of the characterization. Obviously, the risk of modification is still more important when the catalyst samples are not analysed *in situ*. The conditions of sampling, storing and pretreatment before analysis then have to be very carefully selected. It is also advisable to characterize a reasonable number of aged samples, some of them having been operated in similar conditions, others in different conditions.

A multi-technique approach is generally indispensable for characterizing all the catalyst changes undergone during deactivation, which as was already underlined generally, result from several causes. Furthermore, the data, even those originating from well-established methods such as hydrogen titration, used to estimate metal dispersion have to be critically examined; indeed the assumptions on which their interpretation is based are not always verified.

References

[1] Bauer F., Karge H.G., in *Molecular Sieves*, Springer-Verlag, Berlin, Heidelberg, 5 (2006) 249–363.

[2] Guisnet M., Cerqueira H.S., Figueiredo J.L., Ramôa Ribeiro F., in *Desactivação e Regeneração de Catalisadores*, Guisnet M. *et al.* (Eds.), Fundação Calouste Gulbenkian, Lisboa (2008) Chapter 4, 55–77.

[3] Kärger J., Hunger M., Freude D., Pfeifer H., Caro J., Bülow M., Spindler H., *Catal. Today*, 3 (1988) 493–499.

[4] Caro J., Jobic H., Bülow M., Kärger J., Hunger M., *Adv. Catal.*, 39 (1993) 351–361.
[5] Cerqueira H.S., Ayrault P., Datka J., Magnoux P., Guisnet M., *J. Catal.*, 196 (2000) 149–157.
[6] Cerqueira H.S., Ayrault P., Datka J., Guisnet M., *Microporous and Mesoporous Materials*, 38 (2000) 197–205.
[7] Magnoux P., Cartraud P., Mignard S., Guisnet M., *J. Catal.*, 106 (1987) 242–250.
[8] Guisnet M., Magnoux P., *Appl. Catal.*, 54 (1989) 1–27.
[9] Uguina M.A., Serrano D.P., van Grieken R., Ve'nes S., *Applied Catalysis A: General*, 99 (1993) 97–113.
[10] Mignard S., Cartraud P., Magnoux P., Guisnet M., *J. Catal.*, 117 (1989) 503–511.
[11] Magnoux P., Guisnet M., Mignard S., Cartraud P., *J. Catal.*, 117 (1989) 495–502.
[12] Berreghis A., Morin S., Magnoux P., Guisnet, le Chanu V., Kessler H., *J. Chim. Phys.*, 93 (1996) 1525–1542.
[13] Morin S., Ayrault P., Gnep N.S., Guisnet M., *Appl. Catal. A: General*, 166 (1998) 281–292.
[14] Vimont A., Marie O., Gilson J.P., Saussey J., Thibault-Strarzyk F., Lavalley J.C., *Stud. Surf. Sci. Catal.*, 126 (1999) 147–154.
[15] Laforge S., Martin D., Paillaud J.L., Guisnet M., *J. Catal.*, 220 (2003) 92–103.
[16] Karge H.G., Boldingh E.P., *Catal. Today*, 3 (1988) 379–386.
[17] Andy P., Gnep N.S., Guisnet M., Benazzi E., Travers C., *J. Catal.*, 173 (1998) 322–332.
[18] van Donk S., Bus E., Broersma A., Bitter J.H., de Jong K.P., *J. Catal.*, 212 (2002) 86–93.
[19] Ménorval B., Ayrault P., Gnep N.S., Guisnet M., *J. Catal.*, 230 (2005) 38–51.
[20] Bergeret G., Gallezot P., *Handbook of Heterogeneous Catalysis*, Ertl G., Knözinger H., Weitkamp J., Schuit G.C.A. (Eds.), Wiley-VCH Verlag, Section 3.1 (2008), 738–765.
[21] Ji Y., van der Eerden A.M.J., Koot V., Kooyman P.J., Meeldijk J.D., Weckhuysen B.M, Koningsberger D.C., *J. Catal.*, 234 (2005) 376–384.
[22] Matsui T., Harada M., Ichihashi Y., Bando K.K., Matsubayashi N., Toba M., Yoshimura Y., *Appl. Catal., A: General*, 286 (2005) 249–257.
[23] O'Rear D.J., Löffler D.G., Boudart M., *J. Catal.*, 121 (1990) 131–140.
[24] Jacobs G., Ghadiali F., Pisanu A., Borgna A., Alvrez W.E., Resasco D.E., *Appl. Catal. A: General*, 188 (1999) 79–98.
[25] Gallezot P., Leclercq C., Guisnet M., Magnoux M., *J. Catal.*, 114 (1988) 100–111.
[26] van Donk S., de Groot F.M.F., Stephan O., Bitter J.H., de Jong K.P., *Chem. Eur. J.*, 9 (2003) 3106–3111.

Part III: Deactivation Mechanisms

Chapter 6

POISONING OF ZEOLITE CATALYSTS

M. Guisnet

6.1. Introduction

Whatever their nature: acid, base, metal, etc., the active sites of zeolite catalysts can be poisoned during the reactions with, as a consequence, a decrease in activity and often in selectivity. As underscored in Chapter 1, this deactivation results from the strong chemisorption on the catalytic sites of species, generally present as impurities in the feed but which sometimes can also be desired or secondary products. The significance of the poisoning effect depends on the relative strength of chemisorption of these species and of the reactant molecules [1–3]. When desired products compete for adsorption with reactant molecules, the reaction is self-inhibited, i.e. its rate decreases with increasing conversion. This self-inhibition which is frequently observed in the synthesis of very polar products such as organic chemicals is presented in Chapter 17. Furthermore, the secondary carbonaceous products (coke) can act as poisons of the active sites, competing with reactant molecules for chemisorption on the active sites. They can also block the access to micropores, making their catalytic sites apparently inactive. These modes of deactivation by coking — site poisoning and pore blockage – are presented in Chapter 7.

Therefore, only poisoning by feed impurities will be considered in this chapter, the focus being placed on the most common poisons of acid and noble metal zeolite catalysts, i.e. nitrogen-containing bases and sulphur-containing compounds respectively. Three main parameters [1] will be used to characterize the poisoning and deactivation of these catalysts. The first one, the toxicity of the poison (*Tox*), is defined by the number of active sites deactivated per molecule of poison chemisorbed. As the number of molecules of poison introduced on the catalyst is much easier to establish, the toxicity is sometimes given per molecule introduced, which is quite similar in the case of irreversible adsorption under the operating conditions. The second parameter, called tolerance, is the catalyst activity at saturation coverage: A_{sat}; in many cases, the tolerance value is close to zero. The third one, the selectivity of poisoning, will be defined in the text.

6.2. Deactivating Effects of Poisons

6.2.1. *Site poisoning*

The primary effect of poison species is to prevent the access of reactant molecules to the active sites on which they are chemisorbed. The proportion of poisoned sites is determined by the competition between reactant (A) and poison (P) molecules, and hence depends on their concentrations (or pressures) and on their relative chemisorption strengths [1, 4].

Consider the simple catalytic reaction: A $\rightarrow$ products with the surface reaction as the rate-determining step (which is often the case) and a catalyst with active sites having identical properties. The fraction θ_{P} of sites occupied by the poison molecules can be calculated using a Langmuir isotherm:

$$\theta_{\text{P}} = \frac{\lambda_{\text{P}} \cdot p_{\text{P}}}{1 + \lambda_{\text{A}} \cdot p_{\text{A}} + \lambda_{\text{P}} \cdot p_{\text{P}}}. \tag{6.1}$$

where λ_{A} and λ_{P} are the equilibrium chemisorption constants for A and P, and p_{A} and p_{P}, their partial pressures.

With structure-insensitive reactions [5], i.e. for which the rate is proportional to the concentration of active sites, the catalytic activity remaining is proportional to $1-\theta_{\text{P}}$, the fraction of unblocked sites (non-selective poisoning), then to:

$$\frac{1 + \lambda_{\text{A}} \cdot p_{\text{A}}}{1 + \lambda_{\text{A}} \cdot p_{\text{A}} + \lambda_{\text{P}} \cdot p_{\text{P}}}. \tag{6.2}$$

This equation shows that the deactivating effect of P is larger than the values of the equilibrium chemisorption constant and the partial pressure of the poison will be higher than that of the reactant.

The susceptibility to poisoning of structure-sensitive reactions (i.e. reactions which require several sites for their catalysis [5]) will be greater than that of structure-insensitive reactions ("selective" poisoning). This selective poisoning also occurs when the active sites of the catalyst are not identical, which is often the case for the protonic sites of the acid

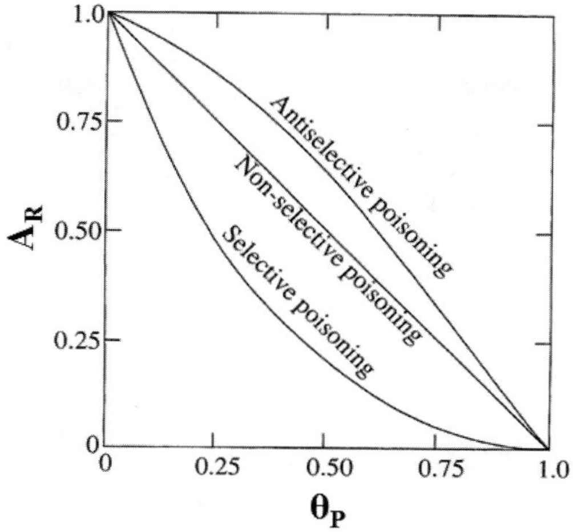

Fig. 6.1. Selective, non-selective and anti-selective poisoning: Effect of the fraction of sites occupied by poison molecules (θ_P) on the catalyst activity (A_R) [1].

zeolite catalysts. Indeed, very often, there is a relatively large distribution in strength (hence in activity) of these sites (Chapter 2). The poison molecules preferentially adsorb and block the stronger, hence more active acid sites, with as a consequence a very significant initial decrease in activity with θ_P. This selective poisoning also appears when the acid active sites can be in different locations, e.g. in micropores with different pore openings, hence with different accessibility by the poison and reactant molecules. Selective and non-selective poisoning are shown in Fig. 6.1. Another possibility, called anti-selective poisoning (Fig. 6.1), can also occur when zeolites have two (or more) micropore systems with different opening sizes, both being accessible to reactant molecules but only one to poison molecules.

It should be underscored that the catalyst activity versus θ_P patterns illustrated in Fig. 6.1 are based on two assumptions: monofunctional catalysis with surface reaction controlling and uniform distribution of the poison in the catalyst bed. With bifunctional catalysis processes, the situation is more complex as shown here in the example of n-hexane hydroisomerization over PtHMOR catalysts [6]. This isomerization involves three successive reaction steps (Fig. 6.2a): 1) dehydrogenation over Pt of n-hexane (nC_6) into hexenes ($nC_6^=$); 2) isomerization over the acid sites of n-hexenes into isohexenes ($iC_6^=$); 3) hydrogenation of isohexenes into isohexanes (i-C_6). Figure 6.2b shows for two series of catalysts differing by the mordenite component — HMOR1 or HMOR2 (the first one having

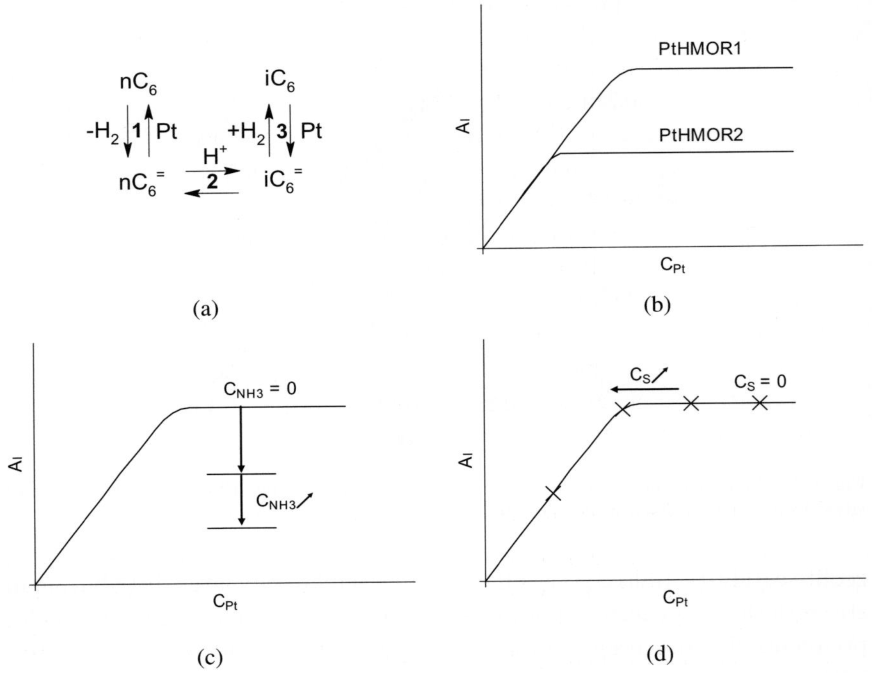

Fig. 6.2. n-Hexane hydroisomerization over PtHMOR catalysts: (a) reaction scheme; (b) influence of the concentration of accessible Pt atoms (C_{Pt}) on A_{I}, the isomerizing activity; (c) effect of poisoning with NH_3 on A_{I}; (d) effect of poisoning with H_2S on A_{I} [7].

more protonic sites than the second) — the change in the isomerization activity A_{I} with C_{Pt}, the concentration of Pt accessible sites. As could be expected, there is first an increase followed by a plateau, the first part of the curve meaning that $n\mathrm{C}_6$ isomerization is limited by the reactions over Pt (1 and 3), the second part by the reaction over the acid sites (2), which is the case under the conditions of the industrial process. The addition in the feed of a basic poison (NH_3) results in an activity decrease (Fig. 6.2c) in agreement with an acid-controlling reaction on the catalyst used while the addition of H_2S has only an effect when the added amount is enough for the reactions over Pt to become rate-controlling (Fig. 6.2d).

As the desired reaction and poisoning by feed impurities occur in parallel, the catalyst deactivation depends on the chosen reactor. With a fixed-bed reactor, poison molecules chemisorb preferentially in the first part of the bed and the front of poison, hence the catalyst deactivation, spreads gradually to the exit of the bed. In contrast, with a fluidized reactor, the concentration of the chemisorbed poison, hence the degree of deactivation, will be independent of the catalyst location. The deactivation curves could be similar only if the active sites are identical (non-selective poisoning). In the case of acid zeolites with protonic sites of different strength, the strongest sites which are the most active will be poisoned in a short time in the fluidized reactor and progressively from the entrance to the exit of the fixed-bed reactor. Therefore, deactivation will be faster with the first reactor type. To extrapolate the results from the laboratory to the industrial scale, it is then essential to operate with the same type of reactor.

Furthermore, when the diffusion of the poison molecules in the pores of the catalyst particles or of the zeolite crystal is slower than their chemisorption, poison molecules will be essentially located near the outer surface (shell poisoning). If the desired reaction is also diffusion-limited, a pronounced deactivating effect will be observed [8].

6.2.2. *Other effects of poisons*

The catalyst deactivation is not always only due to the mere coverage of active sites by molecular poison P. The chemisorbed P molecules often have additional effects, electronic or geometric ones. Thus the deactivation caused by sulphur poisoning of metals may be accentuated namely by [1]:

- Physical blockage of several adsorption-reaction sites per strongly adsorbed S atom;
- Electronic modification of the nearest and even next nearest neighbour metal atoms with as a consequence modification of their chemisorption properties;
- Restructuring of the surface by the strongly adsorbed poison possibly with changes in their catalytic properties, especially for structure-sensitive reactions such as hydrogenolysis;
- Inhibition or limitation of surface diffusion of reactant chemisorbed species with decrease in the probability of combination.

6.3. Poisoning of Acid Zeolite Catalysts by Basic Compounds

Organic nitrogen bases and ammonia are common poisons for acid zeolite catalysts, used in particular in the treatment of heavy feedstocks such as FCC and hydrocracking. Indeed, in crude oil, nitrogen is present in high molecular weight compounds containing often other heteroatoms. A typical feedstock of a commercial FCC unit from a European refinery was recently analysed [9]. This feed contained 0.15 wt% nitrogen, with pyrrole derivatives (alkyl-carbazoles, -benzocarbazoles, -indoles) dominating over the more basic pyridine derivatives (alkyl-quinolines and -tetrahydroquinolines) and largely over amino compounds. These organic compounds have not only a poisoning effect but also contribute to coke formation with as consequences an additional deactivating effect and the formation during coke combustion of NOx which are harmful air pollutants (Chapter 12).

There are relatively few fundamental studies of poisoning of acid zeolite catalysts. Moreover they are carried out under conditions very different from those of the FCC riser, in particular using fixed-bed reactors with the drawback of a progressive poisoning (Section 6.2.1) along the bed. However, experiments were carried out in realistic conditions by Fu and Schaffer [10] to determine the poisoning effect of a large variety of nitrogen bases on catalytic cracking. Large differences were found between the nitrogen compounds. More recently, correlations were drawn from the corresponding results between the poisoning power and various features of the nitrogen compounds [11]. The main conclusion was that the deactivating power of the basic molecules was primarily determined by a balance between their bulkiness (or size) and their proton affinity (i.e. their basicity in the gas phase), which is in agreement with the proposals made by other authors. Thus, values of pyridine toxicity (*Tox*) higher than 1 (from 1.3 to 2.9) were found during *n*-heptane cracking at 703 K over a series of HFAU zeolites. Furthermore, *Tox* of basic poisons was shown to increase with their proton affinity. An inductive deactivation of the non-poisoned neighbour protonic sites was proposed to explain the *Tox* values higher than 1 [12]. Other explanations can however be proposed: additional deactivating effect of coke deposits [13], heterogeneous distribution in strength of the protonic acid sites, the strongest, hence the most active, being the first deactivated [14].

Lewis and protonic acid sites are known to chemisorb basic molecules whereas in contrast, only protonic acid sites are active in most of the catalytic reactions. Therefore a decrease in *Tox* could be expected from an increase in the Lewis acidity. The higher toxicity of 2,6-dimethylpyridine compared to 3-methylpyridine and quinoline was recently related to a limitation in the approach of the Lewis acid sites by the molecules of this base owing to steric hindrance. Therefore, in contrast to the other basic compounds, 2,6-dimethylpyridine could only poison the protonic sites [15]. While this proposal deserves confirmation, the effect of the size of basic molecules on *Tox* is well demonstrated by experiments with other zeolites. Thus, bulky nitrogen bases which cannot enter the micropores of the MFI zeolite have a limited poisoning effect, only the outer acid sites being inhibited. This selective poisoning can be used to improve the shape-selective properties, e.g. to favour para isomers [16].

The poisoning effect of bases is also affected by the catalyst composition. This was shown in catalytic cracking of high-nitrogen feedstocks (with 0.3, 0.48 or 0.74 wt% N) at 783 K in a microactivity testing (MAT) unit, i.e. a fixed-bed reactor, over eight experimental FCC catalysts [17]. The effect of the following features of the FCC catalyst was determined: zeolite content, zeolite type (REY or USHY), matrix acidity, surface area and pore size distribution. The main conclusions drawn from the poisoning effect of nitrogen compounds are the following: i) high zeolite-containing catalysts show a smaller decline in activity than low zeolite-containing catalysts; ii) although less active, USY-based catalysts are less affected by poisoning than REY-based catalysts; iii) high acidity, high surface area and broad pore size distribution in the matrix limit the poisoning effect.

The less costly solution to limit the detrimental effects of nitrogen poisons during commercial processes (such as FCC) over acid zeolites is to use nitrogen-resistant catalysts, these catalysts being developed on the basis of the above conclusions. Another way which is generally costly is to pretreat the feed in order to decrease the amount of nitrogen compounds, which can be done by hydrotreatment, selective adsorption, liquid-liquid extraction or acid neutralization [18].

6.4. Poisoning of Noble Metal Zeolite Catalysts by Sulphur Compounds

Pt- or Pd-containing zeolites are commercially used as monofunctional and, more generally, bifunctional catalysts. Thus, light naphtha aromatization which is catalysed by Pt/non-acidic LTL zeolite occurs through metal catalysis whereas hydrocracking, dewaxing, isodewaxing, light alkane isomerization, C_8 aromatic isomerization, etc. are typically bifunctional processes, in which Pt or Pd intervene with their hydrogenating activity [20].

New application fields have recently appeared owing to stricter regulations related to the content in sulphur and aromatic compounds in diesel fuel [21, 22]. Indeed, aromatic hydrogenation, being exothermic, has to be carried out at low temperatures although the Ni-Mo and Co-Mo sulphide catalysts conventionally used in hydrotreatment processes become active only at relatively high temperatures. Therefore, noble metals which are active in hydrogenation at low temperatures could become attractive as hydrogenation catalysts provided their tolerance to sulphur poisoning has been considerably increased. In current processes, multiple catalyst beds are used to achieve deep desulphurization and hydrogenation [23]. Hydrodesulphurization occurs in the first stage over a conventional Ni-Mo or Co-Mo catalyst, followed by gas (H_2S, etc.) removal and finally aromatic hydrogenation over the noble metal catalyst in the bottom bed where the concentration of S compounds is very low.

Under the operating conditions, the organic sulphur compounds are generally transformed into H_2S which converts the noble metal into the sulphided form with a significant decrease in hydrogenating activity. Many factors were shown to affect the hydrogenation activity and the sulphur tolerance of noble metal catalysts. Rabo *et al.* [24] were the first to demonstrate the strong sulphur resistance of noble metals well-dispersed in acid zeolites. This sulphur resistance was related to the electron deficiency of small metal clusters located within the zeolite micropores (e.g. the supercages of FAU zeolites) with the acid sites [25, 26]. As a consequence, there are on the metal atoms fewer electrons available for the charge transfer toward the electronegative S atoms, hence a weaker bond between S and metal. The reverse (a lower S resistance) would occur [27] when the noble metals are supported over alkaline or basic zeolites as is the case in monofunctional aromatization catalysts (Chapter 16). Some other observations are in favour of the positive effect of acidity on the sulphur resistance of noble metals, namely: i) the exchange of the protons of a PdNaHFAU catalyst with Na cations decreases the sulphur resistance of palladium; ii) Pt highly dispersed in a non-acidic NaFAU (Y type) zeolite shows no sulphur resistance. Note however that highly dispersed Pd, Pt and Pt-Pd over supports with little acidity — high-silica FAU zeolite, silica with small mesopores ($\varnothing = 3\,\text{nm}$) — showed relatively high sulphur tolerance [28].

The poisoning effect of sulphur was also shown to be dependent on metal particle size and on the metal (Pt or Pd). Over noble metal acid zeolite catalysts, electronegative S atoms would be more easily removed from very small electron-deficient particles [26]. However, this size effect was also found with Pt and Pd clusters supported on a high-silica FAU zeolite ($Si/Al = 195$), i.e. with very few acid sites [29]. The surface and bulk phase sulphur tolerance of Pt and Pd were specified by means of S elemental

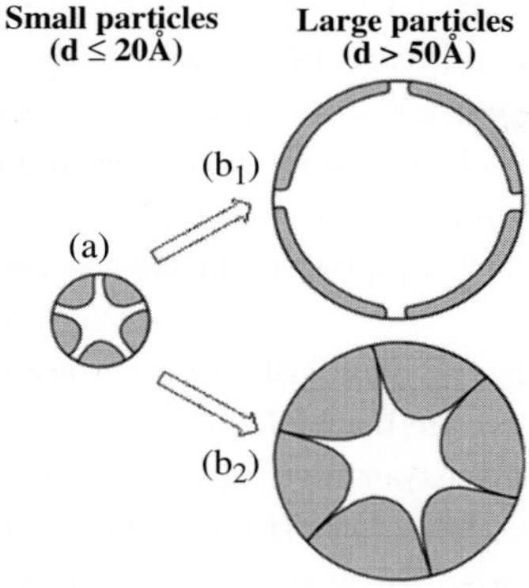

Fig. 6.3. Scheme of the sulphidation patterns of sulphide noble metal (Pt and Pd) clusters: (a) small particles; (b_1 and b_2) large particles of Pt and Pd, respectively. Adapted from [29].

analysis, CO adsorption and EXAFS (extended X-ray adsorption fine structure) [29]. Both Pd and Pt particles with diameters under 2 nm showed the highest surface tolerance, although their bulk phase was subject to penetration by adsorbed sulphur. This suggests that at the surface of these small particles, residual noble metal and sulphide phases coexist (Fig. 6.3a), the residual noble phase being probably responsible for the hydrogenation activity in the presence of sulphur. With increasing particle size, the surface tolerance of Pt decreases gradually, and that of Pd rapidly. For large Pt particles (diameter > 5 nm), the bulk metal structure substantially remains after sulphidation, while for large Pd particles, the PdS grows into the bulk phase as well as on the surface (Fig. 6.3b_2). These different trends of sulphidation were proposed to be due to the stronger affinity of Pt to sulphur and diffusivity of S in the bulk phase [29].

Another way to increase the sulphur tolerance of a noble metal is to modify their electronic properties by association with another metal. Much attention has been paid to the surface tolerance of Pd-Pt/zeolite catalysts and the positive effect of Pd on the hydrogenating activity in the presence of sulphur was demonstrated. This is shown hereafter during tetraline hydrogenation carried out in a high pressure fixed-bed reactor at 553 K over Pt, Pd and Pd-Pt HFAU zeolites, in the absence or presence of dibenzothiophene (500 ppm wt%) [30]. In the absence of S poison, the Pt/HFAU (Si/Al $=$ 340) catalyst was 2.6 and 1.5 times more active than the Pd and bimetallic catalyst, whereas in the presence of poison, there was a remarkable synergistic effect, the maximum in activity and sulphur tolerance being reached at a Pd : Pt mole ratio of 4 : 1. No effect could be observed by using a mixture of monometallic catalysts, which shows that sulphur tolerance requires an alloy (or proximity) of Pt and Pd. Results obtained with Pd-Pt supported on more acidic FAU zeolites suggest a positive effect of acidity on the sulphur tolerance of Pd-Pt [30]. The sulphur tolerance of Pd-Pt supported on HFAU zeolites was confirmed by EXAFS [31].

In a recent work [32], the origin of sulphur tolerance in Pt-Pd bimetallic catalysts was discussed on the basis of theoretical (density functional theory [DFT] computation) and experimental adsorption studies. Compared to monometallic catalysts, the adsorptions of both H_2 and H_2S were found to be enhanced but the adsorption energy of H_2 increased more than that of H_2S, which indicates that on the bimetallic Pd-Pt surface, H_2S adsorption became less favourable with respect to H_2.

The practical problems which appear with acidic supports — poisoning by nitrogen-containing compounds (Section 6.3) of the feed to be hydrogenated with suppression of the sulphur tolerance, catalysis of secondary reactions (e.g. hydrocracking, coke formation, etc.) — make often preferable the other ways to increase the sulphur tolerance of noble metal catalysts: high metal dispersion, bimetallic Pd-Pt catalysts, promoters such as CeO_2, Yb, etc. [33, 34]. In a recent review paper [34], the various factors affecting the sulphidation behaviour and surface tolerance of the bimetallic particles supported over FAU zeolites were discussed in addition to their nitrogen tolerance and resistance to agglomeration which are important for industrial applications.

An original approach to design more sulphur-resistant noble metal catalysts was recently proposed by Song [23] on the basis of two main observations collected in naphthalene hydrogenation in the presence of benzothiophene carried out in a batch reactor at 473 K [35, 36]: i) Pd catalysts (but not Pt catalysts) recovered gradually their activity with increasing residence time, hence the relative amount of H_2S in contact with the catalyst [35]. This observation was interpreted by the existence of two different types of sulphur resistance: tolerance to organic sulphur (I) and to H_2S (II). Note that a greater toxicity of organosulphur compounds compared to H_2S was previously reported [4]; ii) Pd supported over a partially dealuminated HMOR had sulphur tolerance and desulphurization activity greater than when it was supported over a HFAU zeolite [36]. This was related to the presence over the MOR zeolite of two types of micropores — large channels and side pockets with small openings — both

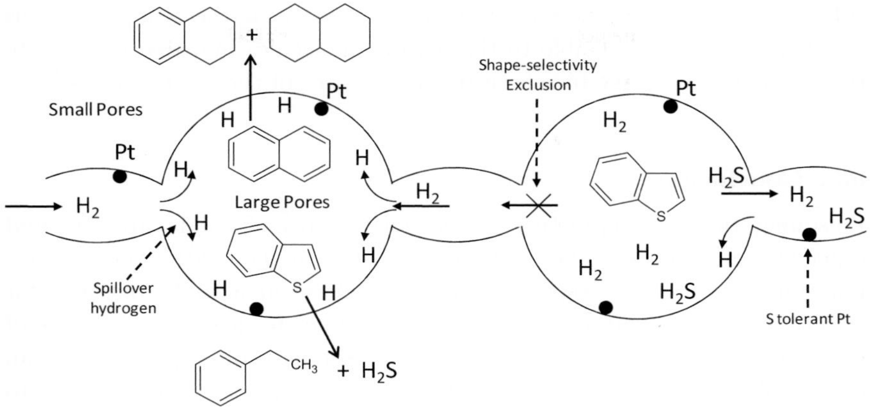

Fig. 6.4. Simplified representation of sulphur-resistant catalysts.

containing Pd particles, those located in the side pockets being protected from type I poisoning by shape-selective exclusion of benzothiophene. These two assumptions were associated with the well-known concept of hydrogen spillover [37] to design sulphur-resistant metal noble catalysts for low-temperature hydrotreatment [23]. A schematic representation of these catalysts is shown in Fig. 6.4.

The proposed catalyst comprises an acidic zeolite with two micropore systems both containing metals, one with pore openings less than ~0.5 nm, the other with larger openings (> 0.6 nm), preferably interconnected. The metal species, especially those in small pores, should have higher H_2S resistance (type II). H_2S and H_2 enter both small and large micropores whereas, because of their bulkiness, the organosulphur molecules can only enter and be desulphurized in the large micropores. Within both types of pores, H_2 molecules are dissociatively adsorbed on the metals, then transported by spillover. The main role of this spillover hydrogen is to recover the metal sites poisoned within the large micropores.

Recent works [38, 39] took inspiration from the model proposed by Song [23] to develop desulphurization noble metal/zeolite catalysts. Efficient monolith Pt catalysts for the deep desulphurization of diesel fuels were designed by Ismaglov *et al.* [38]. These catalysts were constituted by a mixture of a HMFI zeolite with a Ca montmorillonite supporting two types of Pt particles differing by their size and location: i) large particles (2–25 nm) located on the surface of zeolite crystallites and montmorillonite, hence accessible to organosulphur and aromatic molecules but sensitive to sulphur poisoning; and ii) nanoparticles located inside the zeolite channels which would be inaccessible to the organic molecules but able to activate the hydrogen necessary to reactivate the Pt sites of the large particles.

6.5. Conclusion

Poisoning by feed impurities of the active sites of solid catalysts and especially of zeolites is one of the main causes of their deactivation. This poisoning is not only responsible for a decrease in activity but also for selectivity changes, often negative but sometimes positive (e.g. increase of product shape selectivity). Prevention of poisoning is often preferable than curing. When poisoning is due to feed impurities, one natural way is to pretreat the feed so as to eliminate these impurities or at least to limit their concentration. Another way is to design poison-tolerant catalysts.

In certain cases, the activity of zeolite catalysts poisoned by organic species can be recovered by simple flow treatment at temperatures higher than that of the reaction. However, this mode of regeneration is suitable when deactivation is only due to poisoning, i.e. when poisoning is not accompanied or followed by another cause of deactivation. Thus poisoning by sulphur can initiate sintering of noble metals (e.g. in Pt LTL aromatization, Chapter 16). Furthermore, coke formation was shown to occur simultaneously with poisoning of zeolite acid sites by nitrogen organic species, the same likely occurring with poisoning of supported noble metals by organosulphur compounds. In the first case, the catalyst regeneration will require poison desorption then redispersion of the metal species. In the second, poisons and coke could be removed by combustion at high temperatures with harmful SO_2 and NOx products eliminated from the effluent gases (Chapter 12).

References

[1] Bartholomew C.H., *Appl. Catal. A: General*, 212 (2001) 17–60.
[2] Moulijn J.A., van Diepen A.E., Kapteijn F., in *Handbook of Heterogeneous Catalysis*, Ertl G., Knözinger H., Schüth F., Weitkamp J. (Eds.), Wiley-VCH, Weinheim (2008) Chapter 7, 1–18.
[3] Oudar J., Wise H. (Eds.), *Deactivation and Poisoning of Catalysts*, Chemical Industries 20, M. Dekker Inc., New York (1985).
[4] Barbier J., in Ref [3], 109–150.
[5] Boudart M., *Adv. Catal.*, 20 (1969) 153–189.
[6] Ramôa Ribeiro F., Marcilly C., Guisnet M., *J. Catal.*, 78 (1982) 267–274.

[7] Guisnet M., Cerqueira H.S., Figueiredo J.L., Ramôa Ribeiro F., in *Desactivação e Regeneração de Catalisadores*, Guisnet M., Cerqueira H.S., Figueiredo J.L., Ramôa Ribeiro F. (Eds.), Fundação Calouste Gulbenkian, Lisboa (2008) Chapter 5, 81–107.
[8] Wheeler A., *Adv. Catal.*, 3 (1951) 307–342.
[9] Barth J.-O., Jentys A., Lercher J.A., *Ind. Eng. Chem. Res.*, 43 (2004) 2368–2375.
[10] Fu C.M., Schaffer A.M., *Ind. Eng. Chem. Prod. Res. Dev.*, 24 (1985) 68–75.
[11] Ho T.C., Katritzky A.R.J., Cato S., *Ind. Eng. Chem. Res.*, 31 (1992) 1589–1597.
[12] Corma A., Fornés V., Monton J.B., Orchillés A.V., *Ind. Eng. Chem. Res.*, 26 (1987) 882–886.
[13] Caeiro G., Magnoux P., Lopes J.M., Ramôa Ribeiro F., *Appl. Catal., A: General*, 292 (2005) 189–199.
[14] Cerqueira H.S., Ayrault P., Datka J., Magnoux P., Guisnet M., *J. Catal.*, 196 (2000) 149–157.
[15] Caeiro G., Magnoux P., Lopes J.M., Ayrault P., Ramôa Ribeiro F., *J. Catal.*, 249 (2007) 234–243.
[16] Rollmann L.D., *Stud. Surf. Sci. Catal.*, 68 (1991) 791–797.
[17] Scherzer J., McArthur D.P., *Ind. Eng. Chem. Res.*, 27 (1988) 1571–1576.
[18] Cerqueira H.S., Caeiro G., Costa L., Ramôa Ribeiro F., *J. Mol. Catal. A: Chemical*, 292 (2008) 1–13.
[19] Davis R.J., *Heterog. Chem. Rev.*, 1 (1994) 41–53.
[20] Guisnet M., Gilson J.-P., *Zeolites for Cleaner Technologies*, Guisnet M., Gilson J.-P. (Eds.), Imperial College Press, London (2002).
[21] Cooper B.H., Donnis B.B.L., *Appl. Catal. A: General*, 137 (1996) 203–223.
[22] Babich I.V., Moulijn J.A., *Fuel*, 82 (2003) 607–631.
[23] Song C., in *Shape Selective Catalysis*, Song C., Garcès J.M., Sugi Y. (Eds.), ACS Symposium Series 738 (2000) Chapter 27, 381–389.
[24] Rabo J.A., Schomaker V., Pickert P.E., *Proc. 3rd Int. Congr. Catal.*, North Holland Publishing Co., Amsterdam (1964) 1264–1272.
[25] Dalla Betta R.A., Boudart M., *Proc. 5th Int. Congr. Catal.*, North Holland Publishing Co., Amsterdam (1973) 1329–1338.
[26] Gallezot P., *Catal. Rev. Sci. Eng.*, 20 (1979) 121–154.
[27] Besoukhanova C., Guidot J., Barthomeuf D., Breysse M., Bernard J.R., *J. Chem. Soc.*, 77 (1981) 1595–1604.
[28] Matsui T., Harada M., Bando K.K., Toba M., Yoshimura Y., *Appl. Catal. A: General*, 290 (2005) 73–80.
[29] Matsui T., Harada M., Ichihashi Y., Bando K.K., Matsubayashi N., Toba M., Yoshimura Y., *Appl. Catal., A: General*, 286 (2005) 249–257.
[30] Yasuda H., Yoshimura Y., *Catal. Lett.*, 46 (1997) 43–48.
[31] Yasuda H., Matsubayashi N., Sato T., Yoshimura Y., *Catal. Lett.*, 54 (1998) 23–27.
[32] Jiang H., Yang H., Hawkins R., Ring Z., *Catal. Today*, 125 (2007) 282–290.
[33] Yoshimura Y., Yasuda H., Sato T., Kijima N., Kameoka T., *Appl. Catal. A: General*, 207 (2001) 303–307.

[34] Yoshimura Y., Toba M., Matsui T., Harada M., Ichihashi Y., Bando K.K., Yasuda H., Ishihara H., Morita Y., Kameoka T., *Appl. Catal. A: General*, 322 (2007) 152–171.
[35] Lin S.D., Song C., *Catal. Today*, 31 (1996) 93–104.
[36] Song C., Schmitz A.D., *Energy Fuels*, 11 (1997) 656–661.
[37] Pajonk G.M., *Appl. Catal. A: General*, 202 (2000) 157–169.
[38] Ismaglov Z.R., Yashnik S.A., Startsev A.N., Boronin A.I., Stadnichenko A.I., Kriventsov V.V., Kasztelan S., Guillaume D., Makkee M., Moulijn J.A., *Catal. Today*, 144 (2009) 235–250.
[39] Zhou F., Li X., Wang A., Wang L., Yang X., Hu Y., *Catal. Today*, 150 (2010) 218–223.

Chapter 7

MODES OF COKE FORMATION AND DEACTIVATION

M. Guisnet

7.1. Introduction

In most of the heterogeneously catalysed processes, there is a progressive decrease in the catalyst efficiency that is mainly due to the catalytic formation and/or the deposition of bulky organic compounds (coke) which, under the operating conditions, cannot be desorbed from the catalyst [1–3]. It is also the case in zeolite-catalysed processes, the only particularities originating from the location of most of their active sites within micropores [4, 5]. As a consequence, the location of coke components is well specified: the catalytic coke components are essentially formed and trapped within the micropores whereas the bulky feed impurities can only be deposited on the outer surface of the zeolite crystals. An additional remark is that the small size of the micropores limits the growth of catalytic coke molecules, which therefore are not always polyaromatics. It is why, as proposed in Chapter 4, the term coke in inverted commas ("coke") will be used here to designate the non-polyaromatic carbonaceous compounds. Another important remark is that catalytic coke formation can also be responsible for the decrease in efficiency of zeolite adsorbents [6].

Coke has first a direct effect on the active sites, either by poisoning them or by blocking their access by reactant molecules, but it can also have an indirect effect. Indeed, the coke removal needed for the catalyst regeneration requires an oxidative treatment under severe conditions, i.e. high temperatures in the presence of steam with, as consequences, frequent detrimental effects such as dealumination and degradation of the zeolite framework, sintering of supported metals, etc.

This chapter is limited to the description of the main features of zeolite coke — composition, location, etc. — and of its effect on the catalytic properties. The mechanisms involved in coke formation as well as the modes of deactivation that are essential for preventing deactivation by coking (Chapter 11) will be drawn from these data, the examples being chosen in the industrially important fields of acid and bifunctional (metal-acid) catalysis. As shown in Fig. 4.1, coke composition plays a central role in this determination of the modes of coking and deactivation.

7.2. Coke Formation

This paragraph will be essentially limited to the presentation of coke formation involving catalytic steps. However, this does not mean that "coke" resulting from the simple deposition on the outer surface or from the trapping of feed components within micropores cannot be responsible for deactivation. Indeed, as shown in Chapter 13, the first type of coke can participate for a large part in the deactivation of FCC catalysts (Conradson coke). Furthermore, the second type of coke was shown to have a significant deactivating effect on the CaLTA (5A) adsorbent used in the separation by sieving of butane isomers: a very small fraction of the isobutane component that enters the 5A zeolite micropores remains irreversibly trapped within the zeolite cages, causing a progressive blockage of their access [6c].

7.2.1. *General characteristics*

The formation of carbonaceous compounds is primarily a **chemical process**: coke must be a stable reaction product under the operating conditions. Generally, this process is very complex, involving various successive steps (Fig. 7.1). Among these steps, the intramolecular (cyclization) and intermolecular condensation reactions play a key role. These reactions, which are exothermic, are often reversible under the operating conditions; hence the concentration of the condensation products is limited by thermodynamic equilibrium. Moreover, in the absence of the reaction mixture (e.g. in the stripping section of the FCC process), they can be retransformed into the compounds from which they were formed ("reversible coke") [7]. However, these condensation products generally undergo almost irreversible secondary reactions. Thus, coke molecules formed during the catalytic transformation of hydrocarbons are often very stable polyaromatic compounds, which in addition to

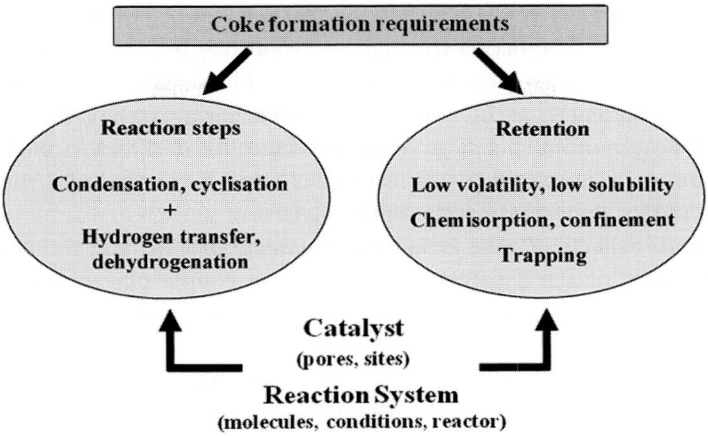

Fig. 7.1. Main chemical and physical requirements for catalytic coke formation [10].

condensation reactions, require for their formation stabilization reactions: dehydrogenation or hydrogen transfer reactions, etc. [8].

Coke formation, although very complex, has the essential features of catalytic processes:

(a) The rate of coking is, like the rate of formation of the desired products, very dependent on the characteristics of the acidic sites and of the reactant molecules;

(b) The formation of coke molecules can occur through different kinetic modes: (i) directly from the reactant molecules in parallel with the formation of the desired products (competitive mode), (ii) from the secondary transformation of the desired products (successive mode) or (iii) from both modes;

(c) Like the formation of desorbed products, coke formation undergoes deactivation by coke (it is an autoinhibited process). However, this deactivation is generally more pronounced than that related to the formation of desorbed products as demonstrated by the decrease with time-on-stream (TOS) of the coking/cracking ratio during *n*-heptane transformation over four zeolites [9].

Another essential particularity of coke is the requirement to be non-desorbed, which means that the formation of coke molecules requires not only chemical steps, but also their **retention** (Fig. 7.1) within the pores or on the outer surface of the catalyst [10]. The retention of coke molecules may be due (i) to their steric blockage (trapping) within the micropores, (ii) to their strong chemisorption generally on the active sites, often coupled with confinement effects and (iii) to their low volatility (gas-phase reaction) or solubility (liquid-phase reaction) (Fig. 7.1). These three possibilities can occur independently or in association. Note however that the first two possibilities are often specific to coke molecules located and formed within the zeolite micropores, while the last one is specific to very bulky molecules formed and/or deposited on the outer surface.

In both aspects of coke formation, chemical process and retention, the characteristics of the catalyst-reaction system couple play a decisive role (Fig. 7.1). Thus, the composition, location of coke and rate of formation are determined by:

- The features of the reaction system: type and rate of the desired and secondary transformations, size and shape of reactant and product molecules, operating conditions (T, pressure, etc.), characteristics of the reactor.
- The features of the catalyst: nature, concentration, strength and location of the active sites, size and shape of pores and openings.

Only the modes of coke formation will be discussed here, the effect of the catalyst-reaction system couple being detailed in Chapter 11 with the aim of establishing guidelines for limiting coke formation.

7.2.2. *Modes of coke formation*

We will show here that the modes of coke formation as well as the origin of retention can be easily drawn from the change in coke composition with time-on-stream (TOS) or duration (batch reactor), hence with coke content, then from data generally accessible in the case of zeolite-catalysed reactions (Chapter 4). Reaction temperature significantly affects the coke composition. At low temperatures ($<473\,K$), the carbonaceous compounds responsible for deactivation are not generally polyaromatic ("coke") whereas at high temperatures ($>623\,K$), polyaromatics can be predominant (coke). As a consequence, both the modes of formation of the carbonaceous compounds as well as the cause of their retention will be different. It is why coke formation will be successively examined in examples of reactions carried out at low ($<473\,K$) and at high temperatures ($>623\,K$). At intermediate temperatures, coke formation could be generally considered to occur through a combination of low and high T modes.

7.2.2.1. *Low-temperature coke*

At low temperatures, the main catalytic applications of acidic zeolites deal with the synthesis of petrochemicals and fine chemicals from highly reactive hydrocarbons (generally in gas phase) or functionalized compounds (often in liquid phase).

Over acidic zeolites, desorbed products and "coke" are competitively formed from highly reactive hydrocarbons such as olefinic, alkyl aromatic and polyaromatic hydrocarbons, oxygenated compounds such as alcohols, aldehydes, ketones, etc. "Coke" and reaction products can even be formed at 373 K from very branched alkanes such as 2,2,4-trimethylpentane. However, in this case, "coke" was shown to result from secondary transformations of isobutene produced with isobutane by cracking (successive or secondary coking) [11]. In contrast, under these mild conditions, n-alkanes, etc. are not reactive enough to be transformed into desorbed products and "coke".

"Coke" formation from propene and toluene was investigated at 393 K over two acidic zeolites: a large-pore zeolite, HFAU (Fig. 2.1a), and an average-pore zeolite, HMFI (Fig. 2.1b) [8]. Whatever the reactant, all the "coke" molecules were soluble in methylene chloride, but only after the acid dissolution of the zeolite matrix, which demonstrates their location within the micropores. Large differences can be observed between "coke" compositions.

Thus, while with both zeolites the "coke" components formed from propene (Table 7.1) are aliphatic and can present a number of C atoms that is a non-multiple of 3, they depend significantly on the zeolite in what concerns the number of C atoms, the number of unsaturations (0 to 4 with FAU against 0 to 2 with MFI) and their degree of branching

Table 7.1. Coke molecules (C_nH_{2n+Z}) formed over large- (HFAU), medium- (HMFI) and small- (LTA) pore zeolites during propene transformation at low temperatures (370–390 K). Adapted from [8].

Coke characteristics	HFAU	HMFI	CaLTA
H/C atomic ratio	1.8	2.0	2.0
n_c	25–40	10–35	6,9,12,15,18,21
Z	−6*,−4*,−2,0,+2	−2*,0*,+2*	−2,0*,+2
n_{I+C}	4,3,2,1,0	2,1,0	2,1*,0
$CH_3/(CH_2 + CH)$	1.5	0.25	1.1
Size (Å)	$(6.0 \times 25 - 40)$	$(4.3 \times 10 - 35)$	$(6.0 \times 9 - 25)$
bp_{760} (K)	625–800	450–760	330–710

n_{I+C} number of unsaturations and cycles; *main values.

(much higher values of the CH3/(CH2 + CH) ratio with FAU, hence more branching).

However, from these "coke" compositions, it could be concluded that on both zeolites, coking occurred through identical schemes (Fig. 7.2) involving five types of reaction: oligomerization and co-oligomerization (1); cracking (2), i.e. the reverse reaction of (1); rearrangement (3); hydrogen transfer (4); and cyclization (5).

The differences in "coke" composition can be related to smaller 3/1 and 4/1 rate ratios over MFI than over FAU, owing to more severe steric constraints at the MFI channel intersections than in the large FAU supercages. Furthermore, it should be underscored that part of "coke" molecules result from reactions (1) only, and hence correspond to "reversible coke" (in particular over MFI). This reversibility was demonstrated by treatment under vacuum [7]. With both zeolites, the "coke" molecules are neither very polar nor very bulky (line 5, Table 7.1). Therefore, their retention within the micropores is due neither to their chemisorption on the acidic sites, nor to their trapping, but rather to their low volatility (boiling point higher than the reaction temperature (line 6, Table 7.1) and probably also to a confinement effect [11].

From toluene, the formation of "coke" was much slower than from propene (at least 20 times). With HMFI, the "coke" components are methyl- diphenyl- and triphenyl-methane (MDPM and MTPM) resulting from a nucleophilic attack of a benzylic carbocation (formed by hydride transfer [HT] from toluene to a pre-existing carbocation or to a protonic site) by a molecule of toluene (T) or of methyldiphenylmethane (Scheme 1). As shown in Scheme 1, all these reactions are reversible.

With HFAU, methyl- diphenyl- and triphenyl-methane were also found in the carbonaceous deposits, but only in small amounts because of their

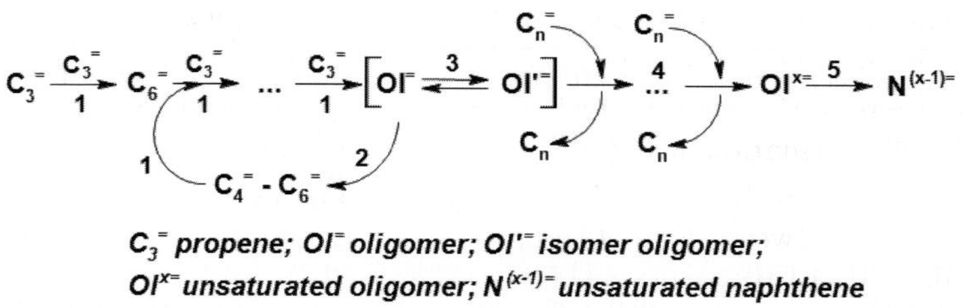

$C_3^=$ propene; $Ol^=$ oligomer; $Ol'^=$ isomer oligomer;
$Ol^{x=}$ unsaturated oligomer; $N^{(x-1)=}$ unsaturated naphthene

Fig. 7.2. Scheme of formation of coke molecules during propene transformation at 293 K over HFAU and HMFI zeolites: oligomerization (1), cracking (2), rearrangement (3), hydrogen transfer (4) and cyclization (5). Adapted from [4].

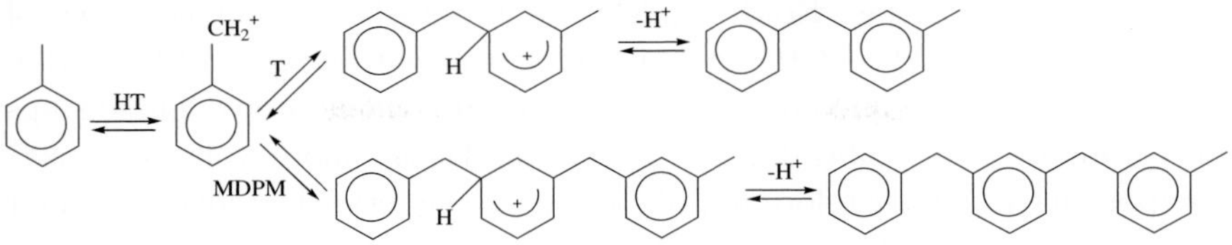

Scheme 1.

transformation into anthracenic and phenanthrenic compounds through cyclization and hydride transfer steps (Fig. 7.6a). Here again, part of "coke", i.e. that which results from condensation reactions only (MDPM and MTPM), is reversible and the differences in coke composition between zeolites can be related to more severe steric constraints at the MFI channel intersections than in the FAU supercages.

Over the MFI zeolite, "coke" formation was also examined at 393 K from an equimolar toluene-propene mixture [12]. The rate of coking was similar to that found from propene but the "coke" composition was very different. The carbonaceous compounds were essentially constituted by mono-, bi-, and tri-isopropyltoluenes resulting from the alkylation of toluene by propene (i.e. by essentially "reversible coke"), with traces of *n*-propyl benzene and of diphenyl methane. The absence of the aliphatic products observed from propene can be explained by an inhibiting effect of the basic toluene molecules with essentially chemisorption of toluene over the protonic sites and preferential attack of the isopropyl cations (in limited amounts) by toluene molecules.

The formation of coke from propene was also investigated over CaLTA (5A) zeolites to simulate their deactivation by alkene impurities during the separation of alkane isomers by molecular sieving [6a]. These small-pore zeolites have large cages with narrow openings (Table 2.2) through which the linear alkane but not the iso alkane molecules can easily enter. They are known to present weak protonic acidic sites that are created by partial hydrolysis of Ca^{2+} cations. Over a 5A sample ($Na_3Ca_{4.5}Al_{12}Si_{12}O_{48}$) at 373 K, coke formation was found to be much slower than over the MFI and FAU zeolites. Another difference is that only part of the carbonaceous compounds (between 20 to 60 wt% depending on the coke content) can be recovered in methylene chloride after acid dissolution of the zeolite matrix, which suggests the elimination during the acid treatment of the aged samples of the most volatile components. GC/MS coupling shows that the recovered "coke" was mainly constituted of olefinic hydrocarbons with

12, 15 or 18 C atoms. However, NMR analysis shows a high degree of branching ($CH_3/CH_2 + CH = 1.1$) in agreement with a mode of formation involving successive oligomerization and rearrangement steps. These steps are probably responsible for the formation of the more volatile "coke" components which therefore are likely C_6 and C_9 branched alkenes. Their branching and their boiling point lower than the reaction temperature are strong complementary arguments in favour of retention by steric blockage of all the "coke" molecules within the 5A zeolite cages.

When noble metals are coupled with acid zeolites, there are often positive changes in activity and selectivity with generally also a significant decrease in coke formation, hence in deactivation. Thus, in *n*-heptane transformation in the presence of hydrogen at 473 K, the introduction of platinum in HFAU zeolites leads to a significant increase in activity, in the selectivity to isomers (at the expense of cracking products) and in the catalytic stability. All these changes can be attributed to a shift from an acidic reaction scheme to a bifunctional one, with as a consequence a decrease of the concentration of coke maker molecules (here, alkenes) in the reaction medium [13]. However, in certain cases, this coupling has a reverse effect. Thus, in toluene hydrogenation at 363 K, the association of Pt or Pd to a HFAU zeolite causes a significant increase in "coke" formation and a very fast initial decrease in the hydrogenation rate. "Coke" was shown to be mainly constituted by C_{14} and C_{21} products resulting from the acid alkylation of toluene molecules by olefinic and dienic intermediates of toluene hydrogenation [14]. This negative effect can be related to the role played by noble metals in the formation of very reactive coke precursors.

The last example deals with the deactivation of acid zeolite catalysts by carbonaceous compounds during liquid-phase synthesis of fine chemicals carried out in a batch reactor. One of the main differences with hydrocarbon reactions is the additional effect of the polarity of these compounds on their retention. The more polar (hydrophilic) the zeolite, the more significant this effect is. Even the desired products can be retained within the zeolite micropores because of the additional effect of solvation by polar zeolites. These product molecules limit or block the access of reactant molecules to the micropores and/or acid sites and hence limit or inhibit the reaction (auto-inhibition). Moreover, their long residence time may favour secondary reactions yielding bulkier and often more polar products which are consequently more strongly retained within the micropores [15]. Thus, for instance, during the acetylation of anisole with acetic anhydride at 363 K, the desired *p*-methoxyacetophenone product (PMA) was shown to be

Table 7.2. Main characteristics of coke formed during low-temperature and high-temperature reactions over acidic zeolites.

Characteristics	Low-temperature coke	High-temperature coke
Composition	Very dependent on the type of reactant	Very dependent on the zeolite pore structure
Location	Within the zeolite micropores	Within the zeolite micropores
Reactions involved	Essentially condensation reactions *to* "reversible coke" plus a limited number of stabilization steps: rearrangement, hydrogen transfer	Condensation plus a large number of stabilization steps: hydrogen transfer, dehydrogenative cyclization and coupling
Cause of retention	Large- and medium-pore zeolites: essentially low volatility, confinement Small-pore zeolites: steric blockage	Steric blockage

strongly retained on the HBEA zeolite catalyst. Moreover, for long reaction times, the micropores were shown to contain di- and tri-acetylated anisole resulting from the acetylation of the PMA side chain [16]. These polar products were also formed in the acetylation of other substrates: veratrole, toluene, benzofurane [15].

The main characteristics of coke formation at low temperatures over acid zeolites are presented in Table 7.2.

7.2.2.2. *High-temperature coke*

7.2.2.2.1. Origin of the retention of coke molecules

At high temperatures (>723 K), i.e. in the temperature range of most refining and petrochemical processes, practically all the hydrocarbons can be transformed over acid zeolites and the desired transformation is always accompanied by coke formation. Like at low reaction temperatures but here only at short time-on-stream (low coke contents), the "coke" molecules are generally soluble in methylene chloride after acid dissolution of the zeolite matrix, which demonstrates their location within the micropores [5a]. It should however be underscored that the higher the reaction temperature, the lower the coke content at which insoluble coke appears.

However, while at low temperatures the composition of "coke" is extremely dependent on the reactant (Table 7.2), this is no longer the case at high temperatures. Thus, at 723 K, at low coke contents

(1.5–3.0 wt%), methylpyrenes are the main coke components resulting from the transformation over HMFI of n-heptane, propene, toluene, or a propene-toluene mixture. At higher contents, large polyaromatic compounds, insoluble in methylene chloride, are also formed [12]. The origin of methylpyrenes retention within the zeolite micropores can be deduced from three main observations [5a, b]: i) their boiling point (~673 K) is lower than the reaction temperature; ii) their molecular size is intermediate between those of the channel intersections and of the channel "diameter"; and iii) their polarity is relatively low. All this suggests that, contrary to what was concluded for low T "coke", the retention of methylpyrenes molecules by the zeolite is due to their steric blockage at the channel intersections.

At the same temperature, the coke composition over a HFAU sample was also independent of the reactant but very different from that found with HMFI. Thus, from various reactants — n-heptane, methylcyclohexane, propene, cyclohexene, toluene — the soluble coke components correspond to relatively bulky polyaromatic compounds with 4 to 7 aromatic rings which can be classified into three main families (A, B, C) with a general formula of C_nH_{2n-26}, C_nH_{2n-32} and C_nH_{2n-36} respectively [8]. Their distribution depends on time-on-stream, hence on coke content: A appears as a primary product, B and C resulting from the secondary transformation of the A compounds. Like over HMFI, highly polyaromatic compounds, insoluble in methylene chloride, are also formed but here they appear from relatively small coke contents (>1.5 wt%). All the soluble coke molecules were shown to have a size intermediate between those of the micropores (i.e. the supercages) and of their apertures, in agreement with retention by steric blockage. Molecular modelling confirms the possibility for soluble coke molecules to be accommodated and sterically blocked within the micropores. Several examples chosen over various zeolites are presented in Fig. 7.3 and Fig. 5.4.

Whereas steric blockage within the zeolite micropores is definitely demonstrated for the molecules of the coke components soluble in CH_2Cl_2, this conclusion deserves to be discussed for the insoluble coke molecules. Indeed, whereas the molecules of soluble coke can be accommodated within the micropores (channels, cages or channel intersections), this cannot be the case for the bulky polyaromatic molecules of insoluble coke. However, as shown in Fig. 7.4 in the example of n-heptane cracking at 723 K over HFAU and HMFI zeolites, insoluble coke generally appears as the expense of soluble coke [5b, c].

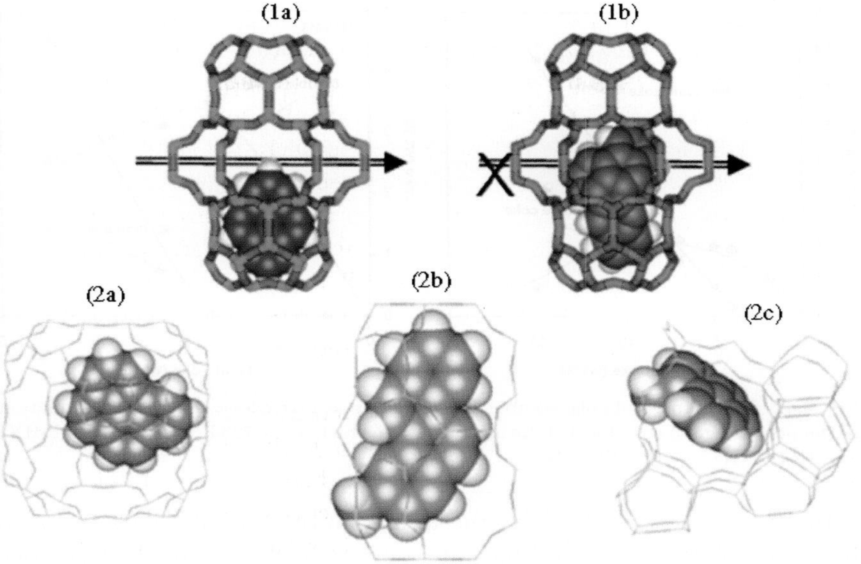

Fig. 7.3. Location of coke molecules within zeolite micropores: Example 1: location of methylpyrenes (1a) and methyldibenzofluoranthene (1b) formed in the supercages of the MWW zeolite during *m*-xylene transformation at 623 K. Example 2: location of coke molecules formed during *n*-heptane transformation at 723 K over three different protonic zeolites: methylcoronene within FAU supercages (2a); methylchrysene within the ERI cages (2b); methylpyrene at the channel intersections of MFI (2c). Adapted from [8].

Therefore, insoluble coke molecules would result from secondary reactions with reactant and/or product molecules of soluble coke molecules trapped within the zeolite micropores. This suggests that at least a part of each molecule of insoluble coke (its "root") is located within the micropores. This hypothesis was confirmed by transmission electron microscopy (TEM) analysis of the coked HFAU and HMFI samples [17] which suggests an overflowing of the insoluble coke molecules onto the outer surface of the zeolite crystals, their root being located within micropores close to the outer surface.

However, in processes operating at high temperatures and/or with heavy feeds such as catalytic cracking (FCC), at least two other modes of formation of insoluble coke can appear: (i) simple deposition on the outer surface of very heavy feed components, (ii) desorption of bulky (but still soluble) polyaromatic molecules from micropores close to the outer surface and then migration in gas or in liquid phase along the catalytic bed

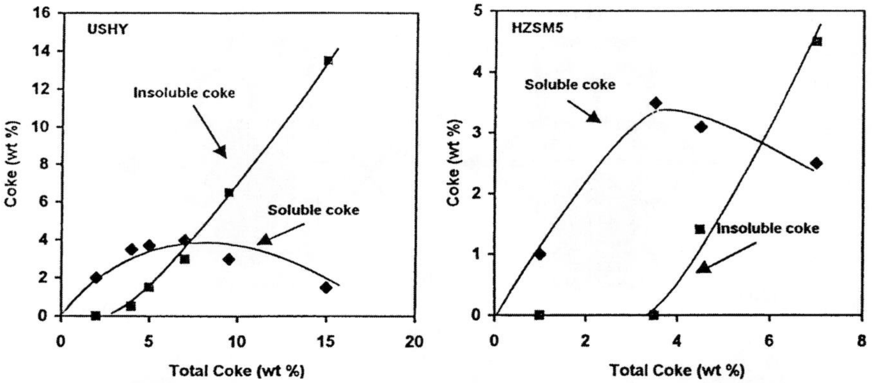

Fig. 7.4. Percentages of coke soluble and insoluble in methylene chloride as a function of the percentage of coke formed during n-heptane cracking at 723 K over HFAU (USHY) and HMFI (HZSM5) zeolites [5c].

with progressive growth up to the formation of molecules heavy enough to remain blocked (coke) on the outer surface of the zeolite. This last one could explain why at very high temperatures (>800 K), only insoluble coke can generally be observed.

Inversely, with some zeolites, insoluble coke molecules can be completely formed and located within the micropore system. This was proposed to occur during n-heptane cracking at 723 K over a MWW zeolite (Fig. 7.5) on the basis of the scheme of coke formation, of physisorption experiments and of molecular modelling [18].

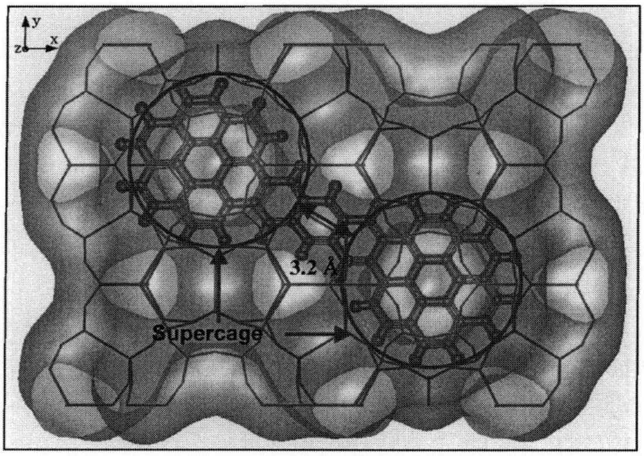

Fig. 7.5. Location of insoluble coke molecules formed by dehydrogenative coupling of two coronene molecules located in adjacent supercages of a MCM22 zeolite [18].

Indeed, insoluble coke was shown to result from transformation of large polyaromatic soluble molecules (from pyrene to coronene) which could be accommodated within the large cylindrical supercages only (7.1 Å Ø 18.4 Å). As the distance between supercages is very limited, the dehydrogenative coupling of part of these molecules located in adjacent supercages was proposed as responsible for this transformation [18].

7.2.2.2.2. Chemical steps involved in coke formation

As was shown in Chapter 4 in the example of the skeletal isomerization of *n*-butene over a HFER zeolite, the scheme of coke formation can be easily deduced from the effect of coke content on the concentration of the various types of coke molecules (Fig. 4.15). From this scheme, the reaction steps involved in their formation, i.e. the mechanism of coke formation, can be specified. However, because of the facile desorption from the zeolite micropores of many intermediates, all the steps of the coke formation scheme cannot be deduced. This is particularly apparent with large-pore zeolites; thus, alkyl cyclopentapyrenes (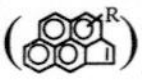), i.e. C_nH_{2n-26}, are the smallest coke molecules formed during *n*-heptane cracking at 723 K and the formation of these complex compounds from the reactant(s) involves a significant number of steps and moreover can occur through different ways. Indeed, generally, several modes of formation of polyaromatic molecules intervene simultaneously, their relative significance depending essentially on the most reactive molecules in acid catalysis present in the reaction medium, i.e. light alkenes, dienes (especially cyclopentadienes), whether alone or in the presence of aromatics [8]. The main modes of formation from hydrocarbon reactants of polyaromatic coke components are presented in Fig. 7.6.

As it was observed at low temperatures, the coupling of metals with acid zeolites can have a positive or a negative effect on coke formation. In alkane transformation at 623 K, the coupling of Pt/Al_2O_3 to HFAU, HMOR and HMFI zeolites was shown to provoke, in the presence of hydrogen, a change from cracking through an acid mechanism to hydroisomerization and to hydrocracking through a bifunctional mechanism. A significant decrease in coke formation and deactivation can be observed, which is essentially due to the much lower concentration of alkene products [19]. In propane aromatization at 803 K, the adding of well-dispersed gallium species to a MFI zeolite has the same effect: change in the reaction mechanism from an acidic to a bifunctional scheme and decrease in coke formation

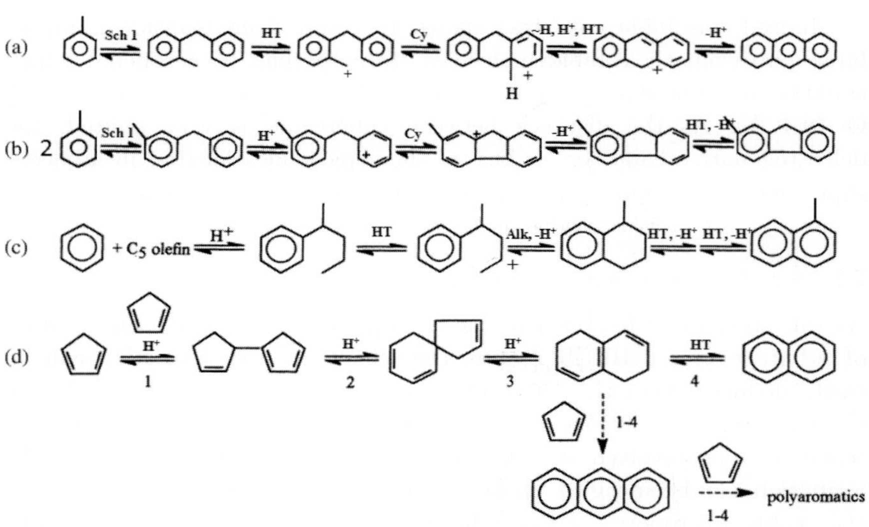

Fig. 7.6. Main modes of formation of polyaromatic coke from benzene (a), from toluene (b), from C_5 alkenes (c) and from cyclopentadiene (d). Adapted from [8].

and deactivation. No change in coke composition could be observed [20]. In toluene disproportionation [21] and transalkylation with heavy aromatics [22], the adding of Pt or Ni to acid zeolites was also responsible for a decrease in coke formation and deactivation. In this case, the positive effect of the metal was not due to a change in the mechanism of the main reactions but to a decrease in the concentration of coke precursors. However, the coupling of Pt to acidic zeolites, indispensable for the catalysis of n-butane transformation into isobutene (dehydroisomerization), leads, as a consequence, to the formation of butadiene, this being the main molecule responsible for the catalyst deactivation due to the formation of coke [23].

7.2.2.3. *Conclusion*

The reaction temperature was shown to have a determining effect on the composition of the carbonaceous compounds responsible for the deactivation of acid zeolite catalysts (coke), hence on the mode of coking and on the cause of coke retention. At low temperatures, "coke" essentially consists of non-polyaromatic compounds that are soluble in organic solvents after mineralization of the zeolite. These compounds, which result from catalytic condensation of reactants and/or products molecules (then rearrangement), present a limited stability, and hence could be

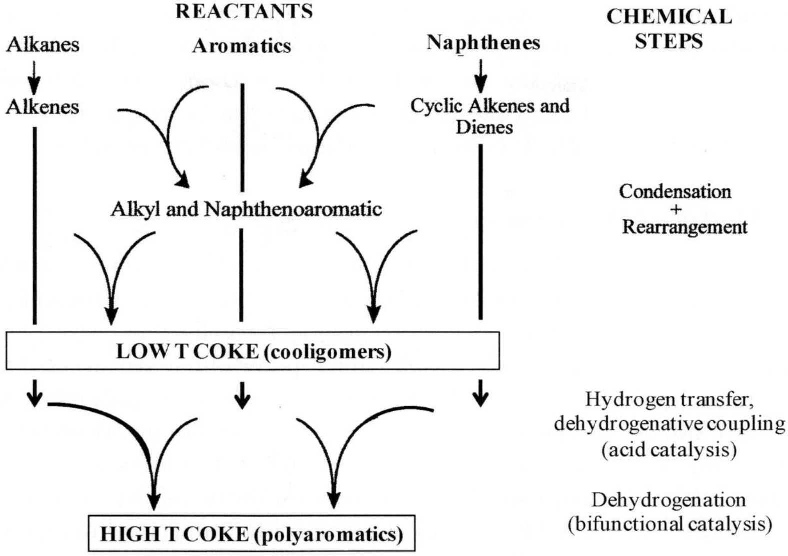

Fig. 7.7. Formation of low- and high-temperature coke from various hydrocarbons over acid and bifunctional metal-acid catalysts: Simplified reaction scheme. Adapted from [8].

reconverted into their precursors, e.g. through thermal treatment under chosen conditions. The "coke" molecules are generally formed within the micropores (channels, cages and channel intersections) in which are located most of the protonic acid sites and retained there owing to their low volatility or solubility (gas- or liquid-phase reactions). A simplified scheme of the formation of low T "coke" from different types of hydrocarbon reactants is presented in the upper part of Fig. 7.7.

At high reaction temperatures, the coke molecules are, like at low temperatures, generally formed and retained within the zeolite micropores. However, the chemical steps involved in their formation as well as the cause of their retention are very different. Alkylpolyaromatics are the main coke components and while their formation requires the condensation and rearrangement steps implied in the formation of low T coke, many additional steps of stabilization, namely hydrogen transfer, dehydrogenative cyclization and dehydrogenative coupling are involved (Fig. 7.7). Moreover, the steric blockage of coke molecules is the main cause of their retention within the zeolite micropores.

As a consequence, the size and shape of the coke molecules are determined by two main characteristics of the zeolite pore system: (i) the size and the shape of channels, cages and channel intersections which determine the shape and the maximum size of the coke molecules which are trapped inside; (ii) the size of the pore apertures which determines the minimum size of the coke molecules which can be trapped. All that demonstrates that **high T coking is a shape-selective process**.

7.3. Modes of Deactivation

It is well known that coke can affect the activity of solid catalysts (including zeolites) in two main ways: poisoning or coverage of the active sites and pore blockage. The effect of these deactivation modes on the active sites, hence on the catalyst activity as well as on the micropore volume still accessible to reactant molecules, are largely different. In the case of site poisoning, only one active site can be generally deactivated per coke molecule. However, the deactivating effect can be more limited when reactant molecules are able to compete with coke molecules for chemisorption on the active sites. It can also be more pronounced for bulky coke molecules if they can interact with neighbouring active sites. Pore blockage generally causes a more significant deactivation. Indeed, a single coke molecule can block the access of reactant molecules to a channel or a cage, hence to all the active sites they contain. The partial limitation of the access of reactant molecules by coke, which has a smaller effect, is sometimes associated with this mode of deactivation. Furthermore, the effect on the micropore volume which can be estimated by physisorption measurements is also more significant with pore blockage than with site poisoning.

7.3.1. *Method developed for determining the mode of deactivation: Simplifying assumptions*

Despite the large differences in the significance of the effect of coke molecules on the active sites and on the accessible pore volume, the deactivation modes are not so easily distinguished. Indeed, serious difficulties generally appear in the quantitative estimation of the number of deactivated sites and of the micropore volume made inaccessible to reactant molecules per coke molecule.

The first one, i.e. the estimation of the concentration of coke molecules which is generally impossible with classical solid catalysts, does not exist with zeolite catalysts, owing to the possible determination of coke composition [5]. The second one concerns the determination of the active sites of the aged catalyst samples, which is not only time-consuming but also imprecise because of the possible desorption of coke molecules by pyridine [23], and in certain cases difficult to relate to the catalytic activity because of the diversity of active sites, e.g. HFAU zeolites can have four bridging OH IR bands sensitive to pyridine or (NH_3) chemisorptions, hence corresponding to four types of acidic protonic sites with different location, strength and hence activity [24]. That is why it is often assumed that all the protonic acid sites present the same activity [5]. With this simplifying assumption, the number of active sites deactivated per coke molecule, sometimes called their toxicity (*Tox*), can easily be drawn from the change in the residual activity $A/A_{\rm f}$ vs. $C_{\rm k}/C_{\rm Af}$, the ratio between the concentrations of coke molecules and of active sites of the fresh zeolite.

The third one deals with the estimation of the micropore volume made inaccessible by coke to the reactant molecules by physical adsorption of inert probe molecules (called $V_{\rm A}$ here, i.e. the volume apparently occupied by coke). Indeed, this estimation can not be very accurate essentially because adsorption measurement is generally implemented at a temperature much lower than that of the reaction. A good way to proceed is to use at least two adsorbate molecules, the first with a size smaller than that of the bulkiest reactant molecule, the second with similar size. This procedure was used in the case of various protonic zeolites coked during *n*-heptane transformation at 723 K: nitrogen and *n*-hexane were chosen as adsorbates, the measurements being carried out at 77 and 273 K, respectively [5a].

7.3.2. *Modes of zeolite deactivation*

The example of the HMFI catalyst was chosen to show that the mode of deactivation can change with the coke content, actually with $C_{\rm k}/C_{\rm Af}$. In the graph of Fig. 7.8a, the toxicity (*Tox*) is given by the slope of the curve. Three domains (A, B, C) can be specified corresponding respectively to average *Tox* values of 0.25, 1 and ~3. In Fig. 7.8b, the ratio between $V_{\rm R}$, the volume really occupied by coke estimated through molecular modelling, and $V_{\rm A}$, deduced from physisorption measurements, was plotted vs. $C_{\rm k}/C_{\rm Af}$. The $V_{\rm R}/V_{\rm A}$ ratio decreases from 1 at low $C_{\rm k}/C_{\rm Af}$ values (domain A) to practically zero at high values (domain C).

From the *Tox* and $V_{\rm R}/V_{\rm A}$ values, the modes of deactivation can be established in each of the domains [5a]:

- In domain A, the $V_{\rm R}/V_{\rm A}$ value of 1 demonstrates that there is no pore blockage. In addition, the *Tox* value of only 0.25, which means that 4 coke molecules are needed to deactivate one acid site, suggests a competition

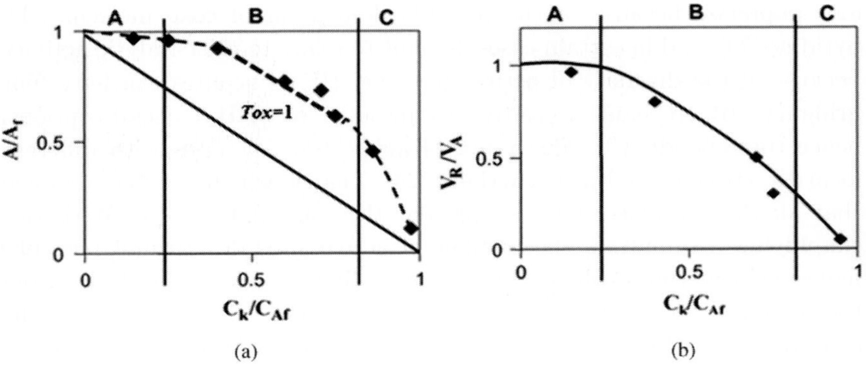

Fig. 7.8. n-Heptane cracking over a HMFI zeolite at 723 K. Plots of the residual activity A/A_{f} (a) and of the ratio between the volumes apparently and really occupied by coke $V_{\mathrm{R}}/V_{\mathrm{A}}$ (b) vs. $C_{\mathrm{k}}/C_{\mathrm{Af}}$, the ratio between the concentrations of coke molecules and of active sites of the fresh zeolite. Adapted from [10].

for chemisorption on the protonic sites between the reactant and the coke molecules. This is not surprising since these coke molecules are only weakly basic.

- In domain B, *Tox* is close to 1, in perfect agreement with deactivation by site poisoning. However, $V_{\mathrm{R}}/V_{\mathrm{A}}$ is lower than 1, which indicates a blockage of the access of adsorbate molecules to part of the microporosity unoccupied by coke molecules. This apparent disagreement can be easily explained by considering that the coke molecules which are trapped at the channel intersections of the MFI zeolite completely block their access, without occupying the totality of their volume (Fig. 7.9b).
- In domain C, both *Tox* and $V_{\mathrm{R}}/V_{\mathrm{A}}$ values are lower than 1, which is typical of deactivation by pore blockage (Fig. 7.9c). This blockage can be related to the presence in coke of insoluble components which overflow onto part of the crystal's outer surface, preventing the access of reactant molecules to the channels.

To examine the effect of the pore structure on the mode of deactivation, the same experiments were carried out over three additional protonic zeolites with initial cracking activity similar to that of the HMFI zeolite: two large-pore zeolites, HFAU and HMOR, and a small-pore zeolite, HERI (Table 2.2). The effect of coke molecules on the residual activity (A/A_{f}) and on the ratio between the micropore volumes really and apparently occupied by coke molecules ($V_{\mathrm{R}}/V_{\mathrm{A}}$) is shown in Fig. 7.10a and b for HFAU and in Fig. 7.11a and b for the other zeolites.

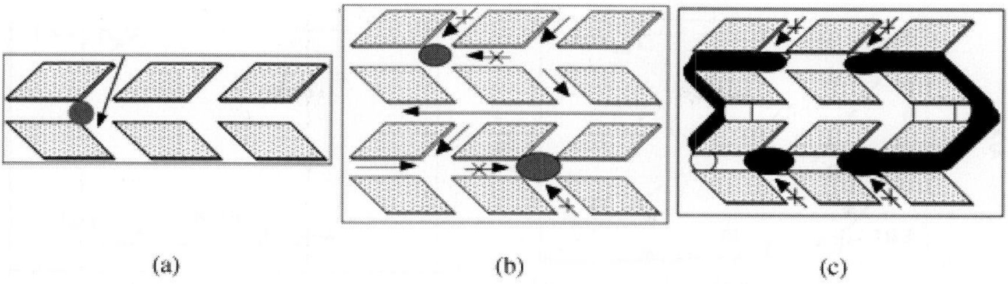

Fig. 7.9. n-Heptane cracking over a HMFI zeolite at 723 K. Modes of deactivation (a and b): site poisoning; (c) pore blockage [10].

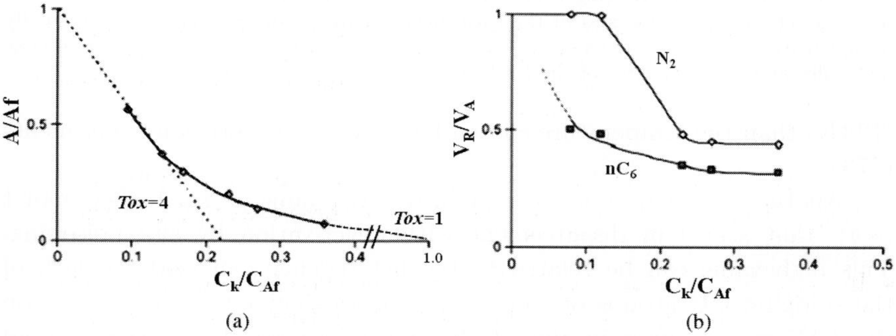

Fig. 7.10. n-Heptane cracking over a HFAU zeolite at 723 K. Plots of the residual activity A/A_{f} (a) and of the ratio between the volumes apparently and really occupied by coke $V_{\mathrm{R}}/V_{\mathrm{A}}$ (b) vs. $C_{\mathrm{k}}/C_{\mathrm{Af}}$, the ratio between the concentrations of coke molecules and of active sites of the fresh zeolite. Adapted from [10].

With HFAU, *Tox* is initially close to 4, which means that apparently one coke molecule would be able to deactivate 4 active sites. However, above a $C_{\mathrm{k}}/C_{\mathrm{Af}}$ value of 0.15, *Tox* decreases, becoming close to 1 (extrapolated value) at complete deactivation of the zeolite (Fig. 7.11a). Furthermore, the effect of coke on the $V_{\mathrm{R}}/V_{\mathrm{A}}$ value depends significantly on the adsorbate molecule (Fig. 7.11b): with nitrogen, $V_{\mathrm{R}}/V_{\mathrm{A}}$ that equals 1 at $C_{\mathrm{k}}/C_{\mathrm{Af}} < 0.15$ progressively decreases up to a value of 0.5; with n-hexane, $V_{\mathrm{R}}/V_{\mathrm{A}}$ close to 1 at $C_{\mathrm{k}}/C_{\mathrm{Af}} = 0$ (extrapolated value) decreases rapidly and then more progressively up to a value of 0.4. From nitrogen adsorption, it could therefore be concluded that for $C_{\mathrm{k}}/C_{\mathrm{Af}} < 0.15$, deactivation is due to site poisoning or coverage. However, this conclusion is only valid if the diffusion within the micropores of the coked HFAU of the n-hexane adsorbate, but not that of the n-heptane reactant, can be considered as limited by coke. This is most likely, the reaction temperature being much higher

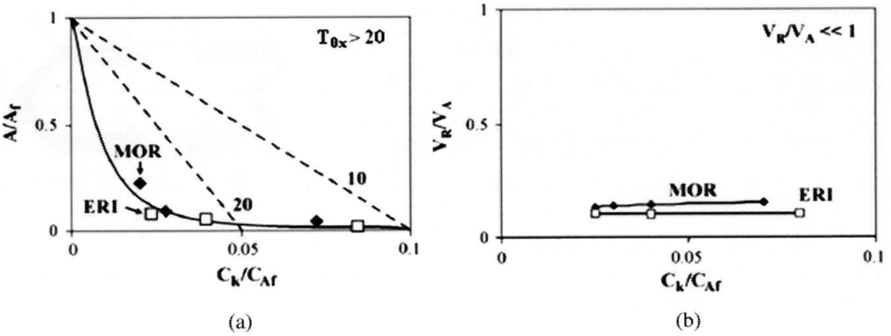

Fig. 7.11. n-Heptane cracking over HMOR and HERI zeolites at 723 K. Plots of the residual activity A/A_{f} (a) and of the ratio between the volumes apparently and really occupied by coke $V_{\mathrm{R}}/V_{\mathrm{A}}$ (b) vs. $C_{\mathrm{k}}/C_{\mathrm{Af}}$, the ratio between the concentrations of coke molecules and of active sites of the fresh zeolite. Adapted from [10].

(723 K) than the temperature chosen for n-hexane adsorption experiments (273 K).

Another observation to be explained is the value of Tox higher than 1 ($\sim$4), that seems in disagreement with deactivation by site poisoning. This high value can be related to the heterogeneity in acid strength of the bridging OH groups of this zeolite, in particular to the presence on the FAU sample used in this study of very strong (hence very active) protonic sites, resulting from interaction with extraframework Al species. IR characterization of the aged zeolite samples confirms this proposal, showing in addition that the first coke molecules interact with the corresponding hydroxyl groups [24], which are therefore preferentially deactivated. This explains why for $C_{\mathrm{k}}/C_{\mathrm{Af}} > 0.15$, a decrease in Tox can be observed, despite the pore blockage shown by adsorption experiments. From the results obtained with HFAU, only modes b (classical site poisoning) and c (pore blockage), described in Fig. 7.9b and c, are clearly demonstrated. It is, however, most likely that at very low coke contents, deactivation of HFAU results from mode a which was observed with HMFI.

The cases of HMOR and HERI are much simpler. Indeed, very high values of Tox (>20) are found for low values of $C_{\mathrm{k}}/C_{\mathrm{Af}}$ (<0.05), which is typical of pore blockage. This conclusion is supported by the low values of $V_{\mathrm{R}}/V_{\mathrm{A}}(\sim 0.1)$ with both zeolites (Fig. 7.12b). It should however be noted that while with HERI these low values are found with both adsorbates (n-hexane and nitrogen), with HMOR the values reported in Fig. 7.11b correspond to n-hexane only. Indeed, with nitrogen, values of $V_{\mathrm{R}}/V_{\mathrm{A}}$ close to 1 can be observed for low $C_{\mathrm{k}}/C_{\mathrm{Af}}$ values, which means that nitrogen molecules can accede to all the micropore volume unoccupied by coke. This observation can be related to the ability of these small molecules (kinetic diameter of 3.6 Å) to reach the large channels by diffusion through the narrow channels. However, V_R/V_A for nitrogen decreases significantly with the increase in coke content and, for $C_k/C_{Af} \geq 0.04$, identical low V_R/V_A values were found with both adsorbates. This blockage of nitrogen access to the pore volume can be related to insoluble coke molecules that overflow onto the crystal's outer surface.

7.4. Conclusion

Three modes of zeolite deactivation by coke formed during high-temperature reactions were demonstrated from the effect of coke molecules on the active sites (i.e. their "toxicity") and on the accessible micropore volume. These three modes are ranked in ascending order of their effect: (i) partial coverage of the active sites of cages or channel intersections by coke molecules; (ii) poisoning (or total coverage) of these active sites; (iii) blockage of the access of reactant molecules to all the active sites of cages, channel intersections or parts of channels in which no coke molecules are located; the coke molecules which block the access are located in other cages, channel intersections, parts of channels. Another possibility shown with MFI (Fig. 7.9c) is that insoluble coke molecules that have their root within the micropores overflow onto the outer surface of the zeolite crystals, blocking one or several pore openings. An additional deactivation mode, similar to pore blockage but in which the access of the active sites is not blocked but simply limited, has also been advanced [5]. The coke content and the zeolite pore structure are the main factors determining the deactivation modes. Thus, with tridimensional zeolites such as MFI and FAU, deactivation occurs successively through modes i, ii and iii with the increase in coke content; with monodimensional zeolites (e.g. MOR) and small-pore zeolites with large cages (e.g. ERI), deactivation occurs only through pore blockage (mode iii).

Finally, it should be underscored that coke does not always have a detrimental effect on the catalytic behaviour of zeolites [25]. Thus, coke can be deposited over zeolites to increase their shape-selective properties, a typical example being the toluene disproportionation process [26]. Furthermore, the relatively simple coke molecules formed within micropores can sometimes be reactive enough to participate in the desired reaction. This possibility will be discussed in Chapters 14, 15 and 16.

References

[1] Figueiredo J.L., in *Progress in Catalyst Deactivation*, NATO ASI Series E, 54, Figueiredo J.L. (Ed.), Martinus Nijhoff Publishers, The Hague (1982).

[2] Oudar J., Wise H. (Eds.), *Deactivation and Poisoning of Catalysts*, Chemical Industries 20, Marcel Dekker Inc., New York and Basel (1985).

[3] Bartholomew C.H., *Appl. Catal. A: General*, 212 (2001) 17–60.

[4] Guisnet M., Cerqueira H.S., Figueiredo J.L., Ramôa Ribeiro F., in *Desactivação e Regeneração de Catalisadores*, Guisnet M. *et al.* (Eds.), Fundação Calouste Gulbenkian, Lisboa (2008).

[5] (a) Guisnet M., Magnoux P., *Appl. Catal.*, 54 (1989) 1–27; (b) Guisnet M., Magnoux P., *Stud. Surf. Sci. Catal.*, 88 (1994) 53–68; (c) Guisnet M., Magnoux P., Martin D., *Stud. Surf. Sci. Catal.*, 111 (1997) 1–19.

[6] (a) Boucheffa Y., Thomazeau C., Cartraud P., Magnoux P., Guisnet M., *Ind. Eng. Chem. Res.*, 36 (1997) 3198–3204; (b) Misk M., Joly G., Magnoux P., Guisnet M., Jullian S., *Microp. Mesop. Mat.*, 40 (2000) 197–204; (c) Magnoux P., Boucheffa Y., Joly G., Guisnet M., Jullian S., *Oil Gas Sci. Tech.*, 55 (2000) 307–314.

[7] Dimon B., Cartraud P., Magnoux P., Guisnet M., *Appl. Catal. A: General*, 101 (1993) 351–369.

[8] Guisnet M., Magnoux P., *Appl. Catal. A: General*, 212 (2001) 83–96.

[9] Guisnet M., Magnoux P., Canaff C., in *Chemical Reactions in Organic and Inorganic Constrained Systems*, NATO ASI Series C, 165, Setton R. (Ed.), Kluwer, Dordrecht (1985) 131–142.

[10] Guisnet M., Costa L., Ramôa Ribeiro F., *J. Mol. Catal. A: Chemical*, 305 (2009) 69–83.

[11] Guisnet M., Magnoux P., in *Zeolite Microporous Solids: Synthesis, Structure and Reactivity*, Derouane E.G., Lemos F., Naccache C., Ramôa Ribeiro F. (Eds.), NATO ASI Series C, 352, Kluwer, Dordrecht (1992) 457–474.

[12] Magnoux P., Machado F., Guisnet M., in *New Frontiers in Catalysis, Proceedings of the 10th International Congress on Catalysis*, Guczi L. *et al.* (Eds.), Akademiaí Kiado, Budapest (1993) 435–446.

[13] Guisnet M., Alvarez F., Giannetto G., Perot G., *Catal. Today*, 1 (1987) 415–433.

[14] Chupin J., Gnep N.S., Lacombe S., Guisnet M., *Appl. Catal. A: General*, 206 (2001) 43–56.

[15] Guisnet M., Guidotti M., in *Catalysts for Fine Chemicals Synthesis*, Microporous and Mesoporous Solid Catalysts 4, Derouane E.G. (Ed.), John Wiley & Sons Ltd, Chichester, 2 (2006) 39–67.

[16] Rohan D., Canaff C., Fromentin E., Guisnet M., *J. Catal.*, 177 (1998) 296–305.

[17] Gallezot P., Leclercq C., Guisnet M., Magnoux P., *J. Catal.*, 114 (1988) 100–111.

[18] Besset E., Meloni D., Martin D., Guisnet M., Schreyeck L., *Stud. Surf. Sci. Catal.*, 126 (1999) 171–178.

[19] Perot G., Hilaireau P., Guisnet M., *Proceedings of the 6th International Zeolite Conference*, Olson D.H., Bisio A. (Eds.), Butterworths (1984) 427–434.
[20] Guisnet M., Gnep N. S., *Catal. Today*, 31 (1996) 275–292.
[21] Gnep N.S., Martin de Armando M.L., Marcilly C., Ha B.H., Guisnet M., *Stud. Surf. Sci. Catal.*, 6 (1980) 79–89.
[22] Tsai T.-C., Chen W.-H., Liu S.-B., Tsai C.-H., Wang I., *Catal. Today*, 73 (2002) 39–47.
[23] Pirngruber G.D., Seshan K., Lercher J.A., *J. Catal.*, 190 (2000) 338–351.
[24] Cerqueira H.S., Ayrault P., Datka J., Guisnet M., *Microp. and Mesop. Mat.*, 38 (2000) 197–205.
[25] Guisnet M., *J. Mol. Catal. A: Chemical*, 182–183 (2002) 367–382.
[26] Chen N.Y., Garwood W.E., Dwyer F.G., *Shape Selective Catalysis in Industrial Applications*, Chemical Industries 36, Marcel Dekker, New York (1989).

Chapter 8

THERMAL ALTERATIONS OF ZEOLITE CATALYSTS

M. Guisnet, H.S. Cerqueira, and F. Ramôa Ribeiro

8.1. Introduction

High temperatures induce chemical and structural alterations of solid catalysts which can significantly affect their activity and/or selectivity [1, 2]. These alterations can occur at all stages of the catalyst life, not only during the catalytic transformation (hot spots in exothermic processes, etc.) and the regeneration via coke combustion, but also during the pretreatment steps (calcination, reduction, etc.) [2]. They can result from three processes: (i) chemical transformations of catalytic phases into inactive phases, which can occur with or without change in the global composition of the catalyst [3]; (ii) loss of the active surface area of the support or of the bulk catalyst due to their collapse; (iii) loss of the active surface area due to crystallite growth or agglomeration of the supported catalytic phase. The two last processes are typically referred to as sintering. In general, sintering processes and chemical alterations are irreversible or difficult to reverse, and can be more easily prevented than cured [1].

Three examples are presented hereafter to describe the chemical and structural alterations undergone by zeolite catalysts, their deactivating effect as well as the ways to prevent, limit and cure this deactivation. The first example deals with the chemical transformation of a catalytic phase into a non-catalytic phase, the two others concern respectively the sintering of zeolites and of noble metals supported on these molecular sieves.

8.2. Transformation of Active Phases During the Catalytic Destruction of Dichloromethane (DCM)

The catalytic destruction of this toxic volatile organic compound was carried out by oxidation on a series of MFAU catalysts (where M was an alkali cation) under the following conditions: fixed-bed reactor, 1000 ppm of DCM in wet air, high space velocity ($20{,}000\,h^{-1}$), temperature (T) between 493 and 723 K [4a, b, c]. The FAU zeolites with Si/Al ratio of 1.3 (X type) or of 2.45 (Y) were used alone or in intimate mixture with an equal amount of 0.4 wt% Pt/SiO_2. With all the catalysts, a rapid deactivation could be observed followed by a plateau in activity. Thus, on a NaY zeolite at 573 K, the DCM conversion (X_{DCM}) decreased from 80% at 5 min time-on-stream (TOS) to 61% at 15 min, then remained practically constant (e.g. 58% after 3 days). The adding of Pt/SiO_2 to NaY had practically no effect on X_{DCM} but caused a drastic change in the product distribution (Fig. 8.1). However, with all samples, CO_2, HCl and formaldehyde appear in the reaction products in relative amounts depending on T and on TOS but CO was only formed with pure NaY. Neither organic products other than formaldehyde nor Cl_2 and $COCl_2$ could be detected in the effluents. Note that the HCl/(formaldehyde + CO + CO_2) molar ratio was always equal to 2 on the stabilized catalysts (as expected from DCM composition) but lower than 2 in the initial period of deactivation. This latter observation shows that part of the DCM chlorine atoms was retained on the catalysts.

Schemes for DCM transformation over the stabilized NaY and NaY-Pt/SiO_2 mixture catalysts can be proposed from the change with T of the product yields (Fig. 8.1). On NaY, there are three apparent successive

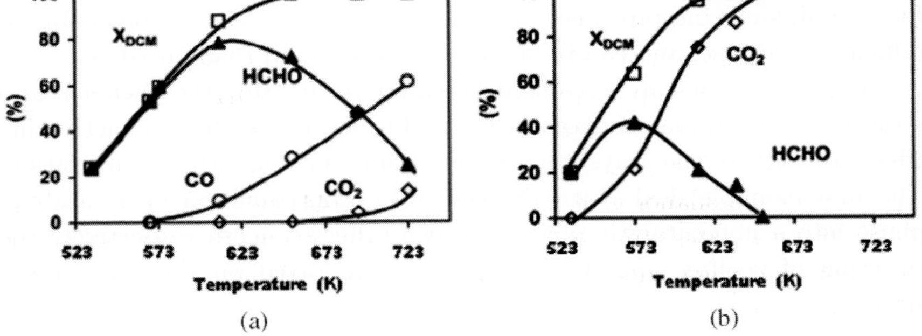

Fig. 8.1. Conversion of dichloromethane (X_{DCM}) and yields in the reaction products vs. temperature over stabilized samples of (a) NaY (140 mg) and (b) NaY + Pt/SiO_2 (140 + 140 mg). Adapted from [4a].

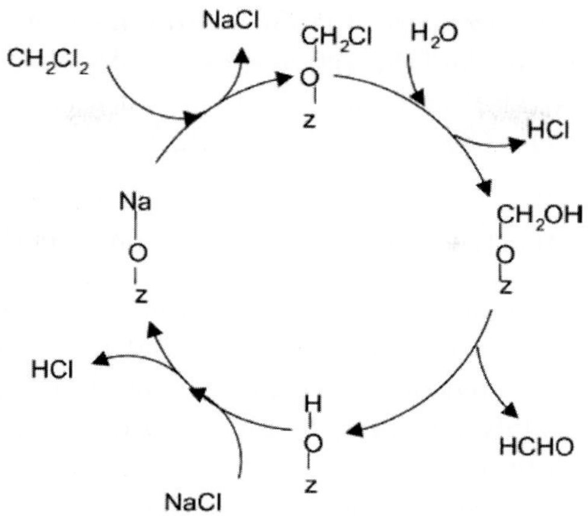

Fig. 8.2. Catalytic cycle of dichloromethane (CH_2Cl_2) hydrolysis over a NaY zeolite sample [4a].

steps: DCM hydrolysis into formaldehyde and HCl, partial oxidation of formaldehyde into CO, then oxidation of CO into CO_2, whereas with the mixture, the two oxidation steps cannot be discriminated. The very close X_{DCM} values on the two types of catalyst show that DCM hydrolysis is the controlling step of DCM transformation [4a]. The catalytic cycle in Fig. 8.2 which involves well-known adsorbed intermediates was proposed to account for the selective and stable hydrolysis of DCM hydrolysis DCM [4a].

To identify the cause(s) of the initial loss in the hydrolysis activity, the physicochemical properties of the NaY sample were determined as a function of TOS. Table 8.1 shows that, in agreement with the initial deactivation, the catalyst undergoes various changes in the first 15 min of reaction: (i) retention of a small amount of coke and of chlorine and removal of Na; (ii) elimination of the strongest sites for DMC adsorption evidenced by calorimetry; (iii) formation of bridging OH groups (IR band at 3649 cm^{-1}) and creation of protonic acid sites demonstrated by pyridine chemisorption experiments. In contrast, neither the accessible micropore volume nor the silanol groups (IR band at 3744 cm^{-1}) are affected and no Lewis acid sites are created. This totally excludes deactivation by micropore blockage, modification of the outer surface or collapse of the zeolite framework.

All the alterations of NaY (except coke deposits) can be simply explained by considering that on the fresh catalyst, step 4 of the catalytic cycle (Fig. 8.2) is strongly displaced toward the formation of bridging OH groups. As a consequence, part of DCM hydrolysis becomes stoichiometric with transformation of the ONa active sites into bridging OH groups.

This equation also accounts for the values of the molar HCl/ formaldehyde (+CO +CO_2) lower than 2 which were observed during the deactivation period.

Reaction 8.1 $CH_2Cl_2 + H_2O + NaOZ = HCHO + HCl + NaCl + HOZ$

The displacement of step 4 (Fig. 8.2) toward the formation of OH groups can be due to a too low amount of NaCl species in the vicinity of the OH groups created in step 3. Indeed, the decrease in the amount of sodium of the catalyst (−13%) in 2 h indicates the migration of these species along the reactor; and/or to the weak acid strength of the first OH groups which are created and the strong basicity of the first ONa groups that are exchanged (demonstrated by the $\Delta_{ads}H^o$ values in Table 8.1). This latter explanation was confirmed by the absence of deactivation found with a partially exchanged NaY zeolite [4a].

With more basic zeolites, NaX and CsY, the same initial period of deactivation with elimination of the strongest sites for DCM adsorption were observed [4b, c]. However, much higher values of activity (~5 times) could be observed over the stabilized samples, which can be related to the higher strength of the residual active sites for DCM chemisorption.

It should be underscored that the initial period with fast loss in activity observed during DCM transformation can be found with many types of catalyst and for deactivation causes other than chemical alterations.

Table 8.1. Influence of DCM transformation in the presence of water on the physicochemical properties of NaY. Adapted from [4a].

	A	B	C	D	E
Reaction time (h)	0	0.09	0.25	2	72
wt% C	0	0.15	n.d.	0.21	0.11
wt% Na	10.5	n.d.	n.d.	9.15	8.4
wt% Cl	0	0.7	n.d.	1.4	1.1
$\boldsymbol{\Delta_{ads}H^o}$ **(kJmol^{-1})**	−72 → −54	n.d.	n.d.	n.d.	−50
***A*3749 cm^{-1}**	0.26	0.39	0.34	0.34	0.35
***A*3644 cm^{-1}**	0	0.19	0.59	0.92	0.25
***A*3617 cm^{-1}**	0	0	0	0	0.04
$\boldsymbol{C_B}$ **(μmolg^{-1})**	0	72	103	107	64
$\boldsymbol{C_L}$ **(μmolg^{-1})**	0	0	0	0	0

A: integrated absorbance.
C_B, C_L: concentrations of Brønsted and Lewis acid sites.

Generally, this initial deactivation is simply considered as an adjustment of the catalyst features to the operating conditions. However, the deactivation phenomena which occur during this period are, though much more pronounced, often similar to those responsible for the subsequent slower loss in activity, and hence deserve to be carefully examined.

8.3. Structural Alteration of Acid Zeolite Catalysts

Zeolite catalysts can be chemically modified in order to exploit their unique shape-selective properties. In some cases, the ideal catalyst is the one where the active sites are exclusively located on (or close to) the external surface. On the other hand, some reactions are favoured inside the micropores, requiring a passivation of the zeolite's external surface.

Selective hydrodewaxing of long-chain alkanes is a classical example of the lock-and-key mechanism, where only the sites located close to the external surface of ZSM-22 and of a silico-aluminate-phosphate (SAPO-11) material are desired [5]. Toluene disproportionation is a reaction where *p*-selectivity can be increased due to product shape selectivity. This can be accomplished with passivation of the external surface of an intermediate-pore zeolite such as ZSM-5. Modifying ZSM-5 with antimony oxide by solid-state reaction, a strong interaction with the hydroxyl groups of the zeolite, i.e. silanol and bridging hydroxyl groups located at the external zeolite surface, and with extraframework aluminum oxide is created, resulting in an enhancement of diffusion constraints and thus an increase in *p*-xylene selectivity [6]. A similar effect can be obtained through partial replacement of zeolite protons with indium. In_2O_3 can be exchanged with the NH_4-form of the zeolites in a different In: Al molar ratio. Working with FAU, MFI, MWW and BEA zeolites, it has been found that this modification largely depends on zeolite pore structure [7]. Large-pore systems favour coke formation and thus the improved selectivity is accompanied by an activity loss.

A key parameter of zeolite catalysts is their thermal stability. This stability depends on the framework structure and composition, type of compensation cation(s) and may be influenced by the presence of contaminant metals. Thermal treatments may be responsible for various structural changes [8, 9], including: (i) cell volume contraction due to the removal of water and/or organic template molecules, (ii) transformations into a more metastable phase, (iii) structural collapse, (iv) amorphization and (v) negative thermal expansion.

For low values of their framework Si/Al ratio, the thermal stability of H-FAU zeolites can be rather small. Their structure starts experiencing changes at temperatures around 720–820 K. Mesopore molecular sieves tend to present an even lower thermal stability, which severely limits their use in commercial processes.

For commercial purposes, more important than the thermal stability is the zeolite hydrothermal stability, i.e. the resistance to the combined effect of steam and temperature. The reaction between the zeolite hydroxyl groups and steam at high temperatures results in framework dealumination and loss of acid sites. After dealumination, the zeolite, although less acidic, becomes more thermally and hydrothermally stable than the parent catalyst. This is one of the reasons why some zeolites are usually steamed under optimized conditions before use in catalytic processes. Ultrastable FAU (USY) is the name given to a FAU zeolite submitted to this type of treatment.

After dehydroxylation [10], the system ends up in a metastable state; the Al can then be easily released from the zeolite framework, producing extraframework Al (EFAL) species. This phenomenon is amplified by the existence of the next nearest neighbor (NNN) position to Al, which decreases the stability of the initial Al atom [11]. In high Si/Al zeolites, where all the acid sites are isolated, the second step is less favoured. Various types of EFAL species can be present in zeolites dealuminated by steaming, e.g. Al^{3+}, AlO^{+}, $Al(OH)_2^{+}$, $AlO(OH)$, $Al(OH)_3$, $AlO(OH)$ [11, 12].

Steaming is not only important from a structural point of view. It also introduces alterations in the concentration, strength and nature (Brønsted or Lewis) of the acid sites [13] and may lead to an increase of the zeolite catalytic activity. Furthermore, steaming not only dealuminates the zeolite, with effects on the remaining acid site strength and distribution, but also creates mesopores that favour the accessibility of bulky molecules to the active sites [14].

Another possibility to increase the thermal stability of H-FAU zeolites widely used in FCC catalysts formulations is through their exchange with rare-earth (La, Ce and Pr) cations [15]. At moderate exchange levels, the enhanced stability relates to the formation of coordination bonds between rare-earth cations and the framework oxygen atoms.

The presence of contaminant metals such as vanadium and sodium can accelerate the damage of the zeolite structure, especially in the presence of steam at high temperatures. Thus, in the FCC regenerator, V deposited on FCC catalysts is oxidized and hydrolysed into vanadic acid which provokes

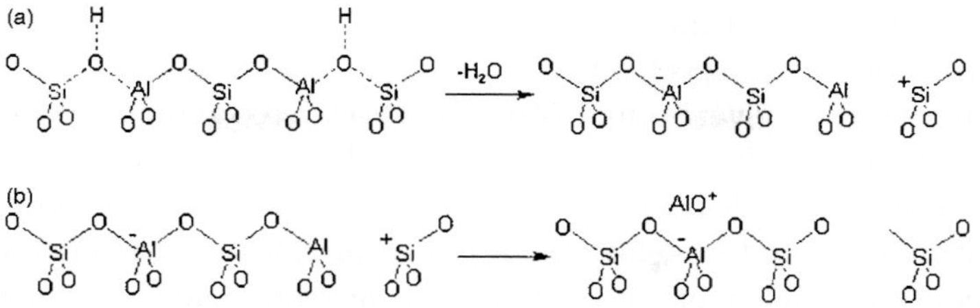

Fig. 8.3. Two-step mechanism for the production of EFAl species. (a) Dehydroxylation and (b) Al segregation. Adapted from Ref. [10].

the destruction of the FAU zeolite framework. One solution to this problem is to use vanadium traps, i.e. solid bases that react with H_3VO_4 to form stable vanadates. The La_2O_3-based traps, which are very effective, can be either introduced as separate particles or directly integrated within the FCC catalysts [16].

8.4. Sintering of Noble Metals Supported on Zeolites

8.4.1. *Mechanisms, kinetics and deactivating effect*

Three mechanisms have been advanced to account for the growth of metal crystallites: (i) crystallite migration, (ii) atom migration and (iii) (at very high temperatures) vapour transport with the possibility of coupling with each other. The first mode of crystallite growth involves migration, collision and coalescence, the second the detachment of atoms from crystallites, their migration over the surface of the support or along micropores, then their capture by crystallites [1].

The loss in the surface area accessible by the reactant molecules due to metal sintering provokes a decrease in the catalyst activity, this decrease depending on the reaction. According to Boudart [17], metal-catalysed reactions can be classified into two types: (i) Structure-insensitive reactions, the rate of which is proportional to the accessible metal area, and hence whose activity of all the accessible metal sites (i.e. their turnover frequency, TOF) can be considered as identical. Therefore, the decrease in activity due to sintering is proportional to the decrease in the accessible metal area. (ii) Structure-sensitive reactions, the rate of which is not proportional to the metal area, which can be related to an effect of the size of metal crystallites on the TOF value. With these reactions, the impact of sintering on the catalyst activity may be either magnified or moderated depending on the positive or negative effect of the crystallite size [1]. Benzene hydrogenation and alkane hydrogenolysis are typical examples of structure-insensitive and -sensitive reactions, respectively.

Sintering rates have been historically correlated by empirical equations such as $-d(D/D_0)/dt = k_s(D/D_0)^n$ where k_s is the sintering rate constant, D_0 the initial metal dispersion and n the sintering order. This order could vary with the catalyst systems from 3 to 15. Moreover, for the same system, it generally depended on the operating conditions and even on time t, which makes it impossible to quantitatively compare sintering kinetics for different catalytic systems. A more satisfying kinetic equation was proposed [18]: $-d(D/D_0)/dt = k_s(D/D_0 - D_{eq}/D_0)^n$ in which the additional term D_{eq}/D_0 accounts for the general observation that D approaches a limiting value with time t. This equation allows a best fit of the sintering rate data with order (n) values of 1 or 2 whatever the t value. It should be underscored that these values could be expected from the fundamental steps involved in sintering modes which are either first- or second-order processes [19].

Analysis of the sintering literature data through this last kinetic equation have shown that the rate of metal sintering depends upon temperature, atmosphere, support and metal generally in this order of decreasing importance [19]. Temperature is the most important parameter with activation energies of 40–150 kJ/mol. Atmosphere exerts also a major influence; thus, in H_2 or N_2 atmosphere, metal sintering is relatively slow whereas in H_2O- or O_2-containing atmospheres it is generally faster. Another key factor is the strength of the metal support interaction; as it can be easily understood, the stronger this interaction, the slower the metal sintering. Furthermore, the mobility of metal atoms on the support can be decreased by promoters (e.g. Ca, Cs) or increased by feed impurities (e.g. sulphur and chlorine) with decrease or increase of the sintering rate. Similarly, the surface mobility of metal particles can be limited by defects or micropores and mesopores of the support [1].

8.4.2. *Sintering of metal supported on zeolites*

Even if few kinetic investigations were carried out over metal-containing zeolite catalysts, similar conclusions can be drawn in what concerns the effect of operating conditions and catalyst features on the sintering rate. An additional factor to be considered is the initial location of the metal atoms or crystallites. Thus, platinum particles of ~3 nm, located on the outer surface of HFAU zeolites, were shown to sinter at temperatures as low as 450 K. Isolated atoms in sodalite cages migrate out as soon they get enough energy to pass through the 0.22 nm apertures, then agglomerate. Pt clusters trapped within the supercages are stable in vacuum up to 1070 K, i.e. they sinter only after the collapse of the zeolite structure [20].

Highly dispersed metal zeolite materials are used in various monofunctional (e.g. C_6–C_7 aromatization) and bifunctional (e.g. alkane hydroisomerization) catalytic commercial processes. As suggested above, metals are preferentially located within the micropores, which limit sintering during the catalyst preparation, pretreatment and use in catalytic processes. The classical method for preparing highly dispersed metals in zeolites involves ion exchange, generally with amino complexes such as $[Pt(NH_3)_4]^{2+}$ (carried out in competition with NH_4^+ ions to obtain a homogeneous macroscopic dispersion) followed by calcination and reduction under optimized conditions [21].

Metal sintering can affect significantly the activity and selectivity of zeolite catalysts. Both changes are generally observed in monofunctional catalysis. In contrast, in bifunctional catalysis processes (e.g. hydroisomerization), the reaction rate is generally controlled by the acid function; hence the loss in activity appears only for a significant decrease in the accessible metal area. In contrast, the effect on selectivity can be immediate, e.g. increase in the yields of the undesired hydrogenolysis products because of creation by sintering of the large ensembles active in this reaction.

8.4.3. *Prevention of metal sintering in zeolite catalysts*

The solutions developed to limit the sintering of metals supported on non-microporous materials can also be applied to zeolite catalysts. A general problem with noble metal zeolite catalysts used in many refining and petrochemical processes originates from the presence of S feed impurities which not only poison the metals but also induce the growth of metal clusters, leading in certain cases to pore blockage. As the ultimate desulphurization of the feed is too costly, solutions to improve the resistance to sintering of supported metals were actively researched. Various promoters were proposed, e.g. in aromatization over PtKLTL catalysts, thulium which acts as a getter for sulphur [22], Ni cations which act as anchoring sites for Pt particles [23], etc. (Chapter 16). Fe ions and/or Fe oxo ions of Pd Fe/MFI were also shown to be very effective chemical anchors of palladium and platinum, limiting significantly the sintering process [24, 25].

Metal sintering may also essentially be due to the high temperatures to which the catalysts are submitted. Thus, automotive exhaust catalysts, typically Pt, Rh and Pd on a suitable support (generally γ alumina), are exposed to extreme temperatures (as high as 1270 K). The sintering of Pt particles is one of the main causes for catalyst deterioration. The susceptibility of Pt to sintering was recently shown to be significantly reduced by substituting the alumina by a MFI zeolite with a high Si/Al ratio (900) and presenting a significant intercrystalline mesoporosity (average mesopore Ø of ~20 nm). Pt particles fixed within these mesopores were found to be more resistant to sintering during high-temperature aging tests than those fixed on γ alumina: from the fresh to the aged samples, their average size increased from 3 to 17 nm (MFI) against 1 to 45 nm (Al_2O_3). This higher resistance was ascribed to mechanical constraints exerted by the mesopore structure on the Pt particles [26]. A similar limiting effect of mesopores created via the so-called carbon-templating procedure was demonstrated during the hydroconversion of *n*-heptane at 573 K in the presence of 100 ppm of S (ex 2-methylthiophene) over bifunctional Pd/HMFI catalysts [27]. Thus, on this hierarchical micro-mesoporous catalyst, the Pd dispersion passes from 47% to 14% after a 5 h reaction, while on a microporous Pt/HMFI of similar composition it decreases more significantly (from 19% to 3.5%). The inhibition of S-induced migration onto the external zeolite surface (then sintering) of Pd particles when confined within the mesopores was proposed to account for this higher sulphur resistance [28].

8.4.4. *Redispersion of metals in zeolite catalysts*

As redispersion is the reverse of metal sintering, the well-known principles of metal dispersion in zeolites [21] may constitute a solid basis for designing the redispersion treatments. It should however be underscored that metal redispersion in aged zeolite catalysts will be generally more complex for the following reasons: (i) sintering of the active metal species is often accompanied by their migration to the outer surface of the zeolite crystallite, (ii) generally, deactivation is not only due to metal sintering but to various other causes: metal poisoning, coking, structural zeolite alterations, etc. Therefore, in addition to the redispersion of the active metal species, catalyst regeneration has to address these other deactivation causes. The situation is even more complicated when more than one metal has to be redispersed (e.g. bimetallic catalysts).

Two main strategies can be adopted to break the strong metal-metal bonds in metal agglomerates [21]. One strategy is to reoxidize the metals into ions which migrate to zeolite cages where the density of the negative charge is highest. The first step is the metal reoxidation into oxide particles:

Reaction 8.2 $Pd_n + (n/2)O_2 \rightarrow (PdO)_n$

The second step involves reaction of zeolite protons with the metal oxide:

Reaction 8.3 $(PdO)_n + 2nH^+ \rightarrow nPd^{2+} + nH_2O$

This last step is virtually complete at 673 K for relatively small PdO particles and high proton concentrations in their vicinity. Subsequent gentle reduction results in the formation of small palladium particles in FAU supercages [28]. Large aggregates are oxidized into PdO (Reaction 8.2) but are not completely redispersed into ions (Reaction 8.3) because of the exhaustion of protons close to oxide particles.

In this case, the alternative strategy which was proposed [28] is to treat the partially redispersed Pd HFAU sample with NH_3, which gives rise to the formation of highly mobile $Pd(NH_3)_4^{2+}$ ions and NH_4^+ ions in supercage channels. Calcination regenerates protons and forms smaller PdO species that can transform into Pd+ (Reaction 8.3) and be further reduced into small Pd particles. Note that oxidative treatment is the first step in both strategies. Therefore, simultaneous metal oxidation and coke combustion can be envisaged, provided however this latter reaction does not require too high temperatures which can be made possible by the presence of metals.

8.5. Conclusion

Structural and chemical alterations of zeolite catalysts can affect each of the active (and non-active) components. These alterations are strongly favoured by high temperatures and hence can occur during reactions carried out at high temperatures but also during coke combustion. Their rate is governed by many parameters, temperature being generally the main factor; as a consequence these alterations can occur in all the steps of the catalyst life and especially during regeneration by coke combustion. Atmosphere and feed impurities are also determining factors. In general, these alterations are difficult to reverse. It is therefore preferable to prevent than to cure them. Among the various ways which were developed for this prevention, the appropriate design of the catalysts (active phases, promoters, traps, matrix, etc.) is often the best solution.

References

[1] Bartholomew C.H., *Appl. Catal. A: General*, 212 (2001) 17–60.

[2] Moulijn J.A., van Diepen A.E., Kapteijn F., (a) in *Handbook of Heterogeneous Catalysis*, Ertl G., Knözinger H., Schuth F., Weitkamp J. (Eds.), Wiley-VCH, Weinheim (2008) Chapter 7.1, 1–18; (b) *Appl. Catal. A: General*, 212 (2001) 3–16.

[3] Delmon B., Grange P., in *Progress in Catalyst Deactivation*, Figueiredo J.L. (Ed.), NATO Advanced Study Institutes Series, Series E, 54 (1982) 231–280.

[4] (a) Pinard L., Mijoin J., Magnoux P., Guisnet M., *J. Catal.*, 215 (2003) 234–244; (b) Pinard L., Magnoux P., Ayrault P., Guisnet M., *J. Catal.*, 221 (2004) 662–665; (c) Pinard L., Mijoin J., Magnoux P., Guisnet M., *C. R. Chimie*, 8 (2005) 457–463.

[5] Martens J.A., Souverijns W., Verrelst W., Parton R., Froment G.F., Jacobs P.A., *Angewandte Chemie Int. Ed.*, 34 (1995) 2528–2530.

[6] Zheng S., Jentys A., Lercher J.A., *J. Catal.*, 219 (2003) 310–319.

[7] Mavrodinova V., Popova M., *Catal. Commun.*, 6 (2005) 247–252.

[8] Nock A., Rudham R., *Zeolites*, 7 (1987) 481–484.

[9] Wang Q.L., Giannetto G., Guisnet M., *J. Catal.*, 130 (1991) 471–482.

[10] Kühl G.H., *J. Phys. Chem. Solids*, 38 (1977) 1259–1263.

[11] Chen T., Men A., Sun P., Zhou J., Yuan Z., Guo Z., Wang J., Ding D., Li H., *Catal. Today*, 30 (1996) 189–192.

[12] Elanany M., Koyama M., Kubo M., Broclawik E., Miyamoto A., *Appl. Surf. Sci.*, 246 (2005) 96–101.

[13] Katada N., Kageyama Y., Takahara K., Kanai T., Begum H.A., Niwa M., *J. Mol. Catal. A: Chemical*, 211 (2004) 119–130.

[14] Scherzer J., *Appl. Catal.*, 75 (1991) 1–32.

[15] Lemos F., Ramôa Ribeiro F., Kern M., Giannetto G., Guisnet M., *Appl. Catal.*, 39 (1988) 227–237.

[16] Habib E.T., Zhao X., Cheng W.C., Boock L.T., Gilson, J.-P., in *Zeolites for Cleaner Technologies*, Guisnet M., Gilson J.-P. (Eds.), Imperial College Press, London (2002) Chapter 5, 105–130.

[17] Boudart M., *Adv. Catal.*, 20 (1969) 153–166.

[18] Fuentes G.A., Gamas E.D., *Stud. Surf. Sci. Catal.*, 68 (1991) 637–644.

[19] Bartholomew C.H., *Stud. Surf. Sci. Catal.*, 88 (1994) 1–18.

[20] Gallezot P., *Catal Rev. Sci. Eng.*, 20 (1979) 121–154.

[21] Sachtler W.M.H., Zhang Z., *Adv. Catal.*, 39 (1993) 129–220.

[22] Jacobs G., Ghadiali F., Pisanu A., Padro C.L., Borgna A., Alvarez W.E., Resasco D.E., *J. Catal.*, 191 (2000) 116–127.

[23] Larsen G., Resasco D.E., Durante V.A., Kim J., Haller G.L., *Stud. Surf. Sci. Catal.*, 83 (1994) 321–329.

[24] Wen B., Jia J., Sachtler W.M.H., *J. Phys. Chem. B*, 106 (2002) 7520–7523.

[25] Li X., Iglesia E., *J. Catal.* 255 (2008) 134–137.

[26] Kanazawa Y., *Appl. Catal. B: Environm.*, 65 (2006) 185–190.

[27] Martinez A., Arribas M.A., Derewinski M., Burkat-Dulak A., *Appl. Catal. A: General*, 379 (2010) 188–197.

[28] Feeley O., Sachtler W., *Appl. Catal.*, 67 (1990) 141–150.

Chapter 9

MODELLING OF DEACTIVATION BY COKE FORMATION

M.-F. Reyniers, J.W. Thybaut and G.B. Marin

9.1. Introduction

Many zeolite-catalysed processes are prone to deactivation. In general, there are several possible reasons for catalyst deactivation [1–4]. Most commonly, deactivation is caused by structural and physical changes in the catalyst, such as sintering [3]; by fouling or poisoning with components present in the feed, such as nitrogen-containing components [5]; or by "coke", i.e. by heavy hydrocarbons that result from side reactions and that are retained on the catalyst. Frequently, catalyst deactivation also alters the product distribution even at a given conversion. The activity and selectivity of the catalyst may completely or partially be restored by burning off the coke in a controlled manner.

Two modes of deactivation of zeolites by coke are generally considered [6, 7]: (1) Site coverage consists in the poisoning of acid sites by coke adsorption. Typically, one acid site per coke molecule is poisoned. (2) Pore blockage occurs when active sites are rendered inaccessible to reactants. The deactivating effect is more pronounced than with site coverage since coke can block access of the reactants to more than one active site. A distinction between blockage of the pore entrance, thereby hampering diffusion of reactants into the zeolite structure (external coke), and internal pore blockage has also been made [6, 8, 9]. Catalyst deactivation by site coverage and pore blockage, and in particular the modelling aspects, have been investigated thoroughly by Froment *et al.* [2, 7, 10–13].

Knowledge of the kinetics of both the "main" reactions and the deactivation is required for the design and optimization of zeolite-catalysed industrial processes as well as for the development of new zeolite catalysts. Since the catalytic activity mostly declines with time-on-stream due to coking, the cycle time before regeneration of the catalyst is limited. Therefore, the optimal operating policy of commercial reactors employing a zeolite catalyst that undergoes deactivation relies heavily on an accurate description of the deactivating effect of coke formation on the reaction kinetics.

9.2. Modelling of Zeolite Deactivation by Coke Formation

Different approaches have been developed to describe the effect of coke formation on the catalyst activity due to coke formation. Most frequently, the deactivating effect of coke on the rate of a reaction $A \rightarrow B$, the so-called main reaction, is expressed by means of a deactivation function, Φ_A, relating the reaction rate on the coked catalyst, r_A, to the rate on the fresh catalyst, r_A^o,

$$\Phi_A = \frac{r_A}{r_A^o}. \tag{9.1}$$

In principle, the deactivation function as defined in Eq. 9.1 pertains to the effect of coking on one chemical reaction only. When diffusional limitations are present, a ratio of the effective rates could be taken, which would not, however, correspond to the deactivation function proper [9, 13].

The deactivating effect of coke on the kinetics is often taken into account by a single deactivation function describing the effect of coke on the global catalyst activity, i.e. not distinguishing between the different main reactions [14–16]. However, in most zeolite-catalysed processes, it is often necessary to describe the catalyst deactivation with more than one deactivation function since the effects of coke on the various reactions are not necessarily identical. The deactivating effect of coke on the rate of a particular reaction depends on several factors, such as the number of sites involved in the reaction, the acid strength distribution, site density and texture of the catalyst [11, 17–19]. Accounting for all of these effects on an explicit basis is very difficult because of the complexity of the resulting kinetic model and the number of structural catalyst properties to be taken into account. For that reason, (semi) empirical deactivation functions are often used to model the deactivating effect of coke and changes in the product distribution.

So-called "time-on-stream" models (see Table 9.1), in which the deactivation functions are expressed as a function of time-on-stream

Table 9.1. Frequently used deactivation functions.

$\Phi = f(t)$	$\Phi = f(C_c)$
$1 - \alpha t$	$1 - \alpha C_c$
$\exp(-\alpha t)$	$\exp(-\alpha C_c)$
$\frac{1}{1+\alpha t}$	$(1 - \alpha C_c)^2$
$\alpha t^{-0.5}$	$(1 + \alpha C_c)^{-1}$

t: time-on-stream; C_c: coke content.

without any explicit link to the operating conditions (p_i, T), are frequently used due to their simplicity [2, 7, 13]. The development of "time-on-stream" models does not require the measurement of the amount of coke on the catalyst. Since the deactivation functions are independent of operating conditions, this approach leads to inaccurate predictions of the activity profile along the reactor [13]. More accurate predictions of the activity profile along the reactor can be obtained using more complicated deactivation functions that also depend on operating conditions [20, 21].

It is generally recognized that the catalyst activity depends on its coke content C_c and hence is only implicitly dependent on the time-on-stream t [2, 7, 13]. Therefore, more reliable descriptions of the decay in catalyst activity can be obtained when the deactivation functions are expressed as a function of a variable related to the cause of the deactivation, i.e. the coke deposited on the catalyst [2, 7, 13]. In this approach, a kinetic model for coke formation is needed to express the rate of coke formation as a function of operating conditions while the deactivating effects of coke on the reaction rates are described using empirical deactivation functions depending on the coke content [22]. Some frequently used deactivation functions are presented in Table 9.1.

In general modelling practice, the mechanisms of site coverage and pore blockage are then used in conjunction with a set of hypotheses concerning coke formation and deactivation. Typically, coke growth is considered to occur from a coke precursor adsorbed on an acid site and to proceed independently of the "main" reactions. Coke precursors species are generally defined as species that contribute directly to coke formation [23] or more precisely to the components which are involved as reactants in the fast irreversible step leading to coke. Derouane *et al.* [24] proposed a general scheme for coke formation starting from acyclic species in which oligomers, cyclic species and aromatics are considered to be important:

$$\text{alkanes} \rightarrow \text{alkenes} \rightarrow \text{oligomers} \rightarrow \text{cycloparaffins} \rightarrow \text{aromatics} \rightarrow \text{coke}$$

This general scheme was later refined by Guisnet and Magnoux [25] who identified coke formation with polyaromatics formation:

$$\text{alkanes} \rightarrow \text{alkenes} \rightarrow \text{oligomers} \rightarrow \text{cycloparaffins}$$
$$\rightarrow \text{monoaromatics} \rightarrow \text{polyaromatics}$$

An acid site covered by a growing coke molecule is thus considered to be exclusively engaged in coke formation and therefore less acid sites are available for the "main" reactions. From this it follows that occupation of an acid site with a precursor or a coke molecule necessarily leads to the unavailability of this site for the "main" reactions and hence to a surface with a diminished reactivity for the "main" reactions. From this point of view, catalyst deactivation may also be described using the concept of an effective space velocity as proposed by Dahl and Kolboe [26, 27] and Sapre [28] or as a loss of the effective amount of catalyst in the reactor [29]. The rate at which the effective amount of catalyst decreases is then taken as a direct measure of deactivation by coke formation. These approaches have been applied to model deactivation in methanol conversion over ZSM-5 [28, 29].

Also, precursor and coke molecules are typically considered to be inert molecules with respect to the "main" reactions. It has however been found that coke formation affects both the activity and product distribution due to the interaction of intermediates participating in coke formation with other species. For instance, coke formation has been found to have an activating effect on the conversion of *i*-butenes over zeolites [30, 31]. Clearly, in this case a reliable description of the effect of coke formation on the activity and selectivity requires the involvement of precursor and coke molecules in the conversion reactions to be explicitly accounted for. For this reason, detailed kinetic models for coke formation have been recently developed that explicitly describe the competition between the coking and the "main" reactions [32–35]. In this way, the effect of coking on the activity and product distribution is, at least partially, inherently accounted for in the kinetics, and the traditionally used deactivation functions then account for the decrease in available acid sites due to poisoning and pore blockage [35]. This approach requires a detailed description of the underlying chemistry of the process. The resulting models provide a wealth of detailed information on the effects of the operating conditions of the activity and the selectivity of the catalyst and hence are a powerful tool not only for process optimization but also for the development of new and more effective catalysts [36]. However, the number of elementary reactions involved is gigantic and the development of these models relies on the use of specialized computer algorithms for the construction of the reaction scheme and the kinetic equations. The single-event microkinetic (SEMK) methodology [37] for both the main reactions and coke formation has been applied to describe the effect of deactivation on methanol-to-olefins [32], solid acid alkylation [33] and fluid catalytic cracking [34, 35, 38, 39]. The approach will be illustrated for fluid catalytic cracking, focusing on the kinetic description for coke formation and deactivation.

9.3. Case Study: Fluid Catalytic Cracking

Catalytic cracking of heavy oil fractions is one of the major processes in the refining industry. It produces lighter, more valuable, cracked products such as LPG, gasoline and middle distillates as well as light alkenes such as ethylene, propylene or isobutene from high molecular mass petroleum fractions, i.e. atmospheric and vacuum gas oils, etc. at temperatures in the order of 770–800 K. Industrial cracking catalysts are typically composed of 20–40 wt% of an ultrastable Y zeolite in an amorphous matrix of silica-alumina. During the catalytic cracking process a significant part of the feedstock is converted into carbonaceous deposits that are retained on the catalyst, i.e. coke. In industrial practice, the heat-balanced riser-regenerator technology is used to accommodate the rapid deactivation of the catalyst. The amount of coke on the catalyst thus constitutes an important factor since its combustion in the regenerator provides the necessary heat for feedstock vaporization and for the endothermic cracking reactions [1, 35]. The coke present on the catalyst leaving the industrial riser-reactor may have different origins and can be classified into different categories, e.g. thermal, contaminant, non-stripped hydrocarbons and catalytic coke [3, 40–42]. Catalytic coke is formed in the zeolitic part of the catalyst and contributes to practically 50 wt% of the final coke. Consequently, the accurate prediction of catalytic coke formation is an indispensable part of a simulation model for an industrial riser-reactor.

9.3.1. *Single-event microkinetics for coke formation*

In the single-event microkinetic (SEMK) methodology, the reaction network is generated based on the classical carbenium and carbonium ion chemistry occurring on the acid sites of the zeolite [22]. The main advantage over other models is that it uses feedstock-independent rate coefficients, which is of paramount importance considering the continuous variation in the composition of catalytic cracking feedstocks [43]. The molecules and ions are represented by a vector, i.e. a label which is transformed into a Boolean matrix in order to generate their chemical transformations according to the reaction rules corresponding to the specified elementary steps. Afterwards, the product Boolean matrix is converted back into the corresponding label. This algorithm has been extensively described in the literature [44, 45]. Within a reaction family the rate coefficients are assumed to depend only on the type of carbenium ions involved as reactant and product and on the global symmetry number of the reactant and the transition state species. The global symmetry numbers can be extracted from the rate coefficient and calculated from statistical thermodynamics, leading to the so-called single-event rate coefficient, $\tilde{k}$, which is assumed to depend only on the reaction family and the type of carbenium ions involved. These assumptions drastically reduce the number of parameters required to describe the complex kinetics involved.

During the cracking of hydrocarbons a huge number of reactions take place; however, the occurring elementary steps can be classified into a limited number of reaction families, e.g. (de)protonations, beta-scissions, alkylations, protonated cyclopropane (PCP) isomerizations, etc. Even though coking is a complex process, there are only a limited number of elementary reaction families involved and, at least as important, the same as those involved in the main reactions, i.e. hydride transfers, alkylations, cyclizations and deprotonations [35]. Recognizing that the same elementary steps are involved in coke formation as in the "main" cracking reactions, Moustafa and Froment [34] and Quintana-Solorzana *et al.* [35, 38, 39] have presented SEMK models incorporating the simultaneous occurrence of the cracking and the coke formation reactions. In the model presented by Moustafa and Froment [34] some global reactions are still present, e.g. aromatization, whereas some relevant elementary reactions, e.g. cyclization of acyclic species and nucleus alkylation of polyaromatics, are not considered. In the model presented by Quinata-Solorzana *et al.* [35, 38, 39], coke precursors and coke are entirely described in terms of elementary reactions. Three-ring aromatic molecules, which are phenantrenic and anthracenic in nature, were selected as coke precursors. To illustrate the elementary reactions considered in the formation of coke precursors, a detailed reaction pathway for coke precursor formation from 1-butene is presented in Fig. 9.1.

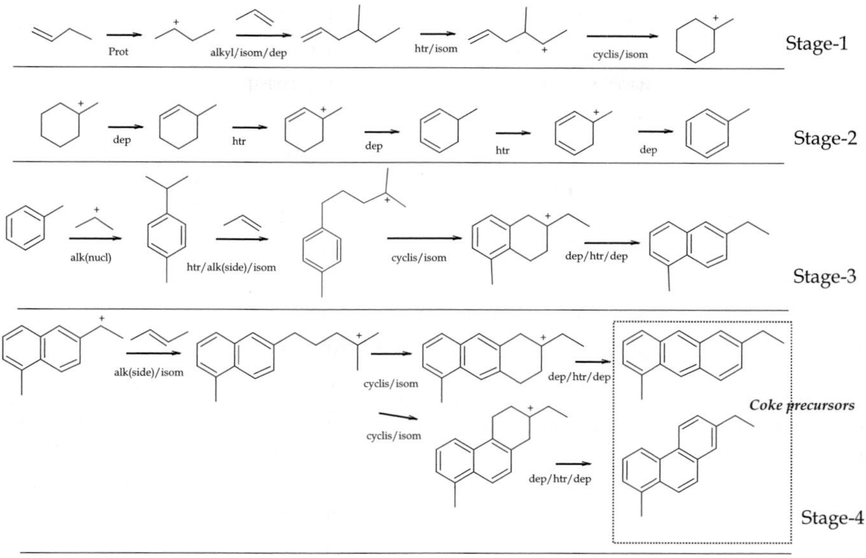

Fig. 9.1. Detailed reaction pathway for the formation of coke precursors starting from 1-butene. To avoid overloading the figure, the reverse reaction steps have been omitted [35].

The rate-determining step in the conversion of coke precursors to coke is considered to be the alkylation of coke precursors that are present in the feedstock or that are formed during the catalytic cracking. As illustrated in Figs. 9.2 and 9.3, both side-chain alkylation and nucleus alkylation of coke precursors are considered in the reaction network.

The number of species in the catalytic reaction network grows exponentially with the carbon number. Current analytical techniques are not capable of giving a detailed composition of complex feedstocks nor of the formed products. Moreover, the solution of the continuity equations for all the components in a reaction mixture leads to excessive calculation times. As a result, a certain degree of lumping is generally required [45]. In accordance with the SEMK methodology, this lumping is performed *a posteriori*, i.e. the fundamental character of the SEMK methodology is preserved and the benefits of a detailed description of all occurring elementary steps are retained. All elementary reactions occurring between the species of the lumps are accounted for via the so-called lumping coefficient. Therefore, this method can be denoted as late lumping or relumping. The reaction rate between lumps is found by the summation of the rates of all elementary reaction steps which transform carbenium

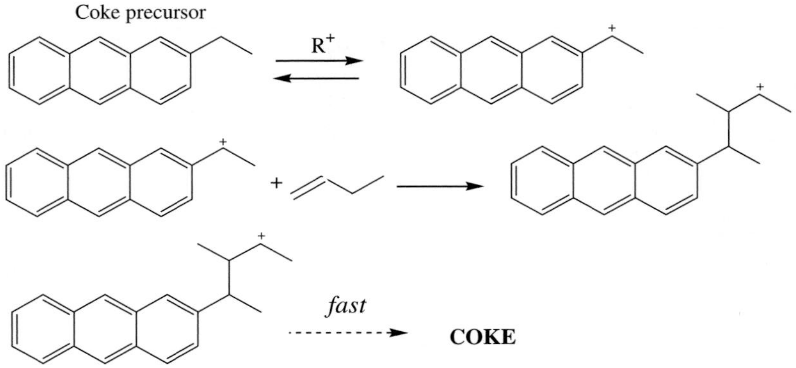

Fig. 9.2. Reaction pathway of coke formation via side-chain alkylation of coke precursors. R^+ is any carbenium ion that can abstract a hydride from a coke precursor [35].

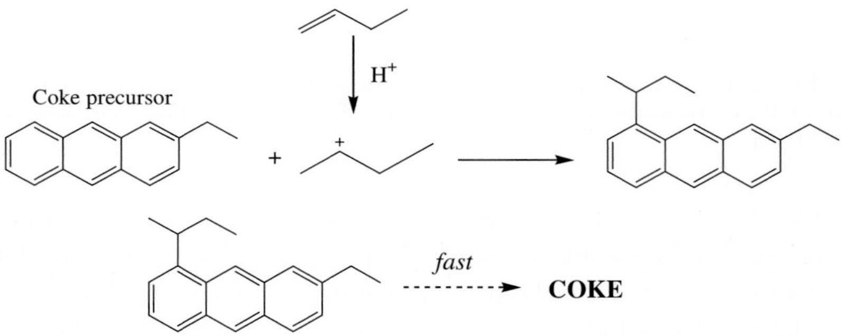

Fig. 9.3. Reaction pathway of coke formation via alkylation of the nucleus part of coke precursors [35].

ions formed out of the reactant lump into carbenium ions which desorb as molecules of the product lump.

In the relumped SEMK, the lump definition is based on analytical capabilities and the chemistry involved. A distinction is made between paraffins, isoparaffins, olefins, isoolefins, mono-, di-, tri- and tetra-naphthenes (cyclic olefins), mono-, di-, tri- and tetra-aromatics (aromatic olefins), naphthen-mono, di- and tri-aromatics per carbon number. Figure 9.4 illustrates the relumping strategy applied to the alkylation reactions for the coke formation from the coke precursors. For instance, Fig. 9.4a shows that all coke precursors in lump L_1 undergo hydride abstraction to produce all corresponding carbenium ions that can be

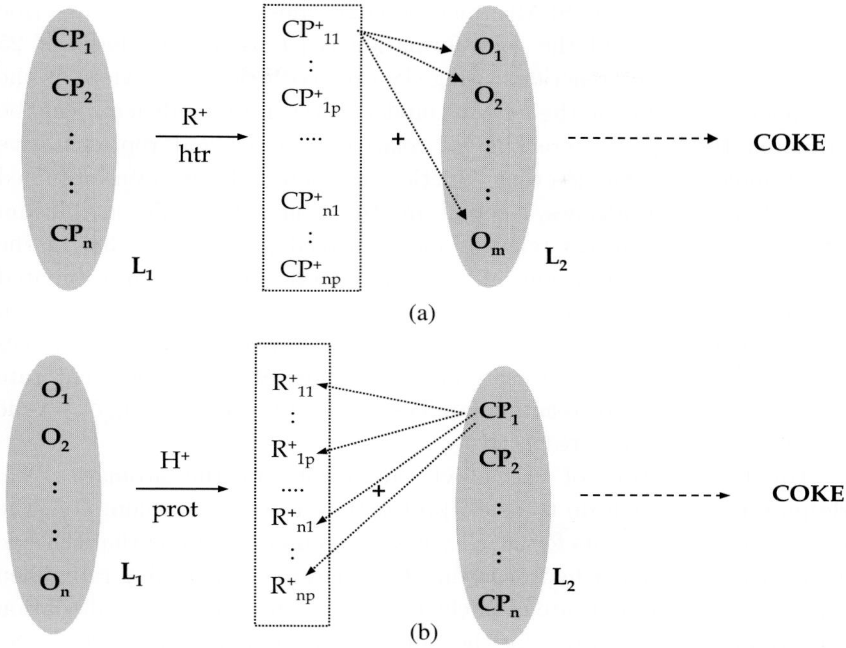

Fig. 9.4. Lumping scheme of the alkylation reactions of coke precursors leading to coke: (a) side-chain alkylation and (b) nucleus alkylation. CP_i are coke precursors, CP^+_{ij} are carbenium ions formed via hydride transfer of coke precursors, O_i are olefins, R^+_{ij} are carbenium ions formed via protonation of olefins and R'^+ is any carbenium ion [35].

alkylated with the olefins contained in lump L_2. Evidently, each carbenium ion formed from coke precursors can react with each olefin in lump L_2. The rate of consumption of the coke precursors contained in lump L_1 is derived from the product of the partial pressure of the reacting lump, the single-event rate coefficient for the elementary step involved and the so-called lumping coefficients. The latter coefficients account for the summations involved in the calculation of the rates of consumption of the species contained in that lump. Assuming a certain composition within a lump, which are typically defined such that an equimolar composition or thermodynamic equilibrium is a good approximation of the reality, these coefficients can be calculated *a priori* and hence the time-consuming summations to account for the entire network do not need to be performed in each single simulation [44]. It should be stressed that the lumping coefficients are independent of the single-event rate coefficients.

Application of the SEMK methodology thus allows us to describe both the cracking and the coke formation reactions on the basis of 25 types of elementary reactions using 48 rate coefficients. In view of the fundamental nature of the SEMK model, these rate coefficients can be estimated through the cracking of relatively simple and representative model molecules. Single-event kinetic parameters have been obtained by cracking (cyclo)alkanes/1-octene mixtures on a RE-USY equilibrium catalyst at temperatures relevant for industrial practice [35, 38] in the presence of coke formation. Activation energies have been estimated via non-isothermal regression while pre-exponential factors have been calculated using transition state theory and statistical thermodynamics [46, 47]. Tables 9.2 and 9.3 present the values of the single-event rate coefficients at 753 K for reaction families involved in the cracking of acyclic and cyclic components, respectively.

For the reaction of the alkyl side chains of the aromatics, i.e. (de)protonation, hydride transfer, protolytic scission, β-scission, etc., the obtained rate coefficients for acyclics have been used assuming that the side chains behave as linear hydrocarbons. Calculation of the coking rates then requires two additional rate coefficients only, i.e. one for nucleus alkylation of (poly)aromatics and one for the side alkylation of (poly)aromatics [38]. At 753 K, The values for these rate coefficients amount to 1.5 10^{-1} kg(kPa kgcat s)$^{-1}$ for nucleus alkylation and 4.7 10^{-3} kg(kPa kgcat s)$^{-1}$ for side-chain alkylation.

Besides giving rise to interactions between the species in the main cracking and in the coking pathways, coke formation also affects the rates of the main cracking reaction families via active site coverage and, in a later

Table 9.2. Single-event rate coefficients computed at 753 K for reactions involved in the cracking of acyclic hydrocarbons [39].

Reaction family	Reaction type			
	p	s		t
Hydride transfer	—	8.3 10^{-8}		9.0 10^{-3}
Protonation	—	2.0 10^{-6}		4.3 10^{-5}
Deprotonation[a]	1.8 10^{7}	5.8 10^{0}		3.3 10^{-3}
Protolytic scission	1.4 10^{-8}	9.2 10^{-10}		3.2 10^{-10}
	(s, s)	(s, t)	(t, s)	(t, t)
PCP-isomerization[a]	1.3 10^{0}	3.2 10^{4}	2.3 10^{-1}	2.3 10^{-3}
β–scission[a]	8.6 10^{-5}	2.1 10^{1}	5.8 10^{-2}	3.9 10^{-5}

p: primary, s: secondary, t: tertiary carbenium ions.
The units of the rate coefficients are: [a]kmol(kg_{cat} s)$^{-1}$ or kmol(kPa kg_{cat} s)$^{-1}$.

Table 9.3. Single-event rate coefficients calculated at 753 K for the reactions involved in the cracking of cyclic hydrocarbons [39].

Reaction family	Reaction type			
	p	s		t
Hydride transfer	—	$8.2\ 10^{-9}$		$2.6\ 10^{-5}$
Protonation	—	$2.3\ 10^{-7}$		$7.9\ 10^{-5}$
Deprotonation[a]	—	$1.1\ 10^{1}$		$3.9\ 10^{-2}$
Exo-protolytic scission[a]	$4.6\ 10^{-8}$	$2.9\ 10^{-8}$		$5.3\ 10^{-11}$
Protonation cycloalkadienes/aromatics	—	$1.1\ 10^{-5}$		$3.4\ 10^{-8}$
Deprotonation cycloalk(adi)enes[a]	—	$5.9\ 10^{4}$		$4.7\ 10^{-8}$
Hydride transfer alkenes	—		$2.1\ 10^{-8}$	
Hydride transfer cycloalk(adi)enes	—		$1.6\ 10^{-4}$	
Disproportionation	—		$1.2\ 10^{-1}$	
	(s, s)	(s, t)	(t, s)	(t, t)
PCP-isomerization[a]	$4.6\ 10^{2}$	$1.0\ 10^{0}$	$2.6\ 10^{-6}$	$3.6\ 10^{-3}$
Endo-β-scission[a]	$1.3\ 10^{-6}$	$1.4\ 10^{-2}$	$1.6\ 10^{-1}$	$5.7\ 10^{-4}$
Exo-β-scission[a]	$5.4\ 10^{0}$	$5.7\ 10^{4}$	$1.7\ 10^{-1}$	$1.6\ 10^{-6}$

p: primary, s: secondary, t: tertiary carbenium ions.
The units of the rate coefficients are: [a]$\mathrm{kmol(kg_{cat}\ s)^{-1}}$ or $\mathrm{kmol(kPa\ kg_{cat}\ s)^{-1}}$.

stage, pore blockage. To account for the deactivation by coke deposition, the total concentration of acid sites is not adjusted. Instead, the decrease in available sites has been accounted for through a deactivation function, Φ, mainly to allow a different deactivation effect for the various reaction families in the model. The reaction rate on a catalyst with coke content C_c is calculated from the reaction rate on a fresh catalyst from:

$$r_\mathrm{i} = r_\mathrm{i}^\mathrm{o}\Phi. \tag{9.2}$$

Because different reaction families require sites with different acid strengths, the deactivating effect of coke can be reaction-family dependent [22]. An exponential dependence on the coke content is put forward because of its general character in describing the loss of catalytic activity [48, 49] (see also Table 9.1):

$$\Phi_\mathrm{k} = \exp(-\alpha_\mathrm{k} C_\mathrm{c}), \tag{9.3}$$

in which α_k is the so-called deactivation constant. Values for α_k per reaction family were taken from Beirnaert *et al.* [22], cf. Table 9.4.

9.3.2. *Industrial riser simulation*

The cracking of a partially hydrotreated vacuum gas oil (VGO) consisting of 26 wt% alkanes, 18 wt% aromatics and 56 wt% cycloalkanes, with an average

Table 9.4. Deactivation constants α_k for the various reaction types and coke formation utilized during the simulations [22].

Reaction	α_k, $kg_{cat}(kg_{coke})^{-1}$
Hydride transfers	0.653
Protonations and PCP-isomerizations	0.148
Deprotonations (p) and (s)	0.031
Deprotonation (t)	0.127
β−scissions (s,s), (s,t) and (t,s)	0.825
β−scission (t,t)	0.407
Protolytic scission	0.445
Coking	1.951

Table 9.5. Riser dimensions, operating conditions and properties of the partially hydrotreated gas oil and the catalyst used in the simulations [39].

Parameter	Value
Dimensions	
Riser length, m	30.0
Riser internal diameter, m	0.94
Operating conditions	
Gas oil mass flow rate, kg s^{-1}(MBPD)	52.6 (33.0)
Cat-to-oil ratio, $kg_{cat}(kg_{feed})^{-1}$	6.0
Feed preheating temperature, K	519
Temperature of the regenerated catalyst, K	1003
Riser total pressure, kPa	193
Catalyst properties	
Catalyst density, kg m^{-3}	1300
Catalyst specific heat capacity, J(kg K)$^{-1}$	1003
Catalyst average particle diameter, μm	75
Feed	
Viscosity[a], Ns m^{-2}	1.1 10^{-5}
Enthalpy of vaporization, kJ(kg)$^{-1}$	150
Average molecular mass, kg(kmol)$^{-1}$	316

[a]Assumed average viscosity of the main hydrocarbons in the gas phase.

molecular mass of 316 g mol^{-1}, a density of 853.2 kg m^{-3}, a mean average boiling point of 685 K and a carbon to hydrogen ratio of 6.18 kg(kg)$^{-1}$ has been simulated in an industrial riser reactor (see Table 9.5). A tubular one-dimensional reactor where axial effective diffusion is neglected, i.e. ideal plug flow is assumed for both gas and solid phase, was used in the simulations.

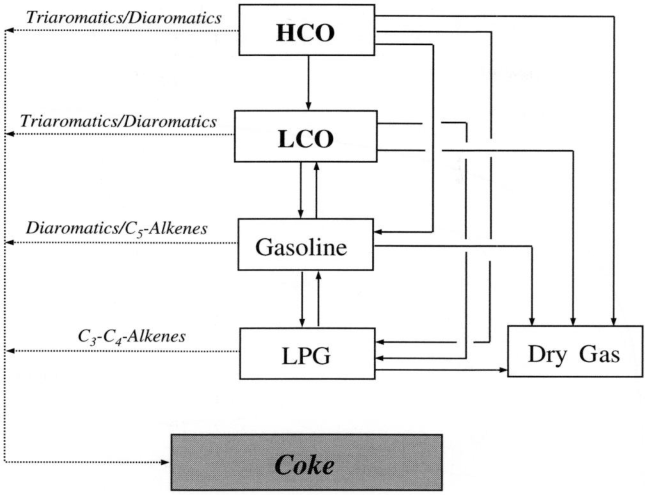

Fig. 9.5. Reaction routes for producing the various cracking products including coke, in terms of the cuts typically utilized in a refinery. Coking pathways are denoted with dashed arrows and hydrocarbon types that are involved in coking from the different cuts are given in italics [39].

Figure 9.5 gives a general overview of the different cracking and coking pathways in terms of the cracked products typically used in a refinery during the conversion of the feed. Cracked hydrocarbon products are grouped according to their boiling point and/or carbon number in the following cuts: dry gas (C_1–C_2), LPG (C_3–C_4), gasoline (C_5–C_{12}), light cyclic oil (LCO, C_{13}–C_{20}) and heavy cyclic oil (HCO, C_{21+}). Coke deposited on the catalyst closes the mass balance between these product fractions and the feed. The simulation results are expressed in terms of the yield of these hydrocarbons cuts. Figure 9.6 illustrates the yield profiles of the cracked products as a function of the riser axial position obtained from simulations performed at the operating conditions displayed in Table 9.5. The observed profiles are in good agreement with those reported in literature [50–52].

The coke yield profile along the riser obtained in the simulation is presented in Fig. 9.6. A moderate to rapid increase in coke yield is observed during the first metres of the riser, which is generally accepted for industrial operation and is typically reported in the literature [52, 53]. A more pronounced increase in coke yield would have been expected when processing a feed with a higher content of (poly)aromatics, e.g. a virgin gas oil. About 60% of the coke observed at the riser outlet is formed in

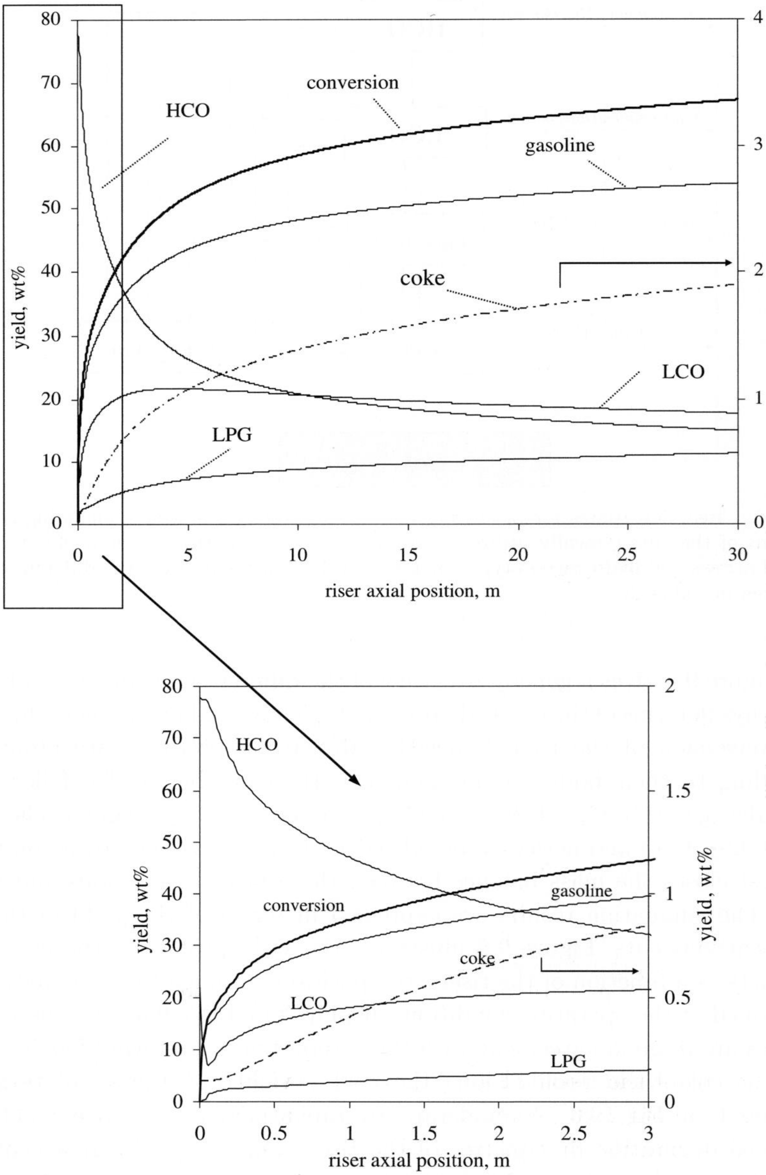

Fig. 9.6. Product yields and feed conversion on mass basis versus axial riser coordinate out of the catalytic cracking of a partially hydrotreated VGO [39].

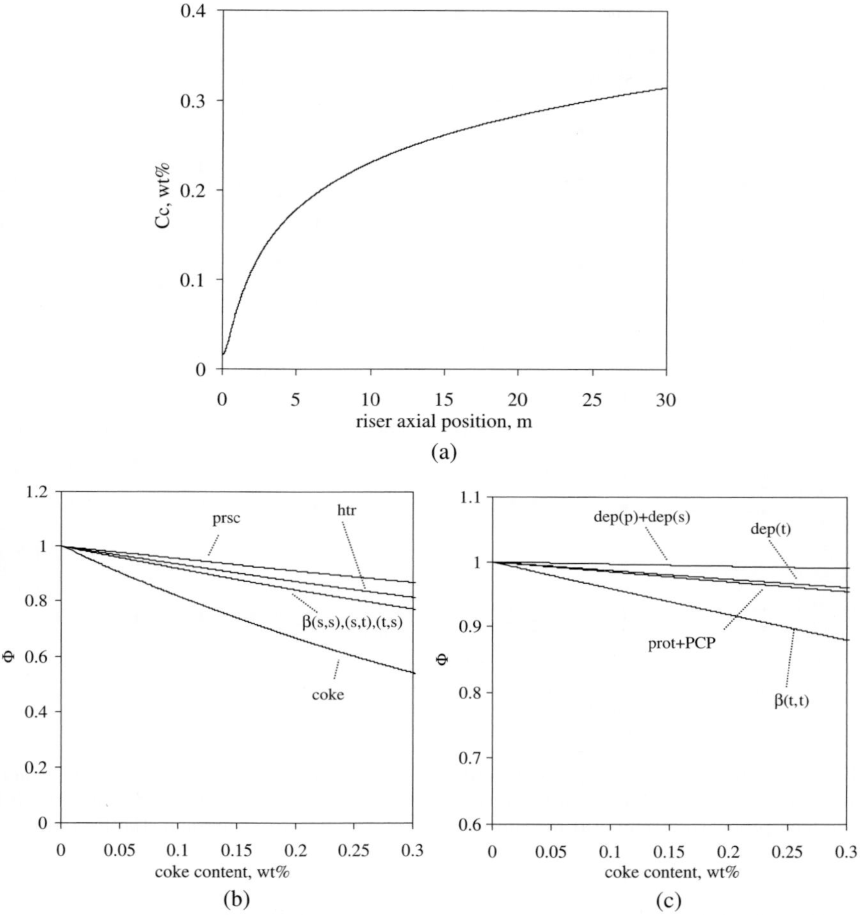

Fig. 9.7. (a) Coke content profile and (b and c) simulated deactivation functions of different elementary reaction families as a function of the catalyst coke content along the axial riser position in the catalytic cracking of a partially hydrotreated VGO [39].

the first 2 m of the riser. The simulated coke yield at the top of the riser amounts to about 2 wt% corresponding to a coke content of 0.32 wt% at the catalyst-to-oil ratio investigated (see Fig. 9.7a). Compared with typical coke yields for conventional feeds, i.e. 3–6 wt%, corresponding to coke contents from 0.6 to 1.4 wt% at a catalyst-to-oil ratio equal to 6.0 kg_{cat}/kg_{feed}, the presented results are somewhat lower [54]. This can be attributed to the fact that the contribution of thermal coke has not been taken into account

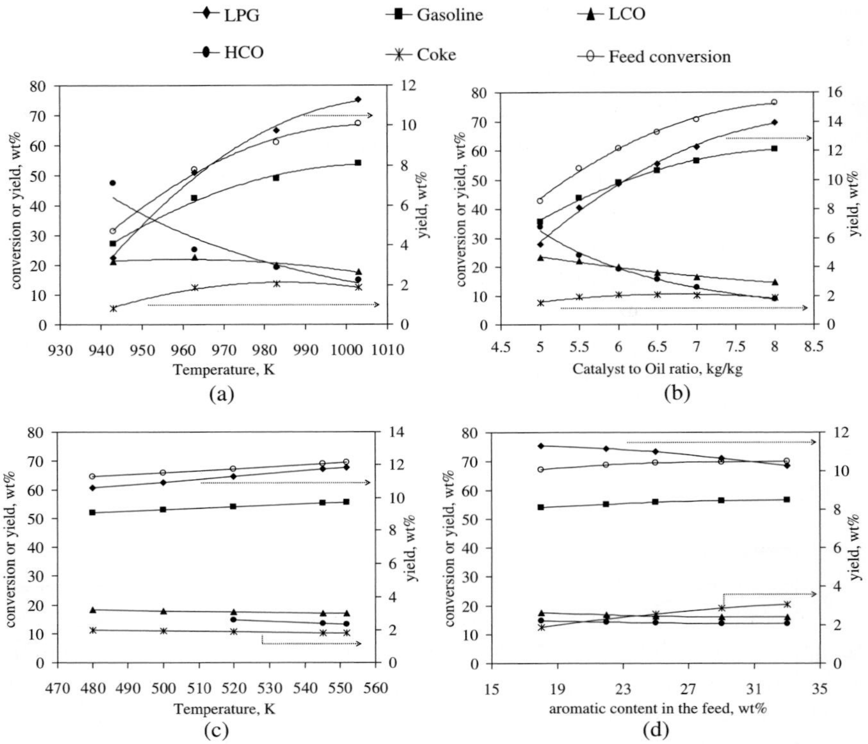

Fig. 9.8. Simulated product yields and feed conversion (wt%) at the riser outlet when varying (a) the temperature of the regenerated catalyst, (b) the catalyst-to-oil ratio, (c) the feed preheating temperature and (d) the total aromatics content in the feed [39].

in the simulation. Moreover, the partially hydrotreated VGO considered in the simulation has a relatively low concentration of polyaromatics and this typically results in lower coke yields.

The deactivation effect of coke on the different types of elementary reaction families including coke formation is presented in Figs. 9.7b and c as a function of coke content. Clearly, the coke formation rate is more strongly deactivated than any other elementary reaction family, which is directly related to the corresponding value of the deactivation coefficient (see Table 9.4). Within the considered reaction families, the susceptibility to deactivation decreases in the following order: beta scission (s, s), (s, t) and (t, s) modes, hydride transfer and protolytic scission. Deprotonation and PCP-isomerization are deactivated to an even lesser extent. In general, at the coke levels considered in this simulation, a moderate deactivation effect is observed.

The model also allows us to explore the effect of relevant variables in industrial operation, i.e. temperature of the regenerated catalyst, feed preheating temperature, catalyst-to-oil ratio and feed composition on the feed conversion, product distribution and coke deposition. The simulation results are presented in Fig. 9.8.

The feed conversion significantly increases with the regenerated catalyst temperature and the catalyst-to-oil ratio, leading to enhanced gasoline and LPG production. The effect of the feed preheating temperature and aromatics content in the feed is less pronounced. Clearly, the regenerated catalyst temperature and the catalyst-to-oil ratio can be identified as the key operating conditions for optimizing the industrial reactor behaviour.

9.4. Conclusion

Various approaches have been put forward to model zeolite deactivation by coke formation. For optimization of commercial processes, the modelling of deactivation needs to account for extensive details concerning the chemistry of the process, the catalyst and the operating conditions. An accurate prediction of the activity profile along the reactor requires that the effect of the operating conditions on deactivation by coke formation is accounted for and that the deactivation is expressed in terms of the deactivating species and not in terms of time-on-stream. Often, the "coke" deposited on the catalyst cannot be considered inert and the species retained on the catalyst are actively involved in the reaction pathways leading to the observed products. The latter is the case in, for instance, the MTO process and catalytic cracking. In these cases, a detailed model that explicitly describes the interaction between the coking and the "main" reactions is required. In hydrocarbon processing, the same elementary steps are involved in coke formation as in the "main" cracking reactions. Single-event microkinetic modelling then offers a powerful chemical engineering approach to account for the effect of coking on the activity and product distribution. The resulting models provide a wealth of detailed information on the effects of the operating conditions on the activity and the selectivity of the catalyst.

For catalytic cracking, it has been illustrated that SEMK modelling provides a powerful tool to explore the influence of important variables in industrial operation on the feed conversion, the product distribution and the coke deposition and to identify the key operating variables for optimization of the industrial reactor behaviour.

References

[1] Froment G.F., *Stud. Surf. Sci. Catal.*, 111 (1997) 53–68.
[2] Froment G.F., *Catal. Rev.*, 50 (2008) 1–18.
[3] Bartholomew C.H., *Appl. Cat. A: Gen.*, 212 (2001) 17–60.
[4] Roininen J., Alopaeus V., *Ind. Eng. Chem. Res.*, 47 (2008) 8192–8196.
[5] Caeiro G., Magnoux P., Lopes J.M., Lemos F., Ramôa Ribeiro F., *J. Mol. Catal. A: Chemical*, 249 (2006) 149–157.
[6] Guisnet M., Magnoux P., *Stud. Surf. Sci. Catal.*, 88 (1994) 53–68.
[7] Froment G.F., in *Progress in Catalyst Deactivation*, NATO Advanced Study Institute Series E, 54, Figueiredo J.L. (Ed.), Nijhoff, The Hague (1982) 103—126.
[8] Guisnet M., Costa L., Ramôa Ribeiro F., *J. Mol. Catal. A: Chemical*, 305 (2009) 69–83.
[9] Jimenez-Garcia G., Quintana-Solorzano R., Aguilar-Lopez R., Maya-Yescas R., *Intern. J. Chem. Reactor Eng.*, 8 (2010) Note S2.
[10] Nam I.-S., Froment G.F., *J. Catal.*, 108 (1987) 271–282.
[11] Beyne A.O.E., Froment G.F., *Chem. Eng. Sci.*, 45 (1990) 2089–2096.
[12] Froment. G.F., De Meyer J., Derouane E., *J. Catal.*, 124 (1990) 391–400.
[13] Froment G.F., *Stud. Surf. Sci. Catal.*, 68 (1991) 53–83.
[14] Pitault I., Nevicato D., Forissier M., Bernard J.R., *Chem. Eng. Sci.*, 49 (1994) 4249–4262.
[15] Watson B.A., Klein M.T., Harding R.H., *Appl. Catal. A: Gen.*, 160 (1997) 13–39.
[16] Jacob S.M., Gross B., Voltz S.E., Weekman V.W., *AIChE J.*, 22 (1976) 701–713.
[17] Corella J., Menendez M., *Chem. Eng. Sci.*, 41 (1986) 1817–1826.
[18] Beyne A.O.E., Froment G.F., *Chem. Eng. Sci.*, 48 (1993) 503–511.
[19] Chen D., Rebo H.P., Holmen A., *Chem. Eng. Sci.*, 54 (1999) 3465–3473.
[20] Corella J., Asua J.M., *Ind. Eng. Chem. Proc. Des. Dev.*, 21 (1982) 55–61.
[21] Corella J., *Ind. Eng. Chem. Res.*, 43 (2004) 4080–4086.
[22] Beirnaert H.C., Alleman J.R., Marin G.B., *Ind. Eng. Chem. Res.*, 40 (2001) 1337–1347.
[23] Quintana-Solorzano R., Thybaut J., Marin G.B., Lødeng R., Holmen A., in *Proceedings of the 13th International Congress of Catalysis*, Paris, France, 11–16 July 2004 (paper P2-214).
[24] Derouane E.G., *Stud. Surf. Sci. Catal.*, 20 (1984) 221–240.
[25] Guisnet M., Magnoux P., *Appl. Catal. A: Gen.*, 54 (1989) 1–27.
[26] Dahl I.M., Kolboe S., *J. Catal.*, 149 (1994) 458–464.
[27] Dahl I.M., Kolboe S., *J. Catal.*, 161 (1996) 304–309.
[28] Sapre A.V., *Chem. Eng. Sci.*, 52 (1997) 4615–4623.
[29] Janssens T.V.W., *J. Catal.*, 264 (2009) 130–137.
[30] Reyniers M.-F., Beirnaert H., Marin G.B., *Appl. Catal. A: Gen.*, 202 (2000) 49–63.

[31] Reyniers M.-F., Tang Y., Marin G.B., *Appl. Catal. A: Gen.*, 202 (2000) 65–80.
[32] Alwahabi S.M., Froment G.F., *Ind. Eng. Chem. Res.*, 43 (2004) 5098–5111.
[33] Martinis J.M., Froment G.F., *Ind. Eng. Chem. Res.*, 45 (2006) 954–967.
[34] Moustafa T.M., Froment G.F., *Ind. Eng. Chem. Res.*, 42 (2003) 14–25.
[35] Quintana-Solorzano R., Thybaut J.W., Marin G.B., Lødeng R., Holmen A., *Catal. Today*, 107–108 (2005) 619–629.
[36] Thybaut J.W., Choudhary I.R., Denayer J.F., Baron G.V., Jacobs P.A., Martens J.A., Marin G.B., *Top. Cat.*, 52 (2009) 1251–1260.
[37] Froment G.F., *Catalysis Reviews: Science and Engineering*, 47 (2005) 83–124.
[38] Quintana-Solorzano R., Thybaut J.W., Galtier P., Marin G.B., *Catal. Today*, 127 (2007) 17–30.
[39] Quintana-Solorzano R., Thybaut J.W., Galtier P., Marin G.B., *Catal. Today*, 150 (2010) 319–331.
[40] den Hollander M.A., Makkee M., Moulijn J.A., *Catal. Today*, 46 (1998) 27–35.
[41] Soares H.S., Chalbaud E., Fabella E., *Appl. Catal. A: Gen.*, 164 (1997) 35–55.
[42] Wojciechowsky B.W., Corma A., *Catalytic Cracking: Catalysts Chemistry and Kinetics*, Marcel Dekker, New York (1986) 99–125.
[43] Yaluris G., Rekoske J.E., Aparicio L.M., Madon R.J., Dumesic J.A., *J. Catal.*, 153 (1995) 54–64.
[44] Feng W., Vynckier E., Froment G.F., *Ind. Eng. Chem. Res.*, 32 (1993) 2997–3005.
[45] Vynckier E., Froment G.F., in *Kinetic and Thermodynamic Lumping of Multicomponent Mixtures*, Vol. 10, Astaritra G., Sandler S.I. (Eds.), Elsevier Science, Amsterdam (1991) 131–161.
[46] Dumesic J.A., Rudd F.D., Aparicio L.M., Rekoske J.E., Trevino A.A., *The Microkinetics of Heterogeneous Catalysis*, ACS Professional Reference Book, Washington, DC (1993).
[47] Martens G.G., Marin G.B., Martens J.A., Jacobs P., Baron G.V., *J. Catal.*, 195 (2000) 253–267.
[48] Marin G.B., Froment G.F., *Chem. Eng. Sci.*, 37 (1982) 759–773.
[49] Jimenez-Garcıa G., Aguilar-Lopez R., Leon-Becerril E., Maya-Yescas R., *Ind. Eng. Chem. Res.*, 48 (2009) 1220–1227.
[50] Souza A., Vargas J.V.C., Von Meien O.F., Martignoni W., Amico S.C., *AIChE J.*, 52 (2006) 1895–1905.
[51] Han I.S., Riggs J.B., Chung C.B., *Chem. Eng. Proc.*, 43 (2004) 1063–1084.
[52] Van Landeghem F., Nevicato D., Pitault I., Forissier M., Turlier P., Derouin C., Bernard J.R., *Appl. Cat. A: Gen.*, 138 (1996) 381–405.
[53] Theologos K.N., Nikou I.D., Lygeros A.I., Markatos N.C., *AIChE J.*, 43 (1997) 486–494.
[54] Sadeghbeigi R., *Fluid Catalytic Cracking Handbook*, 2nd Edition, Gulf Professional Publishing, USA (2000).

Part IV: Prevention of Deactivation and Optimal Regeneration

Chapter 10

SELECTION OF PROCESS AND MODE OF OPERATION

J.-F. Joly, E. Sanchez and K. Surla

10.1. Introduction

The aim of this chapter is to illustrate some general principles that help the selection of the most appropriate reactor technology and process configuration, in order to take maximum benefit of the catalyst. This chapter also highlights the industrial management of deactivation, which is a major issue to be addressed when improving catalyst formulation or developing a new process. In refining and petrochemical industries, units are often characterized by very high feedstock capacities; 50 t/h of hydrocarbon feedstock is quite common today. Therefore, for obvious economic reasons, management of catalyst deactivation is a major concern for operators.

Zeolites are used in petrochemical processes such as aromatic transformation (C_8 aromatic cut isomerization, disproportionation, transalkylation, benzene alkylation with ethylene or propylene), olefins production from methanol (methanol-to-olefins process). More recently, new processes have been announced for isobutane/olefin alkylation to produce high-quality motor gasoline. Zeolites are also largely used in oil refining with at least one zeolite-catalysed process per distillation cut. In addition, the refining industry has to take into account the continuous evolution of motor fuel specifications, available crude oils and environmental constraints.

Over the next 30 years, the most rapid increase in energy demand is expected to come from the transport sector, +2.1%/yr versus 1.7%/yr for total demand. In 2003, the transport sector already consumed 1,500 Mtoe worldwide. At present, this sector relies almost exclusively on petroleum products. Indeed, petroleum-based motor fuels account for more than 95% of the energy used in road transport worldwide. The fuel quality has already been substantially improved. In addition to the specifications already scheduled for implementation, new ones may be enforced between now and 2020 to account for EU air quality targets and/or new modes of combustion in spark ignition and diesel engines. Catalysts and technologies have been improved over the past 30 years in order to meet the demand of greater quantities of higher-quality transportation fuels.

10.2. Evolution of Refining and Petrochemical Industries

With respect to industrial management of zeolite-based catalysts, three major tendencies of refining and petrochemical industries have to be taken into account.

10.2.1. *Fuel specifications*

Regulations, that mandate gasoline reformulation to reduce motor vehicle emissions, have forced refiners to change their strategy with respect to gasoline composition:

- Elimination of lead;
- Reduced benzene (1 vol.%) and aromatic (35%) contents;
- Olefin content reduction (to reach 10 vol.%);
- Increased amount of oxygen (minimum oxygen content: 2 wt%);
- RVP reduction of the gasoline (6.5–8 psi);
- For gasoil, sulphur content as low as 10 ppm wt.

In order to produce high-quality fuels (ultra low sulphur, low aromatic contents), more severe operating conditions have to be applied with as a consequence higher deactivation rates. In addition, as crude oils with low sulphur and heavy metal contents are very expensive, cheaper crudes are more often used, which makes the conversion into desired valuable products more difficult. Conversion processes, such as fluid catalytic cracking, are strongly affected by the presence of metals in these crudes.

10.2.2. *Substitution of highly corrosive liquid catalysts by solid catalysts*

There is a general trend in refining and petrochemical industries to replace liquid acids by heterogeneous acid solids. This has been successfully done on an industrial scale in some petrochemical processes, e.g. ethylbenzene and cumene production in which zeolite-based catalysts are now used.

In the refining industry, the isobutane/olefin alkylation process, which produces a high-quality gasoline cut, uses huge amounts of highly corrosive acids as catalysts (HF or H_2SO_4). Depending on the amount of impurities in the olefin feedstock (water, diolefins, oxygenates, sulphur), catalyst consumption ranges from 0.6 to 0.8 kg HF/t of alkylate and 70 to 103 kg H_2SO_4/t of alkylate.

However, the increase in alkylation capacity poses the question of the replacement of these conventional liquid acids by more environmentally friendly solid acid catalysts. A great research effort has been made over the last decades in order to reach this objective and many solids, among them zeolites, have been proposed as alkylation catalysts. Nevertheless, side reactions such as olefin oligomerization lead to rapid deposition of unsaturated compounds of high molecular weight which remain adsorbed on the active acid sites, causing a rapid deactivation of the catalyst. Therefore frequent regeneration of the catalyst is mandatory and specific technologies have been envisaged to develop a reliable process.

10.2.3. *New technologies for petrochemicals (mainly propylene)*

Propylene is a major petrochemical component used as a raw material for the synthesis of a variety of polymers and other chemical intermediates. It is produced primarily as a by-product of fluid catalytic cracking (FCC) and of ethylene production by steam cracking of hydrocarbon feedstocks, primarily the naphtha cut.

New technologies have been developed in order to open new routes for propylene production. Some of these technologies use zeolite-based catalysts, such as the MTO (methanol-to-olefins) process and deep catalytic cracking which is a petrochemical extension of FCC for propylene and light iso-olefin production. More recently, processes based on the selective catalytic cracking of C_4/C_5 olefins have been developed.

All these processes have in common the use of high reaction temperatures owing to thermodynamic considerations, i.e. olefins being favoured at high temperatures. These processes use zeolite-based catalysts, and management of deactivation is fundamental because high reaction temperatures and the presence of olefins significantly favour deactivation processes. Specific reactor technologies have thus been considered.

10.3. Heterogeneous Catalytic Reactor Technologies

The main reactors technologies can be classified as follows [1, 2]:

- Fixed-bed reactors: the catalyst beads or extrudates are stacked in an immobile catalytic bed with the reactant fluid(s) flowing through the interstices. Generally, a downflow mode is adopted.
- Moving-bed reactors: the catalyst beads are stacked and slowly move downward by gravity. Such reactors are generally operated in gas phase.
- Expanded or ebullated beds: the catalyst, in the form of beads or extrudates, is kept in suspension by a rising stream of reactants.
- Fluidized beds: the fluids and catalyst particles are swept along in the same direction and react in a specific reaction zone often in gas phase.
- Mechanically stirred and bubble column slurry reactors: such reactors offer significant advantages compared with other types of multiphase reactors. They use small catalyst particles which favour intraparticle transfer and a good reactor temperature control. It is also possible to replace in continuous a part of the deactivated catalyst with fresh catalyst, like in fluidized-bed reactors.

The choice of a reactor technology depends on the reaction thermodynamics and kinetics, heat and mass transfer, and deactivation rate. An additional parameter, the number of phases, must also be considered. Thus, in refining processes, the catalytic reaction system can be mono-, bi- or triphasic (hydrocarbon gas or liquid, hydrogen and solid catalyst).

10.3.1. *Single-phase catalytic reactors*

10.3.1.1. *Fixed-bed reactors*

Fixed-bed reactors consist of a compact fixed mass of catalyst particles loaded in a vessel. The appropriate particles are either cylindrically shaped extrudates, multi-lobed extrudates or beads with equivalent diameters ranging from 1 to 5 mm.

The simplest fixed-bed reactor technology is of the "axial flow" type depicted in Fig. 10.1, generally operated in adiabatic conditions. The catalyst is placed in a layer of variable thickness in the cylindrical part of the reactor. At the reactor base, there is a collector for the reaction products as well as a nozzle for draining out the catalyst. When for any reason it is necessary to decrease the pressure drop through the catalyst layer, a radial flow fixed-bed could be used.

Inside such an adiabatic catalytic bed, the temperature varies in relation with the reactant conversion. For exothermic reactions, the fluid

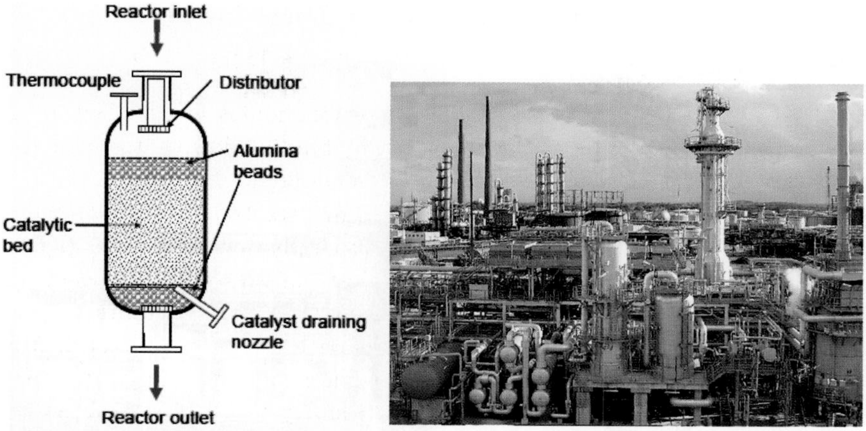

Fig. 10.1. Schematic cross-section and photo of a cylindrical fixed-bed catalytic reactor with an axial flow (photo: copyright Axens).

temperature between catalytic sections has to be readjusted either by thermal exchange or by injecting a cooling fluid. This cooling fluid may be either a gas or a liquid and is preferably one of the reactants. Inversely, for endothermic reactions, the fluid phase must be heated, which is made by using either exchangers or furnaces between catalytic beds, depending on the operating temperature range.

For systems with reaction heats that are too high, adiabatic operation is impossible. The solution generally adopted is to remove or add heat in the midst of the catalytic zone with integrated exchangers (internal exchange surfaces or multitubular reactor).

10.3.1.2. *Moving-bed reactors*

In moving-bed catalytic reactors, the catalyst grains flow by gravity. Moving-bed reactors are associated with two possible types of fluid injection distribution: axially when the catalytic bed is mobile with respect to the reactor wall and radially when the solid is confined between two vertical concentric grids enclosing the catalyst in an annular space. In addition, the reactant flow and moving bed can be in cocurrent, countercurrent (rarely used) or crosscurrent modes (Fig. 10.2).

The moving bed offers all of the features of a fixed bed, in particular the advantages of a gradient reactor. It can achieve high conversions and good selectivity. Moreover, the moving bed offers the possibility of withdrawing the catalyst continuously or intermittently in order to perform

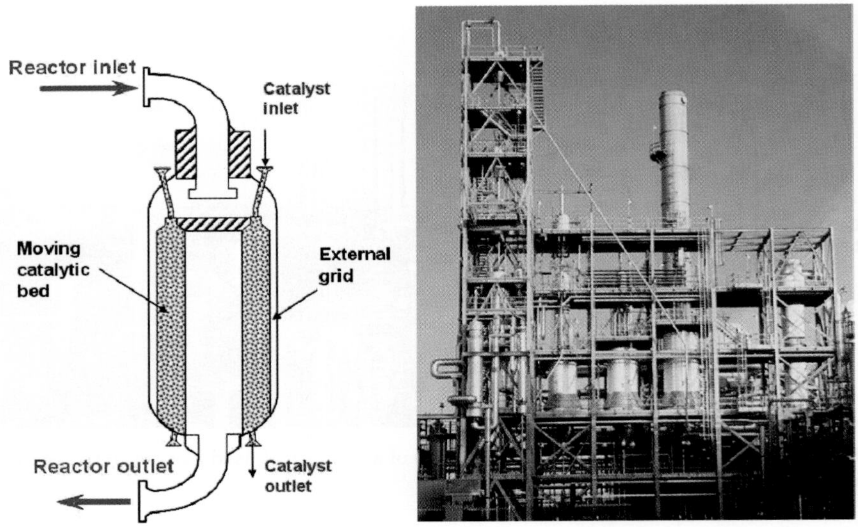

Fig. 10.2. Schematic cross-section and photo of a moving bed catalytic reactor.

steady-state operation. The deactivated catalyst is then sent to *in situ* or *ex situ* regeneration section facilities for reactivation/rejuvenation and then reintroduced back into the loop. The moving bed is well adapted to catalysts with a short lifetime, which hence must be frequently regenerated. With this technology, bead particles must be used and the catalyst properties like density or mechanical strength have to be carefully selected in order to avoid any reactor pinning effect. Moving beds are used in catalytic naphtha reforming and more recently isomerization of pentene.

10.3.1.3. *Fluidized-bed reactors*

In a fluidized-bed system, solid particles are held in suspension or are transported by the flow of the carrier fluid. This fluid is generally a gas, sometimes a liquid. Fluid catalytic cracking (FCC) is the most representative example of gas/solid fluidized beds. A schematic cross-section of an FCC unit is reported in Fig. 10.3.

The main advantages of fluidized-bed reactors are:

- Catalyst can be withdrawn from the reactor and reintroduced continuously. This is extremely efficient when the catalyst is rapidly deactivated;

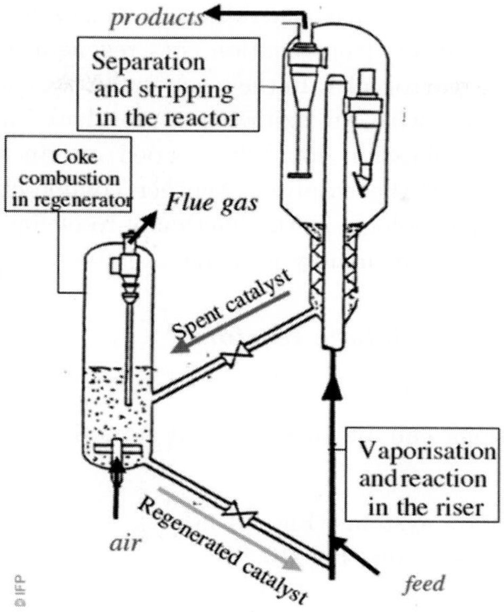

Fig. 10.3. Schematic cross-section of a fluid catalytic cracking unit (source: IFP).

- Elimination of any intra-particular diffusion due to the small particle average diameter;
- Very low temperature gradients;
- Elimination of radial gradients;
- Attenuation of axial gradients;
- Minimization of the risks of hot spots and thermal instability. The fluidized bed is hence ideal for exothermic reactions.

As in any design, the fluidized-bed reactor presents however some drawbacks, namely:

- The bed expansion causes an increase in the reactor volume for the same catalyst weight compared to a fixed bed;
- High fluid velocity is required to fluidize the catalyst;
- A large disengagement zone and special device for recycling the fines are necessary.

Fluidized-bed technology (transported bed) is preferred when the two following conditions are satisfied: very fast reaction rate (reaction time of several seconds) and rapid deactivation of the catalyst.

The deactivated catalyst is generally sent to a dedicated regeneration facility for reactivation (rejuvenation or regeneration) and further reintroduced at the reactor inlet. In the regenerator, the catalyst is generally brought into contact with an oxidizing atmosphere to burn the coke. If necessary, the combustion could be carried out in several stages to limit thermal aging of the catalyst. Another commercial application of the fluidized-bed technology is the methanol-to-olefins process for the production of light olefins, mainly propene.

10.3.2. *Multiphase catalytic reactors*

10.3.2.1. *Fixed-bed reactors*

Three configurations can be distinguished according to gas and liquid phases flows:

- cocurrent downflow systems (Fig. 10.4a)
- cocurrent upflow systems (Fig. 10.4b)
- counter current systems (Fig. 10.4c)

Cocurrent downflow, which is the easiest to implement, is the most commonly used technology. The performance of two-phase downflow reactors depends mainly on the distribution of the phases at the top of the catalyst bed. Due to its industrial importance, this configuration has been extensively investigated.

The cocurrent upflow mode, in which liquid and gas phases pass through the catalyst bed from the bottom to the top of the reactor, is rarely used at the industrial scale because of induced catalyst movement that leads to unacceptable catalyst attrition rates. However, it is often used in small pilot plant units (reactors with less than 1 litre of catalyst).

The countercurrent flow mode is rarely used industrially, but can present advantages for reversible reactions since it allows the thermodynamic equilibrium to be displaced towards the desired products.

10.3.2.2. *Ebullated-bed and slurry reactors*

Ebullated-bed and slurry reactors (gas/liquid/solid) present advantages similar to those of fluidized-bed reactors. The slurry reactors are characterized by a continuous liquid phase with dispersed gas bubbles. Two different technologies can be used: mechanical agitation or bubble column. Ebullated-bed reactors are similar to slurry reactors, except that here, the liquid phase ensures the internal mixing and maintains the catalyst

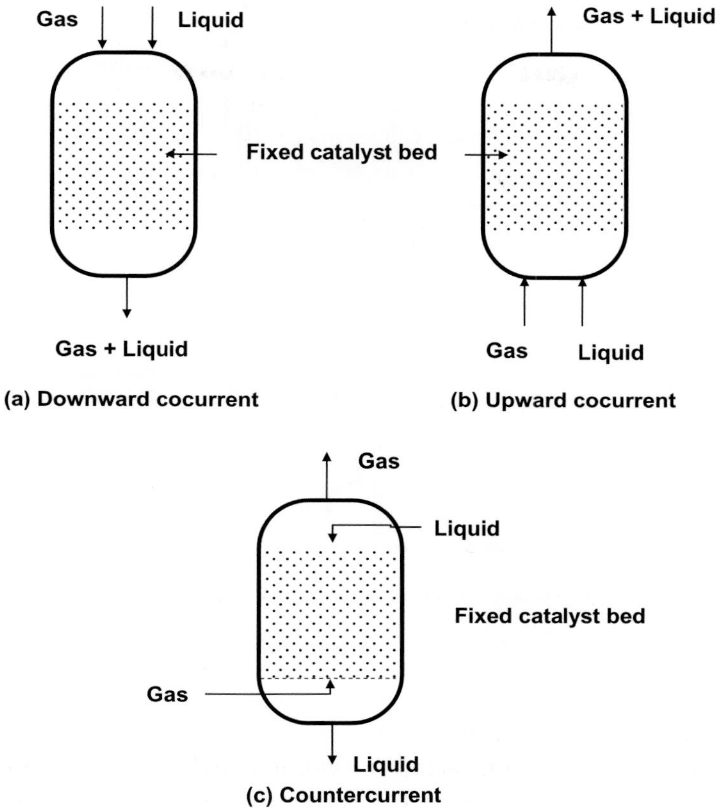

Fig. 10.4. Various types of catalytic reactors with fixed bed and two-phase flow.

in suspension. The catalyst particle size is larger than that of slurry catalyst.

When the catalyst deactivation is too fast, fixed beds cannot be economically considered, and ebullated bed or slurry reactors are alternative choices. Fixed-bed technologies have limitations for handling feeds with high contaminant levels, while ebullated-bed or slurry technologies are more suitable for these feeds and when moderate to high conversions are desired.

An example of an ebullated-bed reactor is shown in Fig. 10.5.

In ebullated beds, the liquid and the gas phase pass through the reactor in an upflow mode and keep the solid catalyst in suspension. The catalyst consists of extrudates with sizes ranging from 1 to 5 mm. Both fluid phases leave the reactor without carrying over the solid, which remains inside the

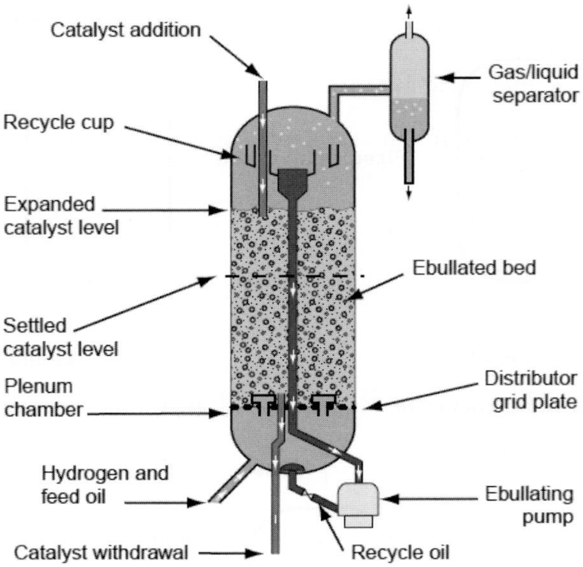

Fig. 10.5. Simplified flow scheme of an ebullated-bed reactor (source: Axens).

reactor. The potential advantages of the ebullated bed compared with the fixed bed are:

- good temperature control for highly exothermic reactions;
- the possibility of withdrawing spent catalyst, batchwise, during operation and of introducing fresh catalyst. The catalyst makeup is mainly governed by catalytic performance and economic criteria.

10.4. Selection of Optimal Reactor Technology

Reactor design engineering is based on thermodynamics, chemical kinetics, heat and mass transfer and economics. It aims to properly select and design the most appropriate process scheme.

10.4.1. *Process design best practices and operation*

In the field of process development, deactivation is taken into account during the selection of the process scheme, using the following guidelines:

Feed purification. Two situations can be considered: coke precursor impurities can be first converted in dedicated reactor vessels (for example selective hydrogenation of di-olefins), and fouling materials can preferably be adsorbed onto guard beds. Generally, guard beds are operated in a swing mode as the fouling material accumulation leads to catalyst deactivation and also to a regular increase in reactor pressure drop.

Catalyst optimization. The design of the catalyst in terms of strength, distribution of the active acid sites and pore size/shape should be adapted to limit to a certain extent the formation of coke or coke precursors while maintaining a sufficient activity level. The aim is to tailor the catalyst properties (active phase, texture, particle size) to obtain a stable catalyst under the operating conditions of the process.

Catalyst loading strategy. Specific catalyst loading can be considered, via a multilayer of catalysts with different active phases and/or average particle diameters. In general, the upper layers of the bed are loaded with large extrudates or beads, acting as filters, in order to limit reactor plugging by fouling materials.

Reactor technologies. This point will be discussed in detail in Section 10.4.2.

Optimizing process operating conditions (start, middle and end of run). Start-of-run conditions are chosen in order to maximize the unit performances. After this initial period, the effects of deactivation are visible; therefore the operator adjusts some operating parameters to maintain the performances as close as possible to the initial ones. This can be achieved by several approaches depending on the reaction mechanisms and thermodynamics. Usually, reactor temperature is regularly increased to compensate for the loss of initial activity; the end-of-run temperature is often limited by metallurgical constraints or by an unacceptable loss of product quality or an extensive coke formation rate. In some cases, the hydrogen partial pressure can also be adjusted, in order to maintain the hydrogenation activity (for example in xylene isomerization processes). This is only possible when the unit design has sufficient flexibility with respect to pressure management (reactor design, compressor, etc.) [3, 4].

10.4.2. *Selection of reactor technology*

Figure 10.6 proposes a pathway for the selection of the optimum chemical reactor design technology and the corresponding catalyst shape.

When the catalyst life cycle is shorter, about six months, dedicated facilities for on-site regeneration/rejuvenation are considered, particularly

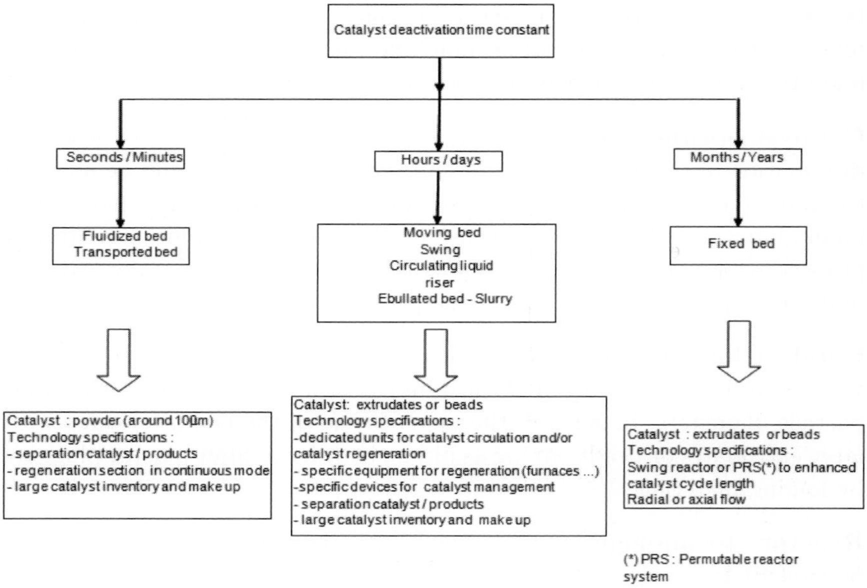

Fig. 10.6. Diagram for process technology selection.

if the catalyst is expensive. The unit operation is usually called "semi-regenerative mode": the catalyst is not unloaded, it is regenerated in the reactor at the expense of a temporary loss of production. To avoid this drawback, several swing reactors can be used. The facilities required for carrying out the regeneration, such as compressors for inert gas circulation and oxygen, may be a permanent part of the unit or be installed on a temporary basis. The operator could decide to have a spare load of fresh catalyst to limit the downtime of the plant during the catalyst reactivation step.

Deactivation rate: few weeks or less

When the catalyst life cycle is shorter, i.e. less than a few weeks, the frequency of regeneration/rejuvenation leads to dedicated permanent installation facilities for the catalyst regeneration/rejuvenation which are considered as an integral part of the process. The catalyst can either remain within the same reactor which switches between operation and regeneration/rejuvenation modes (swing operation) or be transported batchwise or continuously to a separated vessel and then back to the reactor.

In a swing mode type process, the number of reactors is selected in order to ensure an almost uninterrupted product production. If the catalyst reactivation is carried out in a separate vessel, the capacity of this vessel is designed to match the deactivation rate, and the catalyst shape has been suited to the technology in order for a smooth, trouble-free, catalyst transfer.

Compared to swing mode operation, the advantage of the moving-bed technology is the steady-state condition with therefore constant yield and product quality. Moreover, this technology can be applied under more severe operating conditions that enhance yields and product quality but lead to fast deactivation and short cycle lengths. Even with these advantages, the swing mode type process is most often preferred for economic reasons.

In the case of highly exothermic reactions, ebullated-bed and slurry reactor technologies can also be used with the advantage of continuous catalyst replacement and possible regeneration/rejuvenation. The catalyst should offer a high crushing strength and low attrition rate.

Deactivation rate: seconds to minutes

Finally, with still shorter catalyst life cycles, the associated technology is the fluidized bed where the catalyst particles can be moved much faster than catalyst particles in a moving bed.

10.4.3. *Economic criteria consideration*

Aside from scientific considerations, economic constraints have to be taken into account for the evaluation and selection of new optimal industrial process developments. The subject of this section is not to describe the procedure for a complete process economic evaluation [5], but to give some general guidelines.

An estimate of the process design is necessary to obtain the approximate capacities, sizes and types of the basic equipment, like reactors, rotating machines, columns, vessels, heaters, heat exchangers, etc. and the associated erected costs. Main equipment erected costs associated with the cost of commodity materials (structures, insulation, piping, valves, instrumentation, painting, etc.), the cost of indirect construction (tax, insurance, miscellaneous worksite costs, etc.) and unit contingencies (technical and price changes, etc.) permit us to determine the unit ISBL (inside battery limits) which is the main part of the unit CAPEX (capital expenditure). The unit utilities consumption (steam, fuel gas, fuel oil, water, electricity) in association with working capital (precious or semi-precious metals, catalysts or absorbent initial load) determine the unit OPEX (operating expenditure).

Based on these two parameters (OPEX and CAPEX), an estimation of the manufacturing cost of a new product or process and several criteria — the unit pay out time (POT), the internal rate of return (IRR) and the net present value (NPV) — can provide a good indication of the project's economic viability. Usually refining projects are selected when the IRR is over 15% and the POT lower than 3 years. In addition, when different projects are compared, the one with the higher return on investment is considered to be more attractive.

10.4.4. *Main catalytic refining and petrochemical processes*

The main refining and petrochemical processes are reported in Table 10.1. These processes have been developed following the guidelines presented in the previous paragraphs. This table shows in particular the relationship between the time scale of deactivation and the selected technology as well as the relation between the causes of activity loss and the reactivation pathway.

10.5. Case Studies

In this section, two different situations will be analysed, in order to illustrate some of the principles which are presented in Fig. 10.6: (i) liquid-phase alkylation at low temperatures of isobutane with light olefins on zeolite-based catalysts which corresponds to relatively slow deactivation and reaction rates and (ii) fluid catalytic cracking which is an example of very fast deactivation and reaction rates (in gas phase at high reaction temperatures).

10.5.1. *Alkylation of isobutane with light olefins*

Isobutane/olefin alkylation process is a very interesting case, since different process schemes and reactor technologies have been proposed in order to cope with the rapid deactivation generally observed in zeolite-based catalytic systems.

Isobutane alkylation with C_3–C_5 olefins produces iC_5–iC_{12} isoparaffins, that are ideal components for the reformulated gasoline pool. Alkylation units produce a gasoline that perfectly meets the above specifications: minimal aromatic and olefin contents (both lower than 0.7 vol.%) and high octane values (MON: 90–94, RON: 93–97), low octane sensitivity [6, 7].

Table 10.1. Relationship between time scale of deactivation and technology selection.

Process	Time scale of deactivation	Operating conditions	Main deactivation mechanism	Technology	Reactivation
Fluid Catalytic Cracking	Milliseconds	525–550°C 1–2 barg	Coke	Transported Bed	Coke Combustion
MTO	Seconds	350–500°C 1– 2 barg	Coke	Fluidized Bed	Coke Combustion
Aliphatic Alkylation	Minutes/Hours	40–90°C 10–30 barg	Unsaturated Oligomers	Liquid Riser or Swing	Rejuvenation
Catalytic Reforming	Days	480–520°C 2–8 barg	Coke	Moving Bed	Coke Combustion
Skeletal Isomerization	Days	400–450°C 2–5 barg	Coke	Moving Bed or Swing Reactor	Coke Combustion
Paraffin Dehydrogenation	Days/Weeks	450–510°C 2–5 barg	Coke	Moving Bed/Fixed Bed	Coke Combustion/ Metal Recovery
Resid Conversion	Month/Years	400–450°C 150–200 barg	Poisoning + Coke	Ebullated Bed/ Slurry/Fixed Bed	
Aromatic Isomerization	Years	375–415°C 8–15 barg	Coke	Fixed Bed	Coke Combustion
Catalytic Reforming	Years	480–520°C 15–20 barg	Coke	Fixed bed	Coke Combustion
Hydrotreating	Years	350°C 30–150 barg	Poisoning + Coke	Fixed Bed/ Moving Bed	Coke Combustion
Hydrocracking	Years	350–420°C 100–200 barg	Poisoning + Coke	Fixed Bed	Coke Combustion

Different processes were developed in the 1930s to carry out the alkylation of isobutane. Concentrated liquid hydrofluorhydric and sulphuric acids are used as catalysts for the alkylation of isobutane with C_3–C_4 olefins, and up to now, only HF and H_2SO_4 catalyst-based processes have been widely used with regard to economic considerations.

10.5.1.1. *Chemistry*

Aliphatic alkylation refers to the reaction of 1 mole of isoparaffin, i.e. isobutane or isopentane, with 1 mole of C_3–C_5 olefin, giving rise to 1 mole of branched isoparaffin. In practice, isobutane is the only isoparaffin used, owing to its higher reactivity. The reaction is the following when the olefin is butene:

$$iC_4H_{10} + C_4H_8 \rightarrow iC_8H_{18}$$

Industrial alkylate compositions are much more complex than expected with regards to the above stoichiometric equation. For instance, skeletal isomerization of C_8 paraffins also takes place, heavy cations (C_{12}–C_{20} species) are formed and then fragment to produce C_5–C_9 cations and olefins. Because of this complex set of reactions, real alkylates are a complex mixture of C_5–C_{12} isoparaffins.

Compared to isobutane, olefins, like butenes, are much more reactive in acid catalysis. Therefore the main secondary reaction is olefin oligomerization, leading to C_{12}–C_{16} oligomers. Weak acid sites catalyse only oligomerization reactions. Hydride transfer which is invoved in alkylation requires a higher acid strength. Oligomerization proceeds on intrinsically weak acid sites of solid acids. These oligomers can undergo consecutive hydrogen transfer reactions leading to the formation of highly unsaturated polymers, with as the main consequence the fast blocking of the zeolite microporosity. In order to limit olefin oligomerization, isobutane is used in large stoichiometric excess. From process considerations, that means that a de-isobutanizer column will be necessary in order to recycle unreacted isobutane to the reaction section.

10.5.1.2. *Process and reactor technology*

To develop such a process is highly demanding. The required high dilution of butene and the limited reaction time call for a special reaction system because of the fast catalyst deactivation. Additionally, a fine dispersion of the catalyst in isobutane must be realized. Therefore the easiest solution, i.e. a single fixed-bed reactor, cannot be used.

The main requirements are the following:

- The acidity of the zeolite-based catalyst must be high enough to allow a low operating temperature (below 50°C) which is a condition to limit oligomerization reactions as well as iC_8–iC_{12} cracking. Alkylate quality decreases when temperature increases.
- The diffusion from the inner acid sites of the zeolite to the external surface of the catalyst must be fast in order to avoid the secondary alkylation of the desired reaction products, i.e. branched iC_8 paraffins.
- Since deactivation is very fast (lifetime of 12 to 24 hours), the catalyst should be frequently regenerated with limited loss of active sites.

Two main reactor technologies have been envisaged, depending on the level of external diffusion limits encountered:

- **Swing fixed-bed reactors:** This is the simplest technology, but presents some major drawbacks: achieving very high isobutane/olefin ratio is difficult, frequent regeneration means the use of several reactors and swing mode operation.
- **Slurry reactors (CSTR type reactors):** This technology has several key advantages: it is possible to obtain very high dilution of the olefin, a very efficient temperature control, and the possibility to continuously remove deactivated catalyst for regeneration. This technology is quite complex because catalyst/liquid-phase separation requires special devices.

Tubular reactors with a high degree of mixing have also been considered. They are similar to ebullated-bed reactors allowing internal liquid recirculation, and could also be operated in swing mode.

Different regeneration strategies have been developed:

- Conventional regeneration by burning off the coke requires a catalyst with high temperature stability, like in the FCC process.
- Rejuvenation by isobutane (eventually in the presence of hydrogen) for oligomer extraction: in the absence of olefin, isobutane can react with iC_{12}^+ carbocations leading to paraffins; however, long contact times are necessary to promote this reaction.
- Rejuvenation/regeneration by hydrogenation and/or mild hydrocracking can be done at moderate temperatures without damaging the zeolite structure, but a noble metal has to be present on the catalyst.
- Some papers report successful regeneration using supercritical conditions, but no industrial scale-up has been announced.

Solid acid alkylation technologies have been announced. These processes are based on solid acids with a regeneration section utilizing purge and hydrogenation/mild hydrocracking strategies. Typical technology is based on multiple reactors, which swing between reaction and regeneration, using mild regeneration at reaction temperature and pressure with hydrogen dissolved in isobutane frequently performed (well before the end of the theoretical catalyst lifetime). When necessary, the catalyst is fully regenerated at 250°C in a stream of gas-phase hydrogen.

This example demonstrates that different regeneration strategies can be developed, in parallel with different reactor technologies. Compared to FCC, alkylation is much more complex since the reaction is thermodynamically favoured at low temperatures (high temperatures for FCC), and the acidity of the zeolites must be kept as high as possible in order to promote alkylation. It is therefore imperative that regeneration conditions should not damage the catalyst (in case of FCC, high-temperature regeneration is possible, since acidity requirements are less severe).

10.5.2. *Fluid catalytic cracking*

Fluid catalytic cracking (FCC) [8] aims to convert high-molecular-weight hydrocarbons characterized by a boiling range above approximately 350°C (typically vacuum gas oil or residue feedstocks alone or in blend), in the presence of a complex catalyst containing a Y-type zeolite (USY, ReHY or ReHY) in a SiAl matrix. The desired product is a gasoline cut (40–60 wt% on feed basis). Cracking reactions involve the rupture of the C–C bond, which is endothermic hence thermodynamically favoured at high temperatures. Cracking is carried out in gas phase in a circulating fluidized bed.

This process operates at low pressure (typically lower than 2–3 bar abs) in order to enhance cracking reactions and at a riser outlet temperature in the range of 525–550°C for conventional feedstocks. This conversion process is often compared to the hydrocracking process, since it converts the same type of feedstock. Both use zeolite-based catalyst, but FCC is operated in the absence of hydrogen, the main consequence being very fast deactivation due to coke build-up on the catalysts. Coke therefore has to be continuously removed by combustion in a dedicated section in order to recover its activity, which explains why a fluidized-bed process has been developed.

The residence time of hydrocarbons in the reactor section is very low. On the highly acidic catalyst, the feed is converted in a few seconds

(1 to 5). The catalyst is in the form of fine fluidized powder (Group A powder according to Geldart classification). At the outlet of the reactor, the adsorbed hydrocarbons are stripped, then the catalyst is sent to the regeneration section operated in a dense fluidized bed. Coke combustion is ensured by air introduced in the regenerator vessel. One of the most important operating principles of FCC is that the heat of coke combustion is used to vaporize the feed and to compensate for the endothermicity of the cracking reactions that occur in the reactor. This energy comes from the regenerator and is transferred to the reaction zone by the solid itself. The catalyst leaves the regeneration section at a temperature level in the range of 700 to 780°C.

FCC units are operated in adiabatic mode and are auto-sufficient in terms of heat. Heat balance which is a specificity of the FCC process is a key characteristic which directly impacts operating conditions and therefore reactor performances.

10.5.2.1. *Reactor and regenerator technologies*

The reactor is a vertical pipe called "riser". The internal diameter of the riser is in the range of 1 to 2.5 metres and its height is about 45 metres depending on unit pressure balance. It is a plug flow reactor in order to realize short contact times. At the inlet of the riser, the hot catalyst coming from the regenerator is mixed with the liquid feed which has been atomized in order to facilitate its contact with the catalyst and therefore to obtain a quick and homogeneous vaporization. The technology for feed atomization is highly important and has a large effect on riser performance. The mass ratio between catalyst and feed called "cat-to-oil" is commonly in the range of 4 to 8. The feed is completely vaporized. Cracking reactions take place in the riser, and a rapid expansion in volume occurs, modifying the catalyst density along the riser and therefore the cracking conditions. At the outlet of the riser, the catalyst is separated from the hydrocarbon phase, in a disengaging zone containing two stages of cyclones (primary and secondary). The hydrocarbon effluents are sent to the fractionation section while the catalyst flows into a stripper zone to desorb adsorbed hydrocarbons before being sent to the regenerator.

Conversion of the feed is typically 75–85% depending on the feed. The main cracking products are LPG (7–20 wt%), the gasoline cut (30–60 wt%) and the LCO cut (10–20 wt%). These cuts are rich in olefins, and since isomerization also occurs in the system, the gasoline cut presents a very high MON and RON (respectively 75–85 and 85–95). In addition to the desired cracking reaction, coke formation occurs, as well as dehydrogenation induced by metal deposit (nickel and vanadium) on the catalyst. The typical coke content of an FCC catalyst at the outlet of the riser is about 1 wt%.

In the regenerator, oxidation of coke to $CO + CO_2$ and water occurs in a dense fluidized bed; particles below 5 μm will be removed by cyclones. When regeneration is operated in a partial combustion mode, flue gas which contains CO is sent and treated in a CO boiler. Flue gas is sent to a waste heat boiler to recover flue gas heat, which is converted into steam and filtered by an electro-precipitator.

The regenerator technology is a dense fluidized bed with a residence time much greater than in the riser (8 minutes compared to 1–5 seconds). Air is used for coke combustion, the air is homogeneously distributed in the regenerator vessel, and the temperature inside the regeneration zone depends on the coke content at the outlet of the riser. Combustion of the coke deposit is completed in the regenerator; the gas leaving the regenerator vessel typically contains 1–2% mol of oxygen (total combustion). The combination of high temperature, residence time and water partial pressure in the regenerator has a negative impact on the structural integrity of the catalyst (hydrothermal aging), the main consequence being a partial dealumination of the zeolite present in the catalyst (gradual extraction of aluminium atoms from the framework, with formation of extraframework species which accumulate on the Y zeolite. The metals (Ni and V) that come from the feed and further deposit on the catalyst surface in the riser have also negative effects. In particular, vanadium causes partial destruction of the zeolite structure. For these reasons, the equilibrium catalyst present in the unit is very different from the fresh catalyst which is periodically added into the FCC unit.

10.5.2.2. *Catalyst management*

Because of catalyst loss by attrition and by the effect of regeneration conditions, the daily consumption of catalysts in typical FCC unit is about 4 to 20 tons. Therefore, there is a dedicated catalyst handling section which usually consists of 2 to 3 catalyst storage hoppers, with at least one being used for the deactivated catalysts purged from the unit and one for fresh catalysts. These vessels are connected to the regenerator. On average, the catalyst inventory of a FCC unit (100 to 300 tons) is renewed every 2 months.

10.6. Conclusion

Zeolite-based catalysts are used in major refining and petrochemical processes. The main reactor technologies that can be used for taking into account the specificities of the catalysed reaction(s) and the lifetime of the catalyst were described in this chapter.

The selection of the most appropriate technology and process configuration was shown to be strongly influenced by the cause(s) and especially the rate of catalyst deactivation and the possibility of regeneration/rejuvenation. Since catalyst deactivation mechanisms and rates can vary widely with feedstock properties and operating conditions, reactor technologies and process schemes will vary accordingly. The choice is usually made by considering the most appropriate way of coping with deactivation phenomena. Therefore a good understanding of the deactivation causes and their effects is essential to adequately choose the reactor technology and the regeneration/rejuvenation strategy.

In addition, the process development is driven by techno-economic optimization and different process schemes can often be envisaged to maximize profit, and to operate without any trouble.

References

[1] Trambouze P., *Chemical Reactors: Design, Engineering, Operation*, Editions Technip and Gulf Publishing Co., New York (1988).

[2] Trambouze P., *Materials and Equipment*, Vol. 4, Editions Technip, Paris (2000) 391–486.

[3] Trimm D.L., in *Handbook of Heterogeneous Catalysis*, Vol. 4, Ertl G., Knözinger H., Weitkamp J., Wiley-VCH, Weinheim (2008).

[4] Sie S.T., *Appl. Catal. A: Gen.*, 212 (2001) 129–151.

[5] Chauvel A., *Manual of Process Economic Evaluation*, Editions Technip, Paris (2003).

[6] Joly J.F., *Aliphatic Alkylation: Conversion Processes*, Vol. 3, Editions Technip, Paris (2001) 257–287.

[7] Joly J.F., Benazzi E., *Arabian Journal for Science and Engineering*, 21 (1996) 313–319.

[8] Bonifay R., Marcilly C., *Catalytic Cracking: Conversion Processes*, Vol. 3, Editions Technip, Paris (2001) 169–223.

Chapter 11

PREVENTION OF ZEOLITE DEACTIVATION BY COKING

M. Guisnet

11.1. Introduction

Whatever the commercial process in which zeolites are used as catalysts, a more or less fast decrease of their activity and often of their selectivity to the desired products can be observed. This deactivation can be due to different phenomena occurring often simultaneously: poisoning of the active sites, blockage of the access of reactant molecules to these sites, framework degradation, sintering of supported metals, etc. As was underscored in the general introduction (Chapter 1), coking is the main cause of zeolite deactivation. Indeed, not only can coke poison the active sites or block their access, the removal of coke required to recover the catalytic properties is generally carried out under very severe conditions with as a consequence irreversible detrimental effects. It is why we have chosen in this chapter to focus on the prevention of zeolite deactivation by coking.

In most processes, the cost of deactivation is very high and mastering the stability of the catalysts has become as essential as controlling their activity and selectivity for the desired reactions. This mastering requires an integrated approach encompassing the catalysts, the reactors and regenerators and the process conditions [1–3]. In Chapter 10, examples were presented showing how to select the process and the mode of operation. In this chapter, rules for selecting and optimizing zeolite catalysts and operating conditions to limit the rate of coking and the deactivating effect of coke will be proposed. For the sake of simplicity, this chapter will be essentially limited to the case of acid-catalysed transformations of hydrocarbons, i.e. of reactions involved in refining and petrochemical processes.

As it was underscored in Chapter 7, the formation of coke molecules requires not only chemical steps but also their retention within the micropores of zeolites or on their outer surface. As a consequence, the features of this unwanted reaction product are very particular in comparison to those of the other products. First, as coke molecules have to be retained on the catalyst under the operating conditions, their formation requires many chemical steps with the possibility of various routes of formation. Therefore coke is generally a complex mixture, composed of a large variety of compounds differing in their chemical nature and their size. Moreover, coking can be considered as a nucleation-growth process [4]: indeed, the first coke molecules that are formed are not inert, and hence undergo many secondary reactions either intramolecular or intermolecular (between them or with reactant or product molecules). Lastly, carbonaceous deposits not only have a deactivating effect on the formation of the desorbed products but also on their own formation: coking is a self-inhibited process. This autoinhibition is at the source of large difficulties in the determination of the actual effect of the catalyst features and operating conditions on the coking rate. These very particular features of the coke product and of the coking process complicate the establishment of models to account for the effect of the reaction system (reactant and product molecules, operating conditions) and of the zeolite catalyst features (active sites, pore structure) on the rate of coking and the deactivating effect of coke.

11.2. Parameters Determining the Rate of Coking

11.2.1. *Influence of the reaction system*

11.2.1.1. *Reactant and coking rate*

On acid catalysts, coking occurs rapidly from certain molecules which can be considered as coke precursors or **coke maker molecules** [5–8]. These molecules are either highly reactive (e.g. alkenes, dienes in acid catalysis) and/or strongly retained on the outer surface or within the zeolite micropores (e.g. polyaromatics). Note that in coke formation, both conditions of reactivity and retention always have to be satisfied, the first playing the determining role for light alkenes and dienes, the second for polyaromatics. Thus, short-chain alkenes and dienes (especially cyclopentadienes) undergo very fast condensation reactions (oligomerization, polymerization), with the formation of heavy and polar products easily retained on the zeolite. These products are reactive enough to undergo intramolecular reactions (e.g. cyclization) and intermolecular reactions (e.g. hydrogen transfer to desorbed products) with formation of coke molecules. On the other hand, polyaromatic molecules which are not very reactive migrate slowly through the acid zeolites because of their bulkiness and basicity, their long contact time with the acid sites favouring their transformation into molecules too bulky to desorb from the zeolite (i.e. coke molecules).

In contrast, with reactants such as alkanes (in particular the linear ones), non-substituted naphthenes, benzene, etc., none of the conditions required for a direct coke formation is satisfied: they are not very reactive in acid catalysis and moreover neither bulky enough nor basic enough to be strongly retained on the zeolite catalyst. Therefore, from these poorly reactive compounds, coking results from the transformation of either feed impurities or of intermediates or products of the desired transformation. A typical example is the isomerization of *n*-alkanes, in which acid zeolite catalysts are deactivated by coke resulting from the transformation of olefinic cracking products. The solution to limit coke formation, hence deactivation, is to reduce the concentration of these coke maker molecules, for instance by operating under hydrogen pressure over bifunctional Pt acid zeolite catalysts. Under these conditions, alkenes appear only as intermediates of the bifunctional process and their concentration in the reaction mixture is strongly limited by thermodynamic equilibrium [9]. The trace amounts of these highly reactive alkene intermediates are sufficient to allow bifunctional isomerization, but not enough to undergo the fast condensation reactions involved in coke formation. The balance between the hydrogenating and acid functions (given by the ratio of the concentrations of accessible Pt atoms and protonic acid sites: $C_{\mathrm{Pt}}/C_{\mathrm{A}}$) was shown to determine the activity, selectivity and stability of Pt HFAU catalysts in *n*-heptane and *n*-decane transformation [10, 11], a perfect stability being obtained for high $C_{\mathrm{Pt}}/C_{\mathrm{A}}$ values, e.g. >0.17 in *n*-decane transformation (Table 11.1). Note that the activity per acid site is then maximal and that the apparent reaction scheme matches the scheme of transformation of decene intermediates. This means that the number of acid sites encountered by the decene intermediates during their diffusion between two Pt sites is such that only one decene reaction step can occur before hydrogenation, which makes impossible the formation of coke [11].

This positive effect of the association of hydrogenating sites with the acidic ones can occur even without a shift from a purely acid to

Table 11.1. Hydroisomerization and hydrocracking of *n*-decane over Pt HFAU catalysts. Influence of the metal/acid balance ($C_{\mathrm{Pt}}/C_{\mathrm{A}}$) on the catalytic properties [11].

$C_{\mathrm{Pt}}/C_{\mathrm{A}}$	≥ 0.17	>0.03 <0.17	<0.03
Activity/acid site	Maximal	Maximal	Low
Stability	Perfect	Average	Low
Reaction scheme	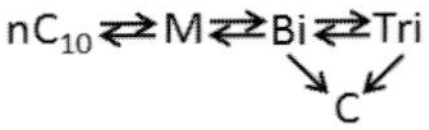	$nC_{10} \rightleftarrows (M,Bi,Tri) \rightarrow C$	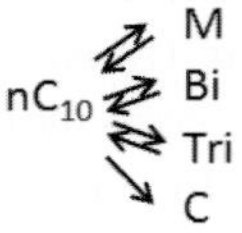

M, Bi, Tri: mono-, bi- and tribranched isomers.

a bifunctional reaction mechanism. Typical examples deal with toluene disproportionation and transmethylation with heavy aromatics [12–15]. In both reactions, a significant decrease in coke formation and an increase in stability could be observed, which was generally ascribed to the hydrogenation of coke precursors. In toluene disproportionation over Ni HMOR catalysts, an increase in hydrogen pressure was shown to result in a simultaneous decrease in the rates of disproportionation and coking, coking being more affected than disproportionation. From this observation, it was concluded [12] that hydrogen activated by metals was able to react with the carbenium ion intermediates involved in both coke formation and toluene disproportionation:

Reaction 11.1 $R^+ + H_2 \rightarrow RH + H^+$

Furthermore, a stabilizing effect of high hydrogen pressure (12 bar) was also shown with Ni-free HMOR samples, probably because of the presence of iron impurities [12].

Benzylic carbenium ions that are involved in both reactions are likely to undergo Reaction 11.1. However, the greater significance of the effect of hydrogen pressure on coke formation suggests that additional intermediates of coking, e.g. those involved in dehydrogenative coupling and cyclization steps (DC and DCy), could also be desorbed through Reaction 11.1. In agreement with this proposal, coke formed during *o*-xylene under hydrogen pressure does not contain CH_2Cl_2 insoluble components whose formation involves DC and DCy steps while insoluble coke was formed under nitrogen [15]. Note that the introduction of Pt on the zeolite used in the transalkylation process provokes a decrease in the purity of the benzene product. To solve this problem, a sulphuration technique was developed to fine-tune the Pt hydrogenating activity while sustaining the stability [13].

11.2.1.2. *Operating conditions and coking rate*

The choice of operating conditions — gas or liquid phase, reactant concentration, pressure, solvent, temperature, etc. — plays a key role in the process efficiency. The increase in catalyst stability is one of the main conditions to be satisfied. However, as many other criteria — economy in terms of atoms and energy of all the steps of the process (reaction, separation, purification, recycling), respect of health and environmental rules — have to be considered, each process constitutes a particular case that demands the implementation of an appropriate solution.

Thus, in the liquid-phase synthesis of fine chemicals, the choice of **solvents** often affects significantly the reaction rate, the product distribution and the stability. In phenol acetylation with phenylacetate carried out in a batch reactor over a HBEA zeolite at 423 K, the polar sulfolane solvent promotes the formation of the para hydroxyacetophenone product, with a para/ortho ratio of 7 compared to 0.8 with dodecane, and also limits the formation of bulky secondary products such as bisphenol A derivatives, hence catalyst deactivation. In the liquid-phase fine chemicals synthesis, the desired products can also behave as coke precursors, simply because they are bulkier and more polar than the reactant molecules [17]. As a consequence, the reactions are limited by product desorption, not only from the active sites but also from the zeolite micropores. Thus, in anisole acetylation with acetic anhydride over a HBEA zeolite carried out in a batch reactor, the desired 4-methoxyacetophenone (4MP) product was shown to preferentially occupy the BEA micropores [17]. This has a dramatic effect on the catalyst stability because (i) acetylation is self-inhibited and (ii) the long contact time of 4MP molecules with acid sites favours their secondary transformation into bulkier and more polar products (coke) which remain trapped within the micropores and block the access to the active sites. By adjusting the operating conditions so as to favour desorption of 4MP molecules — low anisole conversion, large excess of anisole, substitution of the batch reactor by a flow reactor, etc. — the BEA zeolite stability was significantly improved, which allowed the development by Rhodia of an efficient commercial process [19].

Another way to favour desorption of coke precursors is to operate the reaction in a **supercritical medium**. A supercritical fluid is defined as a substance beyond its critical point, i.e. critical temperature T_c and critical pressure p_c, but below the pressure required for condensation into the solid state. Supercritical fluids (SCF) exhibit properties intermediate between those of typical liquids and gases [20–22]: their density, hence

Table 11.2. Comparison of the density ρ, dynamic viscosity η and self-diffusion coefficient D of gases, liquids and supercritical fluids [22].

Property	Gas (at $p_{gas} = 1$ bar)	Supercritical fluid	Liquid
ρ (g dm^{-3})	0.6–2	200–500	600–1600
η (mPa s)	0.01–0.3	0.01–0.03	0.2–3.0
D (10^{-6} m^{2}s^{-1})	10–40	0.07	0.0002–0.002

their solvating power, approaches that of liquids while their viscosity and diffusivity properties are similar to those of gases (Table 11.2).

The SCF properties can be exploited in heterogeneous catalysis [21], in particular for the *in situ* extraction of coke precursors from the zeolite micropores [20, 22]. To discuss this possibility, the effects of the gas, liquid and SC media were compared on the simple case of alkene isomerization over an acid zeolite at low temperatures (e.g. 373 K) for which oligomers are the main by-products [20]. In liquid phase, the oligomers can be solubilized in the reaction medium but because of their bulkiness, diffusion limitations hinder their desorption out of the zeolite micropores whereas in gas phase, oligomers accumulate within the micropores because of their low volatilities. In both media, oligomers are then retained within the micropores and act as coke precursors leading progressively to consolidated coke. This process occurs more slowly in supercritical media which combine gas-like diffusion properties and liquid-like solvent power [20]. A recent review of the applications of zeolite catalysts in supercritical fluids confirms their potential for limiting coke formation and deactivation [22].

Among the zeolite-catalysed reactions tested in supercritical media, the most important is the alkylation of isobutane with n-butenes in which deactivation by coking prevents the industrial application. Two different approaches were followed: the first in which reactants (essentially isobutane which is always in large stoichiometric excess) play the role of SCF, which requires high temperatures (above 408 K, i.e. T_c of isobutane) with as a consequence a low selectivity to alkylate due to undesired cracking reactions; the second with the addition of a supercritical solvent such as CO_2, therefore with the possibility to operate at low temperatures (T_c of $CO_2 = 305$ K) but with low alkylation rate and low selectivity to alkylate to the profit of oligomers, owing to the reactant dilution by CO_2 [22]. In this latter case an increase in stability could be observed [23] while in the first case, the stability improvement was not general [20, 24]. In alkylation experiments carried out at 413 K (without adding CO_2) with a very high isobutane/butene ratio ($\sim$100), the stability as well as the selectivity to alkylates were largely dependent on the zeolite catalyst, with HBEA outperforming HFAU zeolites. This was proposed to be due to differences in crystal sizes (smaller on HBEA) and in the dimensions of HFAU supercages and BEA channels. With HBEA, alkylation and secondary reactions would occur on or near (at the pore mouth) the outer surface, which would facilitate desorption of coke precursors whereas with HFAU, coke precursors being formed within the supercages near or far from the surface could not be easily desorbed.

The variation of the coking rate with **reaction temperature** T depends on the reactant [25]. Thus, over a HFAU zeolite, the rate of coking from n-heptane and toluene increases with T as it is generally found with classical chemical steps. However, from n-heptane, no coking occurs in the absence of cracking, which means that coke results from the secondary transformation of cracking products, actually of the alkene products (coking is a secondary process). Furthermore, from toluene, coke formation occurs at low temperatures without formation of desorbed products. From propene and cyclohexene, the coking rate passes through a minimum with increasing T (Fig. 11.1).

These complex behaviours can be related to the conclusions drawn in Chapter 7 from the large differences in coke composition at low (<473 K) and at high temperatures (>623 K):

- The formation of coke molecules requires both chemical steps and retention within the micropores or on the outer surface. Therefore the

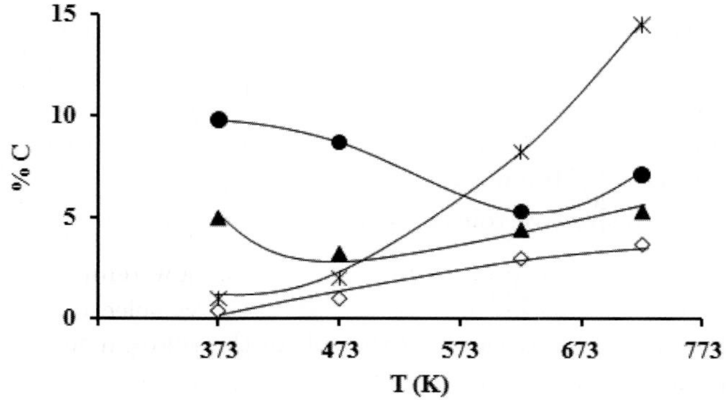

Fig. 11.1. Effect of temperature on coke formation over a HFAU zeolite during the transformation of various hydrocarbon reactants: toluene (*), cyclohexene (•), propene (▲) and n-heptane (◇).

coking rate should depend (i) on the nature, thermodynamics and kinetics of the chemical steps and (ii) on the cause of retention;

- The chemical steps and the cause of coke retention both depend on T. Thus, at low T, coke formation from alkenes involves essentially oligomerization steps (which are exothermic) plus some rearrangement and hydrogen transfer (HT) steps while at high T, coke results from a complex network of condensation and consolidation steps: HT, DC and DCy (dehydrogenative coupling and cyclization). Furthermore, while at low T, coke molecules are generally retained because of their low volatility and strong adsorption, at high T it is because of their steric blockage (trapping) within the micropores.

Because the formation of coke molecules generally requires numerous successive bimolecular reactions, an increase in the coking rate with the **reactant pressure** is to be expected. Moreover, an increase in the selectivity to coke is also likely, the desired reactions involving generally no or only a limited number of bimolecular steps. These increases were observed in *m*-xylene (mx) disproportionation over a HMOR zeolite: the coking rate and the coking/disproportionation rate ratio increased by a factor of about two with the augmentation of p_{mx} from 0.06 to 0.2 bar [26].

11.2.2. *Influence of the zeolite catalyst features*

The main features of acid zeolite catalysts, on the one hand the characteristics of their acidic active sites (generally the protonic ones), on the other hand those of the cages, channels and channel intersections (i.e. of the nanoreactors), play a major role in the various successive chemical steps involved in coke formation, as well as in the retention of precursor and coke molecules. Because of the difficulty in preparing zeolite samples with identical acidities and with different pore structures and *vice versa*, the effect of these parameters is generally difficult to be distinguished and hence quantified [27]. Having said that, some examples are given hereafter, illustrating the respective roles of acidity and of pore structure.

11.2.2.1. *Acidity and coking rate*

The following could be expected of the role of the acidity characteristics on the rate of coke formation (Fig. 11.2):

(i) the stronger the acid sites, the faster the chemical steps and the more pronounced the retention of coke precursors and coke molecules, hence the faster the coking rate; (ii) the higher the density of the acid sites, thus

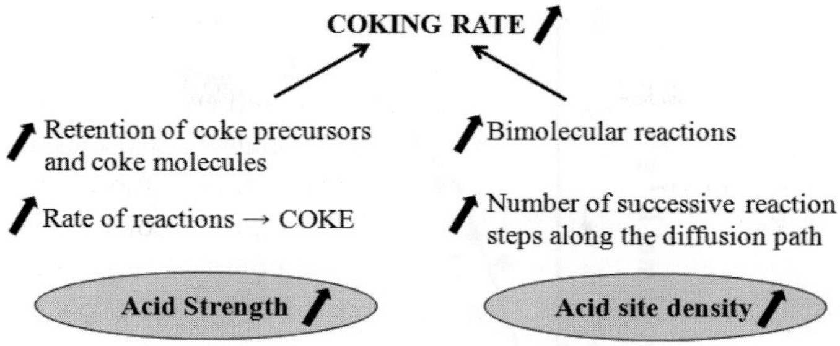

Fig. 11.2. Influence of the acidity characteristics on the rate of coking. Adapted from [8].

the closer these sites are to each other, the larger the number of successive chemical steps undergone by reactant molecules along the diffusion path within the zeolite crystallites and the more favourable the condensation reactions, hence the faster the coking rate.

11.2.2.1.1. Acid strength and coking rate

The effect of acidity on the coking rate was determined by using a series of large-pore zeolites with different framework Si/Al ratios. Figure 11.3 shows the effect of the concentration (C_A) of protonic sites per unit cell of commercial HFAU zeolites (framework Si/Al ratios from 4 to 100) on the initial rate of coke formation (r_k) from propene under the following operating conditions: fixed-bed reactor, 723 K, $p_{\text{propene}} = 30\,\text{kPa}$, $p_{\text{nitrogen}} = 70\,\text{kPa}$, propene flow $= 0.7\,\text{g}\,\text{h}^{-1}\text{g}^{-1}$ zeolite. Note that in this case, coke is an apparent primary product of propene transformation. For low C_A values, r_k is roughly proportional to the density of protonic sites with turnover frequency (TOF) values of $\sim 7\,\text{h}^{-1}$. This quasi-constant value of TOF can be related to the similarity in strength of the protonic sites corresponding to isolated bridging OH groups (Chapter 2). TOF values three times lower were found for C_A values higher than 10, i.e. for the non-isolated, hence weaker, protonic sites. All this suggests that the acid strength plays a determining role in coke formation while the acid site density has no apparent effect.

The positive effect that the acid strength has on coke formation was confirmed during n-heptane cracking at 623 K over two series of dealuminated HFAU samples [29]. The samples of these series differ essentially by the amount of extraframework aluminium species (EFAL), large in series 1, low in series 2. Note that contrary to propene transformation, coke does not result from the direct transformation of the

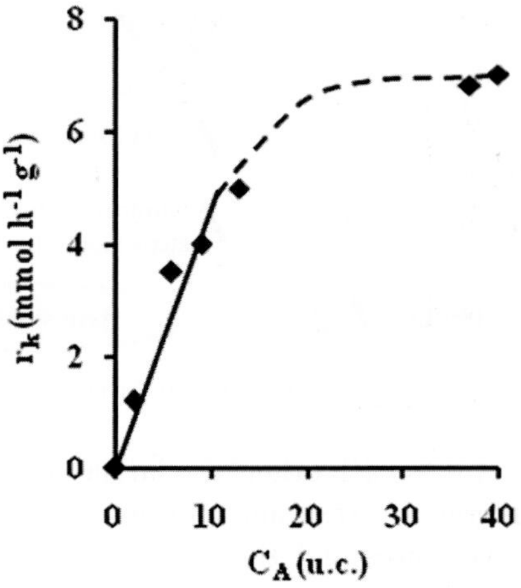

Fig. 11.3. Coke formation from propene at 723 K over a series of commercial HFAU zeolites: Initial rates of coking (r_{k}) as a function of the concentration of protonic sites per unit cell (C_{A}). Adapted from [27].

reactant but from that of olefinic cracking products (coking is consecutive to cracking). Therefore, to discuss the effect of acidity, the coking/cracking rate ratio was substituted for the coking rate. For identical values of C_{A}, this ratio was found to be greater for the samples of series 1 (Fig. 11.4), which suggests that EFAL species have a promoting effect on the selectivity to coke. This promoting effect can be related to the enhancement of the acid strength of protonic sites by their interaction with EFAL species presenting Lewis acidity (Chapter 2).

11.2.2.1.2. Nature of acid sites and coking rate

It should be underscored that most of the authors exclude a direct role of Lewis acid sites in coke formation from hydrocarbon reactants and this is for two main reasons: (i) the active role of protonic acid sites was shown by the good correlations obtained between the rate of most of the reactions and the protonic site concentration but no correlations with the Lewis site concentration were found [30]. It is therefore most likely that protonic sites are the only active sites for the secondary transformations of hydrocarbons into coke; (ii) pyridine chemisorption on aged samples followed by IR spectroscopy shows from very low coke contents a decrease in the number of protonic sites able to retain pyridine adsorbed, which is not the case for the Lewis acid sites (Fig. 5.6).

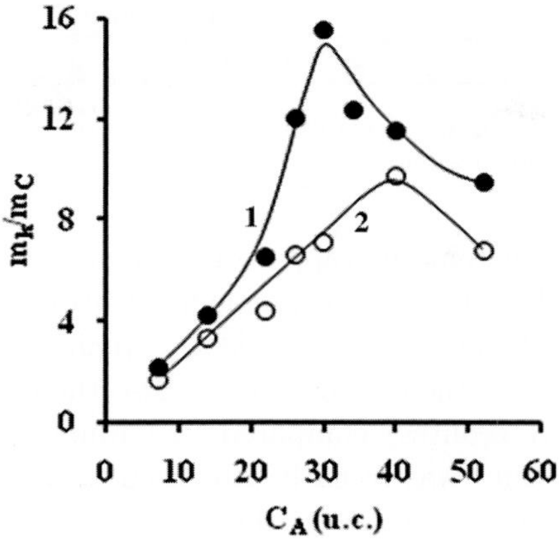

Fig. 11.4. Coke formation during n-heptane cracking at 623 K over two series of dealuminated HFAU samples: 1 and 2, with respectively large and low amounts of extraframework aluminium species. Ratio between the amounts of coke and cracking products after a 70 min reaction ($m_{\mathrm{k}}/m_{\mathrm{C}}$) vs. the concentration of protonic sites per unit cell (C_{A}). Adapted from [27].

11.2.2.1.3. Density of acid sites and coking rate

As shown above, if the effect of the acid strength on the rate of coke formation is well demonstrated, this is not the case for the expected positive effect of the density of the protonic active sites. The existence of this effect can furthermore be contested by the constant TOF values for coking drawn from Figs. 11.3 and 11.4. However, an effect of this parameter was advanced to explain the effect of the balance between hydrogenating and acid functions of bifunctional Pt HFAU catalysts on coke formation and stability during alkane hydroisomerization and hydrocracking. The greater the ratio between the concentrations of accessible Pt and protonic sites ($C_{\mathrm{Pt}}/C_{\mathrm{A}}$), hence the smaller the number of acid sites encountered by the alkene reaction intermediates during their diffusion between two Pt sites, the smaller the number of chemical steps they can undergo, hence the slower the coke formation and the higher the catalyst stability. Above a certain value of $C_{\mathrm{Pt}}/C_{\mathrm{A}}$, the number of chemical steps becomes so small that no coke formation is possible [11].

An effect of the acid site density was also proposed in the case of zeolites presenting non-interconnected channels. Indeed, along these channels, the molecule diffusion is unidirectional. Therefore, the number of successive reaction steps undergone by reactant molecules during their diffusion from the entrance to the exit of the channels, hence the probability of forming

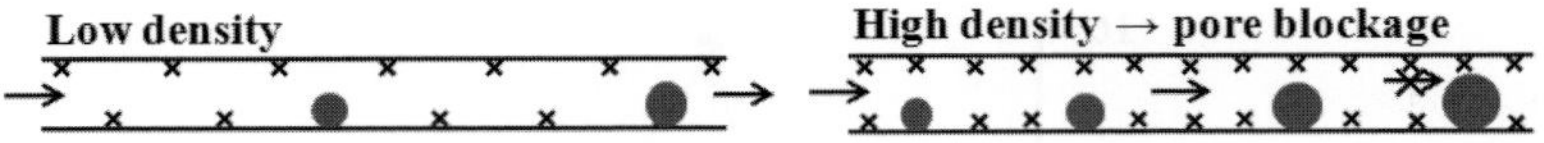

Fig. 11.5. Effect of the acid site density on coke formation in the non-interconnected channels of monodimensional zeolites.

bulky molecules which remain trapped, increases with the density of the protonic acid sites (Fig. 11.5). This particular feature of monodimensional zeolites (called tunnel shape selectivity [31]) explains not only the effect of their acid site density on the coking rate but also their very high selectivity to coke (e.g. HMOR zeolites) compared to tridimensional zeolites. This effect of acid site density was recently advanced to provide an explanation of the unexpected effect of coke on the rate of *n*-butene transformation and on the selectivity to isobutene that was observed over HFER zeolites [32].

11.2.3. *Pore structure and coking rate*

The effect of pore structure was first investigated by Rollmann and Walsh [33] by comparing the transformation at 700 K of an equimolar mixture of *n*-hexane, 3-methylpentane, 2,3-dimethylbutane, benzene and toluene over a large series of zeolites. The size of the pore apertures was found to be the determining parameter, with coke yield decreasing from ~1 g per 100 g of alkane converted with large-pore zeolites to 0.1 and 0.05 with medium- and small-pore zeolites, respectively. This conclusion which excludes the effect of the size and shape of cages, channels and channel intersections in which coke molecules are formed and trapped at high reaction temperatures seems surprising. More recent works lead to different conclusions.

The first deals with coke formation during *n*-heptane cracking at 723 K over four protonic zeolites with different pore structures: HFAU, HMOR, HMFI and HERI [27, 34]. The degree of protonic exchange was chosen so that the initial cracking activities were similar. The initial coking/cracking rate ratio was found to be close to one with HMOR and HERI, four times lower with HFAU and 1000 times lower with HMFI. The high selectivity to coke of the first two zeolites was attributed to the fast blockage of coke precursors in the one-dimensional MOR channels and in the ERI trap cages, i.e. large cages (6.3 Å Ø × 13 Å) with small apertures (3.6 × 5.1 Å^2). This blockage is less easy with the FAU and MFI zeolites which have three-dimensional micropore systems with moreover relatively small differences between the dimensions of cages (FAU) or channel intersections (MFI) and those of the apertures. This difference in size is much smaller with MFI — channel intersections of ∼8.5 Å Ø and pore openings of $5.3 \times 5.6\,\text{Å}^2$ and $5.1 \times 5.5\,\text{Å}^2$ — than with FAU — cages of 13 Å Ø with apertures of 7.4 Å Ø — which explains for a large part the very low coking activity of HMFI. An additional reason, i.e. the low acid site density of the chosen MFI sample, was also suggested [27] but remains to be confirmed (see Section 11.2.2.1).

In the second example, only one zeolite (MWW, also called MCM22) which possesses unique features, i.e. three independent systems with very different characteristics that do not exist in the other molecular sieves, was used. Therefore, the effect of different micropore characteristics on coke formation during catalytic reactions can be established under quite identical conditions. One of the pore systems (Fig. 11.6a) is defined by two-dimensional narrow 10-MR sinusoidal channels of slightly elliptical cross-sections (4.0 × 5.0 Å) which form small cages at their intersections (6.4 × 6.9 Å). The second (Fig. 11.6b and c) is comprised of large cylindrical supercages (10-MR diameter: 7.1 Å Ø, height: 18.4 Å) interconnected through 10-MR windows (4.0 × 5.5 Å). The last one is constituted by large 12-MR cups (diameter: 7.1 Å Ø, depth: 7 Å) on the external surface (Fig. 11.6d), hence by hemi-supercages [35].

The three pore systems contain bridging OH groups, i.e. protonic acid sites, and acid-catalysed reactions were shown to occur in each of them. Generally, the concentration of protonic sites in the sinusoidal channels is slightly greater than in the supercages, both being 2–5 times greater than in the external cups [37].

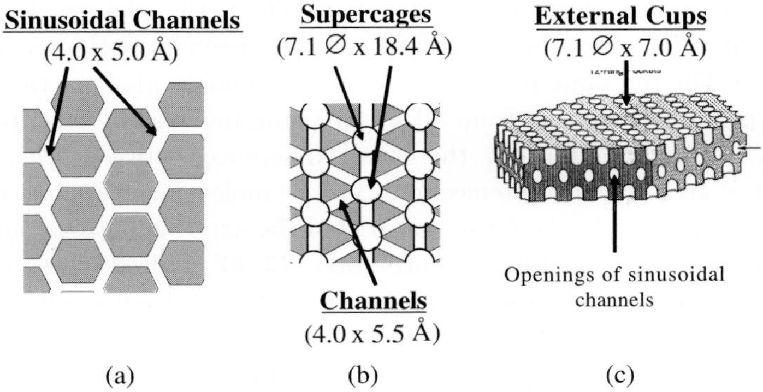

Fig. 11.6. Micropore systems of the MWW (MCM22) zeolite: (a) supercage system, (b) external cups, (c) sinusoidal channels. Adapted from [36].

In order to specify the role of each of the three pore systems, a simple method was developed [37] and applied to the transformation of various reactant molecules — *o*-, *m*-, *p*-xylenes, methyl cyclohexane, *n*-heptane, propene-toluene mixture, etc. — into both desorbed products and carbonaceous compounds (coke). Large differences in turnover frequency (TOF), selectivity and stability were found between the three pore systems. However, whatever the reactant, a rapid formation of carbonaceous products occurs within the large supercages, causing their very fast deactivation. In contrast, there is practically no deactivation of the reactions occurring within the sinusoidal channels and external cups [37–40].

The fast coke formation within the supercages could be expected from the large difference between their dimensions and those of the apertures: supercages can be considered as typical trap cages. Thus, in these large cages, xylene disproportionation, which involves bulky bimolecular intermediates (trimethyl diphenyl methane) that are known as coke precursors, can occur in addition to isomerization. These coke precursors which desorb slowly through the narrow supercage apertures are progressively transformed into methyl polyaromatics (coke) [38]. From methylcyclohexane and *n*-heptane, another mode of coke formation involving the participation of alkene cracking products in the successive formation of aromatic rings becomes predominant [39, 40].

The quasi-impossibility to form coke within the sinusoidal channels and the external cups has different origins. Within the narrow sinusoidal channels, it can be related to the difficulty of transforming reactant molecules into "coke" precursors. Thus from xylene reactants, only the reversible isomerization of meta- to para-xylene can occur (the ortho isomer is unable to desorb from or to enter the channels) and the bimolecular disproportionation with coke precursor formation is sterically inhibited. For the same reason, the secondary bimolecular reactions of the alkene products resulting from alkane cracking involved in coke formation cannot occur. In addition, the small difference between the sizes of apertures and channel intersections makes molecular trapping unlikely. The impossibility of coke formation within the external cups cannot result from steric limitations on coke formation [37–40]. Indeed, these cups are large enough to allow the formation of bulky molecules such as coke precursors. Therefore, it is most likely that it is the second condition for coke formation, i.e. the possibility of retention of coke precursors, which cannot be satisfied. This difficulty of retaining molecules within the outer cups seems highly probable when considering their large diameter and their small depth. Indeed, both these characteristics favour desorption of the primary products by sweeping, thus preventing their secondary transformation.

11.2.4. *Other features of zeolite catalysts affecting the coking rate*

Characteristics of the zeolite catalysts other than acidity and pore structure can significantly affect the rate of coke formation. One of the most important is the size of the zeolite crystals which, as underscored in Chapter 2, plays a determining role in zeolite-catalysed reactions, in particular when reactant and/or product molecules cannot easily diffuse within the micropores because of their bulkiness and/or their polarity: the smaller the crystal size, the greater the turnover frequency value (TOF) of the active sites for the desired reaction. Indeed with large crystals, because of diffusion limitations, the catalytic transformation of bulky molecules can occur only on the active sites close to the outer surface whereas with small crystals, the reaction can occur on all the active sites, including those of the core. Therefore, milder conditions (shorter contact time and/or lower temperature) are suitable to get the same yield as with large crystals, which explains for a large part the decrease in coking rate with decreasing crystal size. However this decrease in coking rate was also observed when the desired reaction was not diffusion limited, i.e. when TOF values were independent of the crystal size [41]. In this case, the increase in coking rate with the crystal size can be related to limitations in the diffusion of coke precursors within the zeolite micropores. With small crystals, these coke precursors formed near the outer surface can easily desorb; with large crystals, they undergo successive secondary reactions with formation of highly polyaromatic molecules (coke) which limit then block the access of reactant molecules to the active sites. This difference in coke composition with crystal size was shown during methylcyclohexane transformation at 723 K over HBEA zeolites, with coke insoluble in methylene chloride appearing with large crystals only [41]. It should however be underscored that in many cases, the demonstration of the effect of crystal size on the coking rate is made difficult by differences in acidity characteristics of the zeolite samples. Furthermore, the positive effect of the decrease in crystal size on the catalyst stability can sometimes be interpreted as a more significant catalytic role of the outer surface with less coke formation [42]. The opposite opinion was however advanced by Ding *et al.* [43] on the basis of the stabilizing effect of a decrease in the outer site concentration of a Mo/HMFI methane aromatization catalyst provoked by a silanation treatment.

The creation of mesopores by post-treatment of the zeolites, which shortens the micropore length, is expected to have a negative effect on coking similar to that caused by the decrease in crystal size. Actually, the creation of mesopores by desilication of mordenite zeolites was shown to significantly decrease coke formation in the liquid-phase alkylation of benzene with ethylene [44].

In most of the commercial zeolite catalysts, the zeolite is associated with a matrix (binder) which has various physical and chemical functions. Indeed the matrix provides the appropriate particle size and shape, improves the resistance to attrition, traps feed contaminants, serves as carrier for promoters, passivators, etc., catalytically pre-convert the molecules too bulky to enter the zeolite micropores and stabilize the zeolite active phase. Part of this stabilizing effect can be due to limitation of the coke formation, probably by trapping of coke precursors by the binder [45, 46].

11.3. Parameters Determining the Deactivating Effect of Coke

The deactivating effect of coke (its toxicity, *Tox*) depends on the same parameters: reaction system (reactant, solvent and product molecules, operating conditions) and zeolite features (characteristics of the active sites, pore structure) that have been shown to determine the coking rate. However, the dependence of the toxicity is more complex than that of the coking rate because often *Tox* depends both directly and indirectly on these parameters. In particular, the composition and location of coke which depend on the same parameters significantly affect the coke toxicity.

11.3.1. *Operating conditions and deactivating effect of coke*

The operating conditions essentially affect indirectly the deactivating effect of coke molecules. Under severe conditions (high temperatures and/or high reactant pressure), the desired reaction is often diffusion limited. As a consequence, this reaction and the secondary ones, including coke formation, will occur essentially within the micropores of the outer part of the zeolite crystals. Coke molecules that are trapped in these micropores will block the diffusion of the reactant molecules towards the crystal core, inducing a high *Tox* value. This so-called pore mouth (or shell) blockage

Table 11.3. Deactivating effect of coke molecules (*Tox*). A: initial values; B: values at complete deactivation [27].

	A	B
HFAU	5	1.3
HMOR	25	7.5
HMFI	0.25	1.0
HERI	30	7.0

demonstrated a long time ago [47] was observed in methylnaphthalene transformation at 623 K over a HFAU zeolite [25].

11.3.2. *Pore structure and deactivating effect of coke*

As shown in Chapter 7, the deactivating effect of coke molecules (quantified by the *Tox* values) depends significantly on the zeolite pore structure and on the coke content, location and composition (which are interconnected characteristics). This can be seen in Table 11.3 which reports the *Tox* values at low coke contents and at complete deactivation in the example chosen in Section 11.2.3 to present the effect of pore structure on the coking rate [27].

With three-dimensional zeolites having no trap cages (i.e. large cages with small apertures), such as MFI and FAU, deactivation that at low coke contents is due to site poisoning progressively evolves to pore blockage with the increase in coke content [27].

Therefore, with zeolites whose acid sites have similar features — essentially strength and accessibility — the greater the coke content, the higher the toxicity of coke molecules. It is what was observed with a HMFI zeolite sample: (i) at low coke contents, *Tox* was smaller than 1, which was explained by a competition between reactant and weakly basic coke molecules for adsorption on the acid sites("partial coverage or poisoning"); (ii) at average coke contents, *Tox* became equal to 1, which is typical of site poisoning; (iii) lastly, at high coke contents, *Tox* was greater than 1, in agreement with pore blockage by very bulky molecules overflowing onto the crystal outer surface.

The situation is slightly different when the zeolite catalyst has acid sites with different features. In Chapter 7, the effect that the heterogeneity in strength of the protonic acid sites has on the coke "toxicity" (*Tox*) was shown in the example of a HFAU zeolite [27]: (i) At low coke contents, the strongest acid sites, which are the most active in both the desired reaction and coke formation, are preferentially deactivated with a consequent apparent *Tox* value of four. Of course, this high value does not mean that one coke molecule is able to deactivate four active sites, but that the TOF value of the deactivated strong acid site is four times higher than the average TOF value. (ii) The increase in coke content causes a decrease in the apparent toxicity, owing to opposite effects of the decrease in strength of the residual acid sites and to the development of pore blockage demonstrated by adsorption experiments, the first parameter having a predominant effect with the chosen HFAU sample.

With zeolites such as MOR, in which the diffusion of organic reactant molecules is unidirectional, and with zeolites such as ERI, which contains trap cages, deactivation is always due to pore blockage with as a consequence a very pronounced deactivating effect of coke molecules ($Tox > 20$) [27]. Indeed, one coke molecule located in a large channel of HMOR is enough to block the access of the reactant molecules to all the acid sites located in this channel (Fig. 11.7b). Similarly, the coke molecules trapped within HERI cages located near the outer surface of the crystals block the access of the reactant molecules to the acid sites located within the inner cages (shell blockage). In both cases, the acid sites made inaccessible to the reactant molecules appear to be deactivated.

11.3.3. *Acidity and deactivating effect of coke*

Besides the pore structure, other features of the zeolite can affect the apparent deactivating effect of coke molecules (*Tox*). The effect of a heterogeneous distribution in strength of the acid sites (that is often found

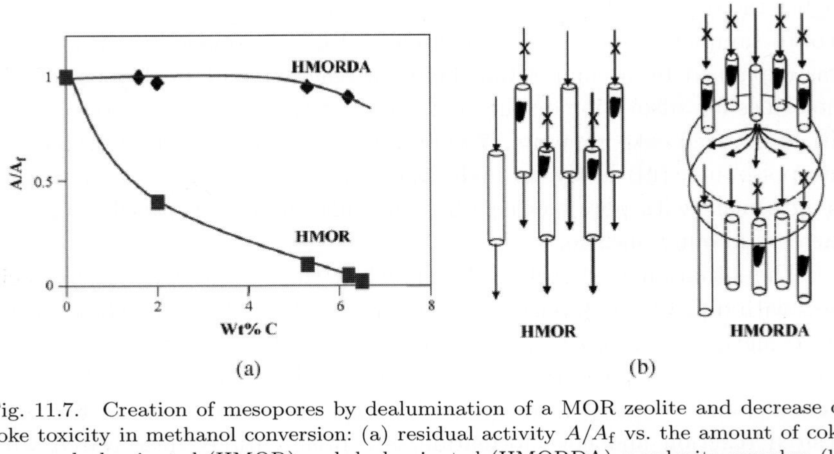

Fig. 11.7. Creation of mesopores by dealumination of a MOR zeolite and decrease of coke toxicity in methanol conversion: (a) residual activity A/A_f vs. the amount of coke for non-dealuminated (HMOR) and dealuminated (HMORDA) mordenite samples; (b) diffusion of reactant molecules in the channels of HMOR and HMORDA samples [8].

with zeolite catalysts) has been underscored in Section 11.3.2. An additional argument in favour of an increasing effect of the acid site strength on *Tox* is provided by the observation that the effect of coke on the reaction rate is all the more pronounced when the reaction demands stronger acid sites. Thus, over a HFAU zeolite, the effect of coke was found to be more pronounced on *n*-hexane cracking which requires very strong acid sites than on *m*-xylene isomerization which is a less demanding reaction [48].

11.3.4. *Other features of zeolite catalysts determining the deactivating effect of coke*

When considering the deactivation of zeolites with a one-dimensional pore system or presenting trap cages, a significant effect of the crystal size could be expected. Thus, the larger the size of MOR crystals, the longer the channels will be and the higher the *Tox* value. Similarly, the larger the size of ERI crystals, the higher the number of inner acid sites apparently deactivated per coke molecule. Therefore, *Tox* can be significantly reduced by operating with small crystal zeolites, e.g. nano-sized zeolites.

Another way to reduce *Tox* is to create intracrystalline mesopores within the zeolite crystals. Figure 11.7a shows that in methanol conversion into hydrocarbons, dealumination of HMOR zeolites decreases more than 10 times the coke toxicity [49]. This significant *Tox* decrease was explained by the creation of inner mesopores which make the pore system quasi-three-dimensional (Fig. 11.7b). Desilication by alkaline treatment is another way to create mesopores in zeolites which presents the advantages of dealumination to preserve the protonic active sites and to lead to zeolite (e.g. mordenite [44]) samples with a superior mesopore surface area.

11.4. Guidelines for Preventing Deactivation

Deactivation by coking depends on both the rate of coke formation and on the deactivating effect of coke, both of which are affected as shown in Sections 11.2 and 11.3 by the features of the reaction system (in particular operating conditions) and of the zeolite catalyst (acidity, pore structure, etc.).

The main rules to follow in the design of zeolite acid catalysts with low sensitivity to deactivation by coking are reported in Fig. 11.8.

Choosing the appropriate pore structure of the zeolite is essential to limit the formation of coke and to minimize its deactivating effect. High rates of coking as well as high "toxicity" of coke molecules are found with zeolites with pore systems either monodimensional or with trap cages

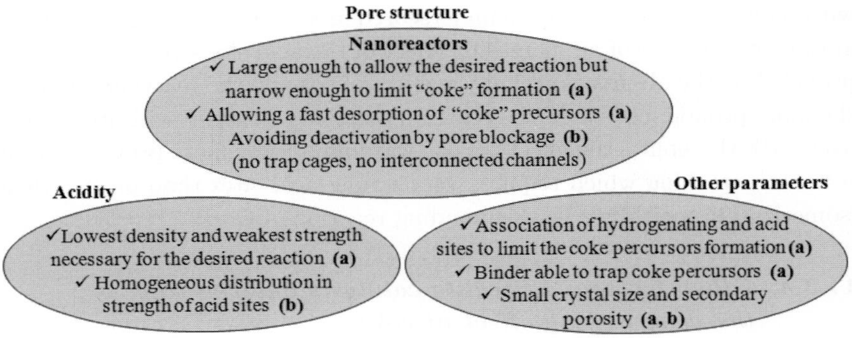

Fig. 11.8. Design of the acid zeolite catalyst for reduced deactivation by coking; (a) limitation of coke formation; (b) reduction of the deactivating effect of coke.

(i.e. large cages with small openings). Indeed, these pore systems favour the retention of coke precursor molecules and deactivation by pore blockage. These drawbacks can however be minimized in different ways, in particular by creating mesopores by post-synthesis treatments (dealumination and preferably desilication).

Fine-tuning the acid properties is no less essential. In particular, the strength of the protonic acid sites should be limited to the values necessary to obtain a sufficiently high rate for the desired reaction. In the case of monodimensional zeolites, great care must be taken in the optimization of the acid site density, a compromise having to be found to obtain sufficiently high rate and selectivity for the desired reaction with relatively slow coke formation.

The operating conditions also have to be judiciously chosen. Thus, operating at the lowest contact time that allows sufficient conversion to the desired product(s) is recommended, especially when coking is consecutive to the desired reaction. Furthermore, too severe conditions of temperature and reactant pressure can provoke a preferential formation of coke on the outer part of the crystals or particle (shell coking) with as a consequence pore blockage, and hence pronounced deactivating effect of coke.

It should be underscored that most of the guidelines to limit the coking rate and the deactivating effect of coke presented in this chapter apply not only to acid zeolite catalysts but also to all the other solid catalysts.

References

[1] Gosselink J.W., Stork W.H., *Ind. Eng. Chem. Res.*, 36 (1997) 3354–3359.
[2] Gosselink J.W., van Veen J.A.R., *Stud. Surf. Sci. Catal.*, 126 (1999) 3–16.

[3] Sie S.T., *Appl. Catal. A: General*, 212 (2001) 129–151.
[4] Guisnet M., Magnoux P., in *Deactivation and testing of Hydrocarbon Processing Catalysts*, O'Connor P., Takatsuka T., Woolery G.L. (Eds.), ACS Symposium Series 534, American Chemical Society: Washington, DC (1996) Chapter 5, 77–90.
[5] Guisnet M., Magnoux P., Moljord K., in *Zeolite Microporous Solids: Synthesis, Structure and Reactivity*, Derouane E.G., Lemos F., Naccache F., Ramôa Ribeiro F. (Eds.), NATO ASI Series C, Kluwer, Dordrecht, 352 (1992) 437–457.
[6] Guisnet M., Magnoux P., *Catal. Today*, 36 (1997) 477–483.
[7] Guisnet M., Cerqueira H.S., Figueiredo J.L., Ramôa Ribeiro F., in *Desactivação e regeneração de catalisadores*, Guisnet M., Cerqueira H.S., Figueiredo J.L., Ramôa Ribeiro F. (Eds.), Fundacão Calouste Gulbenkian, Lisboa (2008) Chapters 3, 4, 6, 17.
[8] Guisnet M., Costa L., Ramôa Ribeiro F., *J. Mol. Catal. A: Chemical*, 305 (2009) 69–83.
[9] Guisnet M., Perot G., in *Zeolite Science and Technology*, Ramôa Ribeiro F. *et al.* (Eds.), NATO ASI Series E, Martinus Nijhoff Publishers, The Hague, 80 (1984) 397–419.
[10] Guisnet M., Alvarez F., Giannetto G., Perot G., *Catal. Today*, 1 (1987) 415–433.
[11] Alvarez F., Ramôa Ribeiro F., Perot G., Thomazeau C., Guisnet M., *J. Catal.*, 162 (1996) 179–189.
[12] Gnep N.S., Martin de Armando M.L., Marcilly C., Ha B.H., Guisnet M., *Stud. Surf. Sci. Catal.*, 6 (1980) 79–89.
[13] Tsai T.-C., Chen W.-H., Liu S.-B., Tsai C.-H., Wang I., *Catal. Today*, 73 (2002) 39–47.
[14] Roldan R., Romero F.J., Jimenez C., Borau V., Marinas J.M., *Appl. Catal. A: General*, 266 (2004) 203–210.
[15] Henriques C.A., Bentes A.M., Magnoux P., Guisnet M., Monteiro J.L.F., *Appl. Catal. A: General*, 166 (1998) 301–309.
[16] Jayat F., Sabater Picot M.J., Rohan D., Guisnet M., *Surf. Sci. Catal.*, 108 (1997) 91–98.
[17] Guisnet M., Guidotti M., in *Catalysts for Fine Chemicals Synthesis*, Microporous and Mesoporous Solid Catalysts 4, Derouane E.G. (Ed.), John Wiley & Sons, Chichester (2006) Chapter 2, 39–67.
[18] Derouane E.G., Dillon C.J., Bethell D., Derouane-Abd Hamid S.B., *J. Catal.*, 187 (1999) 209–218.
[19] Metivier P., in *Fine Chemicals Through Heterogeneous Catalysis*, Sheldon R.A., van Bekkum H. (Eds.), Wiley-VCH, Weinheim (2001) 161–172.
[20] Subramaniam B., Arunajatesan V., Lyon C.J., *Stud. Surf. Sci. Catal.*, 126 (1999) 63–77.
[21] Grunwaldt J.-D., Wandeler R., Baiker A., *Catal. Rev.-Sci. Eng.*, 45 (2003) 1–96.
[22] Gläser R., *Chem. Eng. Technol.*, 30 (2007) 557–568.
[23] Clark M., Subramaniam B., *Ind. Eng. Chem. Res.*, 37 (1998) 1243–1255.

[24] Mota Salinas A.L., Sapaly G., Ben Taarit Y., Vedrine J.C., Essayem N., *Appl. Catal. A: General*, 336 (2008) 61–71.

[25] Guisnet M., Magnoux P., in *Zeolite Microporous Solids: Synthesis, Structure and Reactivity*, Derouane E.G., Lemos F., Naccache F., Ramôa Ribeiro F. (Eds.), NATO ASI Series C, Kluwer Academic Publishers, Dordrecht, 352 (1992) 457–474.

[26] Benamar A., Miloudi C., Gnep N.S., Guisnet M., unpublished results.

[27] Guisnet M., Magnoux P., Martin D., *Stud. Surf. Sci. Catal.*, 111 (1997) 1–19.

[28] Moljord K., Magnoux P., Guisnet M., *Appl. Catal. A: General*, 122 (1995) 21–36.

[29] Wang Q.L., Giannetto G., Guisnet M., *J. Catal.*, 130 (1991) 471–482.

[30] Guisnet M., *Acc. Chem. Res.*, 23 (1990) 392–398.

[31] Guisnet M., Morin S., Gnep N.S., in *Shape-Selective Catalysis*, Song C., Garcés J.M., Sugi Y. (Eds.), ACS Symposium Series 738 (2000) Chapter 24, 334–352.

[32] de Ménorval B., Ayrault P., Gnep N.S., Guisnet M., *J. Catal.*, 230 (2005) 38–51.

[33] Rollmann L.D., Walsh D.E., *J. Catal.*, 56 (1979) 139–140.

[34] Guisnet M., Magnoux P., *Appl. Catal.*, 54 (1989) 1–27.

[35] Leonowicz M.E., Lawton S.L., Partridge R.D., Chen P., Rubin M.K., *Science*, 264 (1994) 1910–1916.

[36] Guisnet M., Ramôa Ribeiro F., *Zeolithes, un nanomonde au service de la catalyse*, EDP Sciences, Les Ullis (2006) Chapter 11, 186–200.

[37] Laforge S., Martin D., Guisnet M., *Appl. Catal. A: General*, 268 (2004) 33–41.

[38] Laforge S., Martin D., Paillaud J.L., Guisnet M., *J. Catal.*, 220 (2003) 92–103.

[39] Matias P., Lopes J.M., Laforge S., Magnoux P., Russo P.A., Ribeiro Carrott M.M.L., Guisnet M., Ramôa Ribeiro F., *J. Catal.*, 259 (2008) 190–202.

[40] Matias P., Lopes J.M., Laforge S., Magnoux P., Guisnet M., Ramôa Ribeiro F., *Appl. Catal. A: General*, 351 (2008) 174–183.

[41] Magnoux P., Rabeharitsara A., Cerqueira H.S., *Appl. Catal. A: General*, 304 (2006) 142–151.

[42] Choi M., Na K., Kim J., Sakamoto Y., Terasaki O., Ryoo R., *Nature*, 461 (2009) 246–249.

[43] Ding W., Meitzner G.D., Iglesia E., *J. Catal.*, 206 (2002) 14–22.

[44] Seitz M., Klemm E., Emig G., *Stud. Surf. Sci. Catal.*, 111 (1997) 1–19.

[45] Fougerit J.M., Gnep N.S., Guisnet M., Amigues P., Duplan J.L., Hughes F., *Stud. Surf. Sci. Catal.*, 84 (1994) 1723–1730.

[46] Misk M., Joly G., Magnoux P., Guisnet M., Jullian S., *Microp. Mesop. Mat.*, 40 (2000) 197–204.

[47] Wheeler A., *Adv. Catal.*, 3 (1951).

[48] Guisnet M., Bourdillon G., Gueguen C., *Zeolites*, 4 (1984) 308–309.

[49] Gnep N.S., Roger P., Cartraud P., Guisnet M., Juguin B., Hamon C., *C. R. Acad. Sci.*, Serie II, 309 (1989) 1743–1746.

Chapter 12

REGENERATION OF COKED ZEOLITE CATALYSTS

M. Guisnet

12.1. Introduction

All catalysts lose over time their activity and/or selectivity, with however the rate and cause(s) of deactivation — poisoning and/or coking and/or physical degradation, etc. — differing significantly from an industrial process to another. Thus, deactivation time can range from some seconds, e.g. in catalytic cracking (FCC), up to several years, e.g. in ammonia synthesis [1–4]. The time scale (tc) of deactivation is the determining factor in the choice of the reactor and mode of regeneration in continuous processes: (i) entrained-flow reactors (riser) with continuous catalyst circulation from the reactor to the regenerator for extremely fast deactivation (tc of seconds); (ii) fluidized-bed and slurry reactors with continuous regeneration for fast deactivation (tc of minutes to days); (iii) moving-bed reactors for continuous regeneration (tc of weeks); (iv) fixed-bed reactors with regeneration while reactor is off-line (tc of months); (v) fixed-bed reactors with *ex situ* regeneration or even no regeneration for very slow deactivation (tc of years) [1, 3].

The mode of regeneration depends naturally on the cause(s) of deactivation. As coking is often the main cause, this chapter is essentially devoted to regeneration by coke removal. The classical oxidative treatment under air flow at high temperatures will be presented, the focus being placed on the reaction steps involved in the combustion of hydrocarbon and nitrogen coke components. The severe conditions of this mode of coke removal have often detrimental effects on the zeolite catalysts such as dealumination and degradation of the framework, or sintering of metal components in the case of bifunctional catalysts, which constitutes a strong incitation to develop milder methods. The milder modes of regeneration presently proposed will be critically reviewed.

12.2. Removal of High-Temperature Coke

12.2.1. *Classical oxidative treatment*

12.2.1.1. *Oxidation of polyaromatic coke*

The combustion under air flow, sometimes with oxygen added, is the common way to remove coke from the aged refining and petrochemical catalysts. The combustion of the polyaromatic coke formed over acid zeolites at high temperatures (>623 K) during the transformation of pure hydrocarbons has been investigated by many authors by operating in fixed-bed reactors at increasing temperature. The first conclusion, drawn from the measurements of the amounts of H_2O, CO_2 and CO produced during the combustion of this polyaromatic coke, was that the oxidation begins with the hydrogen atoms as is shown in Fig. 12.1 [5]. Indeed at low oxidation temperatures (<600 K), water was formed in very large molar amounts while at high temperatures (>773 K) only CO_2 and CO were evolved. Various oxygenated compounds — aldehydes, ketones, anhydrides — were furthermore found in the partially oxidized coke. These intermediates that disappear at higher temperatures with essentially the formation of CO and

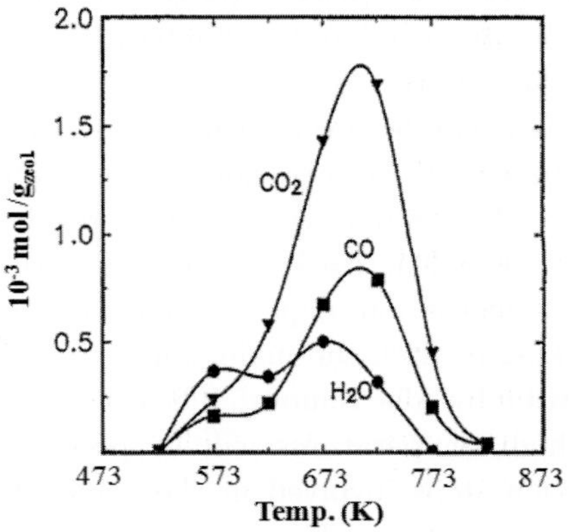

Fig. 12.1. Evolution as a function of the oxidation temperature of water (after substraction of water liberated from the zeolite), carbon monoxide and carbon dioxide during the oxidation of coke (8.5 wt%) deposited during propene transformation at 723 K over a HFAU zeolite [5].

CO_2 could not be identified because of their large diversity and their high molecular weights.

Fortunately, as demonstrated by Moljord *et al.* [6], the oxidation of pyrene molecules located within the supercages of HFAU was shown to occur exactly in the same way as that of HFAU zeolite coke. Note that this could be expected from the previous observation that coke oxidation did not depend on the zeolite coke content, hence on the related coke composition [7]. The composition of the oxygenated intermediates recovered in CH_2Cl_2 after mineralization of the zeolite was established by GC/MS coupling. At low oxidation temperatures (430–520 K), there was formation of ketones with a pyrenic skeleton and of aldehydes, acids and anhydrides with a phenanthrene skeleton, whereas at high temperatures (620–670 K), less bulky oxygenated compounds with the same functional groups appeared. At intermediate temperatures, highly polyaromatic compounds insoluble in CH_2Cl_2 were also present. The following steps are then involved in pyrene (hence polyaromatic coke) oxidation: (i) functionalization of polyaromatics from low T; (ii) condensation of reactant and/or oxidation product molecules with formation of highly condensed polyaromatics at intermediate T; (iii) decarbonylation and decarboxylation of oxygenated compounds at $T > 550$ K.

Coke removal was studied for a number of HFAU zeolites with framework Si/Al ratios from 4 to 100, coked to different levels (from 2 to 12–15 wt%) during propene transformation at 723 K [5]. As shown in Fig. 12.2a, coke content had practically no effect on the oxidation rate, despite the large differences in coke composition — H/C ratios from 0.9 to 0.4 — and location — only within the supercages at low coke contents, with part of the many coke molecules overflowing onto the outer surface at high contents [8]. Note, however a difficult oxidation of coke at very high contents (~20 wt%), because of a complete blockage of the access to the micropores. In contrast, the concentration of framework Al atoms, hence of the protonic sites, had a significant effect on coke oxidation: the higher this concentration, the faster the coke oxidation. Thus, the percentages of coke (or of the pyrene model molecule) that require for combustion temperatures higher than 723 K passed from 10–15% with the more acidic zeolite to 50% with the less acidic zeolite (Fig. 12.2b). This suggests that the protonic sites of the HFAU zeolites play an active role in coke oxidation. The other features of the HFAU samples — concentration of extraframework Al species, small quantities of sodium content (<0.15 wt%) — do not influence coke removal, the CO_2/CO ratio

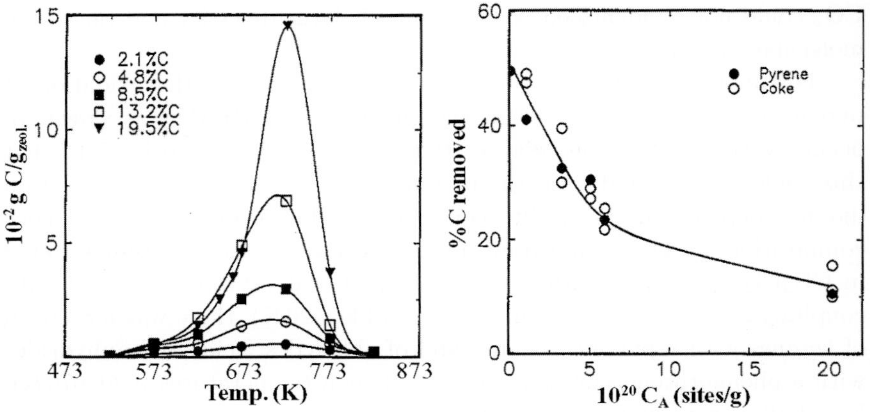

(a) (b)

Fig. 12.2. Oxidation of coked HFAU samples: (a) normalized amounts of carbon eliminated as $CO + CO_2$ for different values of coke contents; (b) % of C eliminated as $CO + CO_2$ above 723 K vs. C_A, the concentration of framework Al atoms (or of protonic sites) of HFAU zeolites. Adapted from [5].

increasing however with the Na content: from 1.4 below 400 ppm to more than 5 at 1400 ppm. The effect on coke removal of the zeolite pore structure was previously examined with three zeolites: HFAU, HMOR and HMFI [7]. While practically no differences were found between the two first zeolites, coke oxidation was much slower on HMFI, which was ascribed to a pore structure effect. However, it seems most likely that the slower oxidation rate on this zeolite originates from the much lower concentration of its protonic sites (~10 times less).

There are various ways to minimize CO production while limiting the adverse effects of the removal of coke on the acidic properties. Combustion promoters (e.g. Pt/Al_2O_3) are frequently used, not only in the refining processes such as the FCC process (Chapter 13), but also in fine chemical processes such as the liquid-phase anisole or veratrole acetylation with acetic anhydride. Another way, developed in FCC units devoted to the treatment of residues in which full combustion would result in excessive regenerator temperatures, is to use a two-stage regenerator. The first stage operates at a temperature just sufficient to allow the combustion of most of the H atoms (which as shown above are preferentially oxidized) and only part of the C atoms, the second stage at higher temperatures completing the combustion. With this double regeneration system, the detrimental effects (framework dealumination and degradation) are limited, owing to the relatively low temperature in the first stage and to the low water partial pressure in the second.

12.2.1.2. *Role of nitrogen species in* NO_x *formation during FCC coke oxidation*

Nitrogen oxides (NO_x) which are harmful air pollutants result mainly from road traffic. Nevertheless, from a typical refinery, 2000 t/year of NO_x (essentially NO: >90%) are released, half of them originating from FCC regenerator flue gases. Regulations limiting the release of NO_x from stationary sources affect the FCC regenerator, which leads refiners to invest in new technologies to reduce the NO levels in the FCC exhaust gases. It should be underscored that the Pt-based combustion promoters used in the last decades to control CO emissions have significantly increased NO emissions [9, 10].

FCC feedstocks contain a variety of aromatic N-heterocyclic compounds. The distribution of a typical feed was reported in Ref. [10]. This feed contained 0.15 wt% nitrogen and 0.075 wt% basic nitrogen. Polyaromatic pyrrole derivatives — alkyl carbazoles, alkylbenzocarbazoles and alkylindoles — were the predominant nitrogen species compounds (60%); their amount was about 3 times higher than that of the 6-ring nitrogen species (pyridine derivatives) — alkylquinolines and alkyltetrahydro quinolines — and much higher than that of amino compounds [10].

In the riser reactor, about 50% of the nitrogen-containing compounds are converted into ammonia while the rest ends up as coke. In the regenerator, approximately 90% of the nitrogen of coke is converted into N_2 while the remaining part is released as NO. Owing to the very low nitrogen contents in coked FCC catalysts (e.g. ~250 ppm), it seems easier for establishing the mode of NO_x formation in the regenerator to operate with model nitrogen compounds. This was done by Babich *et al.* [11]. Three simple representative nitrogen compounds — pyridine, aniline and pyrrole — were transformed into coke at 823 K over an industrial FCC catalyst. After XPS analysis of this coke, its oxidation was monitored by GC and MS analysis of the evolved products. The following conclusions were drawn from this work:

- Whatever the nitrogen precursor, three types of N species — pyridinic, pyrrolic and quaternary nitrogen — are present on the coked catalysts.

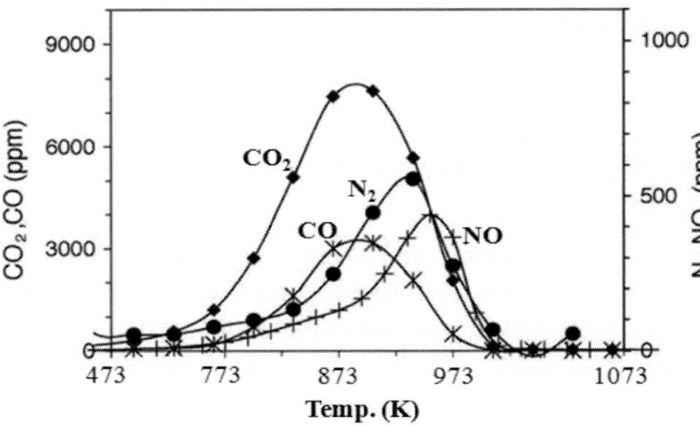

Fig. 12.3. Evolution of gases during temperature-programmed oxidation (heating rate: 10 K/min) of coke formed from pyridine at 823 K over a FCC catalyst (4 h): (♦) CO_2; (*) CO; (•) N_2 and (+) NO [11].

The last ones which are preferentially formed during the first seconds of coking are strongly adsorbed on the catalyst acid sites.

- During temperature-programmed oxidation (TPO) of the coked catalysts, the order of appearance of the gases is always: CO_2 = $CO/N_2/NO$ (Fig. 12.3). The pyrrolic- and pyridinic-type species are easily oxidized with N_2 as the main product owing to the large amount of C or CO still available. The quaternary nitrogen species are the last to be oxidized, the lack of CO resulting in the appearance of NO in the gas phase.

This TPO study of the coke formed from model compounds leads to useful data on the progressive changes of nitrogen-containing coke and on the main released gas products during the FCC catalyst regeneration. It should however be underscored that the nitrogen-containing species of this coke are different from those found in FCC coke which, for a large part, likely results from reactions of the nitrogen-containing molecules of the feed with the predominant hydrocarbon components. Furthermore, under the operating conditions, only very small amounts of NH_3 and HCN which were proposed as intermediates in the transformation of nitrogen coke species [9] were observed.

The formation of these species and their transformation into N_2 and NO_x were addressed in a paper from Barth *et al.* [12] on the basis of vacuum temperature-programmed desorption (TPD) and TPO measurements carried out over coked catalysts obtained from a commercial

FCC unit. HCN and NH_3 were generated by cracking and hydrolysis of coke above 723 K. Water favoured NH_3 formation via hydrolysis of isocyanate intermediates adsorbed on the alumina matrix of the FCC catalyst. Both HCN and NH_3 intermediates can be oxidized into NO at temperatures above 823 K. This reaction was confirmed to be favoured when a Pt combustion promoter was added to the FCC catalyst. It is therefore essential to develop catalytic NO_x reduction additives that could operate in the presence of a CO combustion promoter or better even able to simultaneously reduce CO and NO_x emissions. This explains why the research is so active in this field.

12.2.2. *Alternative ways of regeneration by coke removal*

12.2.2.1. *Reactivation with alternative oxidants*

More expensive oxidizing agents such as ozone and nitrous oxide were also tested for removing coke from zeolites. With ozone-enriched oxygen, coke can be oxidized at low temperatures (400–450 K) with therefore no risk of hydrothermal degradation. However, ozone reactivation has many drawbacks: no possibility to remove coke from the core of extrudates because of its rapid dissociation [13], attack of steel reactor vessels by this strong oxidant.

These inconveniencies do not exist with N_2O, which moreover is cheaper and is even produced as waste in adipic acid preparation. However, relatively high temperatures (700–800 K) are required for coke removal. Therefore, despite the good results generally obtained [14], the substitution of the N_2O treatment for the classical combustion under air seems very unlikely except in processes using N_2O as an oxidizing agent, such as those catalysed by Fe-containing MFI zeolites: propane oxidative dehydrogenation [15], direct benzene oxidation to phenol (hydroxylation) for which a pilot-scale process ("AlphOX^{TM}") has been developed [16], etc.

In this latter case, a comparative regeneration of coked catalysts by treatment with O_2 and N_2O diluted with helium was carried out [17]. N_2O was shown to be more active in coke oxidation than O_2, the treatment with N_2O for 2 h at 708 K removing the same amount of coke as the O_2 treatment for 6 h at 723 K. Figure 12.4a, in which the activity in benzene hydroxylation of three series of samples (coked before (1) and after treatment with O_2 (2) or N_2O (3)) is plotted vs. the coke content, shows that both treatments decrease the deactivating effect of coke, the effect being more pronounced with N_2O than with O_2. Note also that the hydroxylation activity is

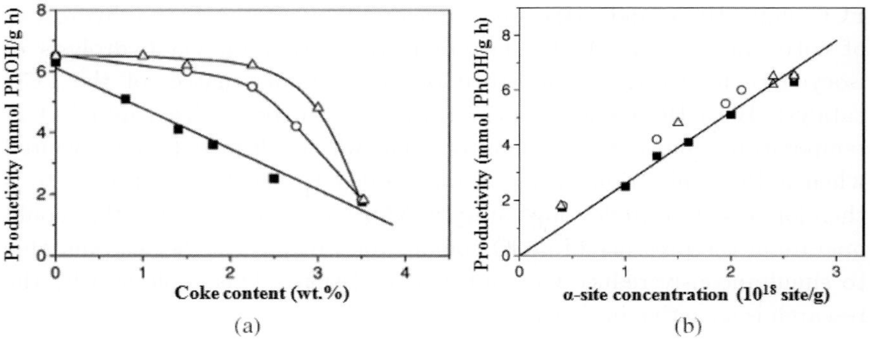

Fig. 12.4. a) Effect of coke on the activity in benzene hydroxylation of the coked catalyst samples (■) and of the samples regenerated with O_2 (○) and with N_2O (△); b) catalytic activity of the samples vs. α-sites [17].

totally recovered after removal of ~35 % of the coke by N_2O, which occurs only after quasi-complete coke removal by O_2. All this suggests that coke elimination by N_2O is preferable.

Extraframework complexes of Fe^{2+} stabilized within the MFI micropores (called α-sites) were demonstrated to be active in benzene hydroxylation [17–19]. The proportionality that was observed between the activity of the 1, 2, 3 samples and their concentration on α-sites (Fig. 12.4b) is a clear demonstration of deactivation of these sites by coke poisoning. A preferential oxidation by N_2O of coke located near the α-sites was proposed to explain the unexpected complete regeneration of the hydroxylation activity before complete coke oxidation.

12.2.2.2. *Non-oxidative reactivation*

Non-oxidative treatments were investigated to regenerate zeolites coked at high temperatures (>623 K). As these treatments are operated under milder conditions (lower temperatures, no water produced), the zeolite degradation is expected to be minimized. Reactivation of HMFI catalysts coked in methanol to gasoline conversion at 640 K was shown to occur after treatment at reaction temperature with hydrogen or with alkanes [20]. The same occurs in acetone transformation into hydrocarbons over HMFI at 673 K [21]. In both cases, sweeping of the catalyst with inert gas was less efficient.

A positive effect of hydrogen sweeping was also found in toluene disproportionation (TDP) [22, 23]. The applicability of this hydrogen

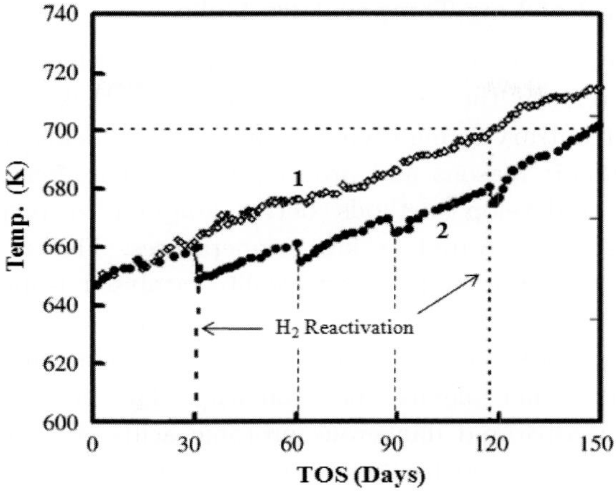

Fig. 12.5. Toluene disproportionation over a HMOR zeolite. Reaction temperature T (K) vs. time-on-stream TOS (days) for normal operation (curve 1) and for operation with periodic reactivations under hydrogen flow (curve 2). Adapted from [23].

reactivation procedure was demonstrated in a long-term aging study with a typical industrial operating mode: TDP is operated at a constant level of toluene conversion (45%) with compensation for the activity loss by rising the reaction temperature (T) (Fig. 12.5). After each hydrogen reactivation treatment, T is reduced to achieve the desired 45% conversion. As shown from the comparison of the T increase with normal operation (curve 1) and with intermediate reactivations (curve 2), a significant activity gain (corresponding to a T decrease higher than 15 K) can be obtained.

The optimal conditions of the reactivation treatment were specified in previous experiments; the activity was shown to be fully restored after sweeping of the catalyst under hydrogen flow for 8 h at a temperature 30 K higher than T. Only partial reactivation was obtained by sweeping under nitrogen flow. Whatever the sweeping gas, a decrease in the content of coke could be observed, e.g. from 12.3 wt% to 9.1 and 10.1 wt%, with hydrogen and nitrogen respectively. To explain the reactivation, it was proposed that sweeping removes preferentially coke molecules with a low aromaticity degree (lower than that of alkyl pyrene), the greater reduction of coke content by hydrogen treatment arising from the higher desorption affinity of hydrogen. Another possibility could be however the hydrogenation of carbocations of coke precursors, that was advanced in Chapter 11 to explain the positive effect of hydrogen pressure on the HMOR stability in TDP.

12.3. Removal of Low-Temperature Coke

In low-temperature processes ($<473\,K$), oxidation which should not be necessary to remove the carbonaceous deposits ("coke") that are not polyaromatic (contrary to high-temperature coke) remains largely applied. However, economic reasons and progress in the characterization of coke have helped to develop methods better adapted to regenerating the zeolite catalysts deactivated at low temperatures. This is shown below in the example of the liquid-phase zeolite-catalysed isobutane- butene alkylation.

Alkylation yields one of the most valuable gasoline cuts: high octane number, with neither olefins nor aromatics. The current technologies use strong, corrosive and dangerous protonic acids (HF and H_2SO_4) as catalysts, which explains why there is a major research effort to substitute them with environmentally friendly catalysts. Large-pore acid zeolites, essentially with FAU and BEA structures, were particularly studied. Unfortunately, these zeolites, as was also the case with most of the other solid catalysts which were tested, suffered from fast deactivation by carbonaceous compounds. Optimizing their physicochemical features and the operating conditions (in particular use of a continuous-flow stirred tank reactor with a high isobutane/butene ratio) allowed a significant improvement of their stability. But despite these advances, frequent regeneration steps are needed to extend the lifetime of zeolite catalysts to an acceptable length.

The composition of the carbonaceous compounds ("coke") responsible for the deactivation of alkylation zeolite catalysts is now well-known [24–26]. On the deactivated catalysts, "coke" was constituted by a complex mixture of highly unsaturated and highly branched species containing cyclic structures with 35–40 C atoms at the maximum. However, there is over time an increase in the molecular size of the carbonaceous compounds and a change in nature: from butene oligomers to species with cycles and with more double bonds and branchings even in the last stage formation of aromatic rings. As a consequence, the mode of deactivation passes from site poisoning to a limited then a complete pore blockage [26].

This change in coke composition with time-on-stream and the related decrease in selectivity to the desired trimethylpentanes suggest that periodic regeneration steps should be preferentially carried out on partially deactivated samples. Moreover, various options could then be considered to substitute the classical high-temperature coke combustion. Milder oxidation methods at low temperatures using air containing ozone or hydrogen peroxide solutions that were tested by Querini [27] did not allow a complete "coke" removal. Regeneration through "coke" hydrocracking gives more promising results. This method was tested by Klingmann *et al.* [28] over 0.4 wt% Pt/La FAU (Y) samples aged during alkylation in a continuous-flow stirred tank reactor at 348 K for different times. As shown by comparison of the product distribution with purely acid zeolite catalysts used in similar conditions, Pt did not affect the alkylation process. Up to 2 h, *n*-butene was totally converted and the "coke" deposited had only a limited effect on the micropore volume accessible to nitrogen adsorbate. After this time, there was a decrease in butene conversion and in the formation of coke with a more pronounced effect on the micropore volume, which suggests a change in the mode of deactivation from site poisoning to pore blockage. Aged zeolite samples were recovered after 3 h of alkylation, just before a fast decrease in *n*-butene conversion and being submitted to regeneration treatment under the following conditions: fixed-bed reactor, hydrogen pressure of 15 bar, $100\,K\,h^{-1}$ increase in temperature up to the final temperature (preferentially 573 K), total regeneration time of ~8 h. This treatment causes a decrease in "coke" content from ~3 wt% to 0.2 wt% and a complete restoration of the alkylation activity and selectivity even after several regeneration steps (Fig. 12.6).

Supercritical fluid (SCF) extraction of "coke" is another possible way to regenerate alkylation zeolite catalysts. It has been previously shown

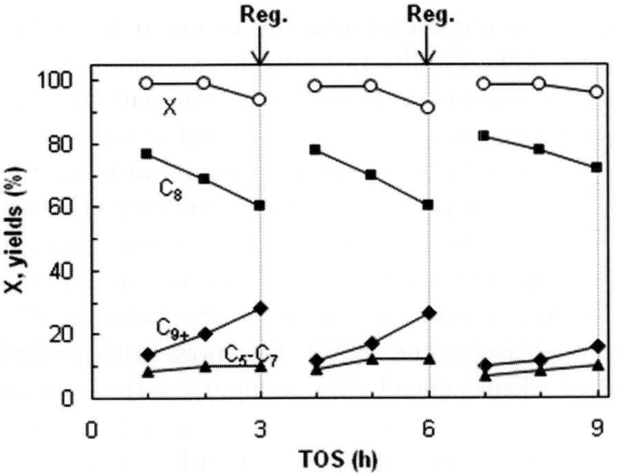

Fig. 12.6. Isobutane-butene alkylation on a 0.4 wt% Pt/La FAU catalyst: butene conversion and product distribution within three alkylation-regeneration cycles vs. time-on-stream (cumulative values) [28].

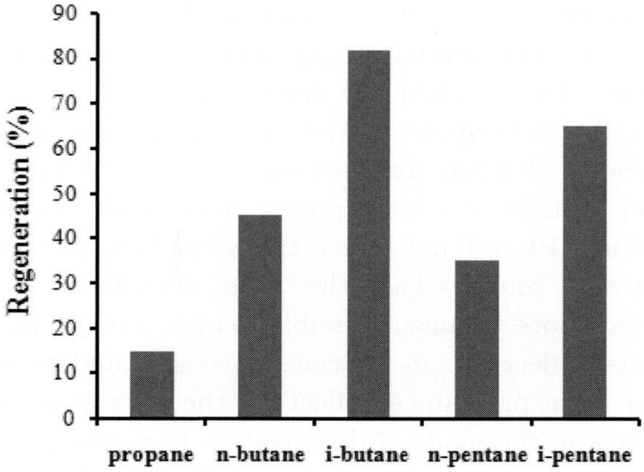

Fig. 12.7. Catalyst reactivation of a HFAU zeolite through supercritical extraction of "coke". Comparison of the effect of various alkanes. Adapted from [29].

(Section 11.2.1.2, Chapter 11) that SCFs could be used for controlling catalyst deactivation by extracting continuously the carbonaceous compounds from the zeolite. However, the benefit was limited, either by the necessity of a very high isobutane/butene ratio (with isobutane used as SCF) or to a low activity and a low selectivity to the desired trimethylpentanes (CO_2 as SCF). More promising results were obtained in the regeneration of aged alkylation catalysts.

A comprehensive study of the extraction of carbonaceous compounds by SCFs was carried out on a totally deactivated HFAU zeolite catalyst in a flow reactor [29]. Fluids including linear and branched alkanes — propane, *n*-butane, *n*-pentane, isobutane and isopentane — were examined individually in reactivation. Figure 12.7 that corresponds to reactivation treatments at a constant pressure of 111 bar and at temperatures between 453 K (C_3, C_4) and 483 K (nC_5) to operate in supercritical conditions shows that the nature of the fluid has a very pronounced effect on the catalyst reactivation (estimated by the ratio between the yields in desired trimethyl pentanes on the regenerated and on the fresh catalysts). The best results were obtained with isobutane (82%), the worst with propane (14%). C_4 and C_5 branched alkanes gave better results than the corresponding linear alkanes.

To explain this large difference between linear and branched alkanes, it was proposed that alkanes with tertiary carbons may easily transfer hydrides to high-molecular-weight carbocations resulting from the adsorption of the unsaturated carbonaceous species. This reaction is easier with branched alkanes than with linear alkanes because of the higher stability of the resulting tertiary carbocations:

e.g. 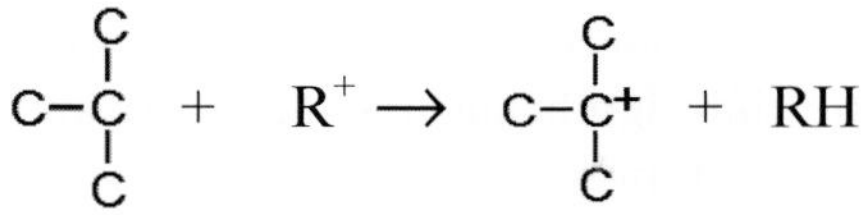

Contrary to the unsaturated coke molecules, the saturated ones (RH) are not chemisorbed on the acid sites, and hence are more easily desorbed from the zeolite micropores. As could be expected from the increase of the size of coke molecules with the extent of deactivation of the alkylation catalyst and with the shift from site poisoning to pore blockage, better results could be obtained by operating the reactivation treatment on partially deactivated zeolite catalysts. Actually, SC regeneration with isobutane was shown to restore 100% of the activity of a partially deactivated HFAU zeolite, and this more than 23 times on the same sample [30].

This removal of low-temperature coke from deactivated zeolite by reaction with a "solvent" was also observed in benzene alkylation with long-chain *n*-alkenes whose products, i.e. linear alkylbenzenes (LAB), are the precursors of biodegradable detergents. A more environmentally friendly process has been developed [31] in which a solid proprietary catalyst was substituted for the previously used hydrofluoric acid. This catalyst is most likely a large-pore zeolite whose applicability to the synthesis of linear alkyl benzenes is well demonstrated [32]. Alkylation tests carried out at 363 K with a model reaction, the liquid-phase alkylation of toluene by 1-dodecene on a HFAU zeolite, shows that catalyst deactivation is due to the formation and trapping of tridodecyltoluenes within the micropores [33, 34]. Various regeneration treatments were tested. Whereas sweeping under nitrogen or decane flow of the coked catalyst at the reaction temperature were totally negative, sweeping with toluene reactant led to complete removal of "coke" with recovery of the activity of the fresh zeolite [34]. The trapped tridodecyltoluenes molecules were shown to react with toluene, leading through transalkylation to mono- and di-dodecyltoluenes that easily desorbed from the zeolite. This regeneration by transalkylation was also successfully applied to benzene alkylation over a BEA zeolite under commercial conditions, the benzene reactant being used for washing the deactivated catalyst [35].

12.4. Conclusion

For a long time, most of the investigations of zeolite catalyst regeneration were done in industrial companies, with as a consequence results considered as proprietary, hence inaccessible. The times are progressively changing. Thus, thanks to the significant technical advances in the methods of characterization, great progress has been made in the knowledge of the chemical steps involved in the removal of most organic coke components by the classical oxidative treatment. This fundamental knowledge has opened the way to more economical and ecological technologies. Another important point concerns the regeneration of zeolite catalysts deactivated during low-temperature reactions. While removing the polyaromatic high-temperature coke demands oxidation treatments, the components of low-temperature "coke" are generally still enough reactive to be eliminated through milder treatments. Here again, good knowledge of coke composition enables us to design and develop economical and ecological regeneration methods.

References

[1] Moulijn J.A., van Diepen A.E., Kapteijn F., Bartholomew C.H., *Appl. Catal. A: General*, 212 (2001) 3–16.
[2] Bartholomew C.H., *Appl. Catal. A: General*, 212 (2001) 17–60.
[3] Sie S.T., *Appl. Catal. A: General*, 212 (2001) 129–151.
[4] Trimm D.L., *Appl. Catal. A: General*, 212 (2001) 153–160.
[5] Moljord K., Magnoux P., Guisnet M., *Appl. Catal. A: General*, 121 (1995) 245–259.
[6] Moljord K., Magnoux P., Guisnet M., *Catal. Lett.*, 28 (1994) 53–59.
[7] Magnoux P., Guisnet M., *Appl. Catal.*, 38 (1988) 341–352.
[8] Gallezot P., Leclercq C., Guisnet M., Magnoux P., *J. Catal.*, 114 (1988) 100–111.
[9] Habib E.T., Zhao X., Yaluris G., Cheng W.C., Boock L.T., Gilson J.P., in *Zeolites for Cleaner Technologies*, Guisnet M., Gilson J.-P. (Eds.), Imperial College Press, London (2002) Chapter 5, 105–130.
[10] Barth J.-O., Jentys A., Lercher J.A., *Ind. Eng. Chem. Res.*, 43 (2004) 2368–2375.
[11] Babich I.V., Seshan K., Lefferts L., *Appl. Catal. B: Environmental*, 59 (2005) 205–211.
[12] Barth J.-O., Jentys A., Lercher J.A., *Ind. Eng. Chem. Res.*, 43 (2004) 3097–3104.
[13] Hutchings G.J., Copperthwaite R.G., Themistocleous T., Foulds G.A., Bielovitch A.S., Loots B.J., Nowitz G., van Eck P., *Appl. Catal.*, 34 (1987) 153–161.
[14] Hutchings G.J., Themistocleous T., Copperthwaite R.G., EP 307239, Pratley Invest. Pty. Ltd and Zeofuels Res. Pty. Ltd (1998).

[15] Sanchez-Galofré O., Segura Y., Perez-Ramirez J., *J. Catal.*, 249 (2007) 123–133.

[16] Parmon V.N., Panov G.I., Uriarte A., Noskov A.S., *Catal. Today*, 100 (2005) 115–131.

[17] Ivanov D.P., Sobolev V.I., Panov G.I., *Appl. Catal. A: General*, 241 (2003) 113–121.

[18] Pirutko L.V., Chernyavsky V.S., Uriarte A., Panov G.I., *Appl. Catal. A: General*, 227 (2002) 143–157.

[19] Pirutko L.V., Chernyavsky V.S., Starokon E.V., Ivanov A.A., Kharitonov A.S., Panov G.I., *Appl. Catal. B: Environmental*, 91 (2009) 174–179.

[20] Bauer F., Ernst H., Geidel E., Schödel R., *J. Catal.*, 164 (1996) 146–151.

[21] Aguayo A.T., Gayubo A.G., Erena J., Atutxa A., Bilbao J., *Ind. Eng. Chem. Res.*, 42 (2003) 3914–3921.

[22] Gnep N.S., Martin de Armando M.L., Guisnet M., *React. Kinet. Catalysis Lett.*, 13 (1980) 183–189.

[23] Tsai T.-C., *Appl. Catal. A: General*, 301 (2006) 292–298.

[24] Pater J. ,Cardona F., Canaff C., Gnep N.S., Szabo G., Guisnet M., *Ind. Eng. Chem. Res.*, 38 (1999) 3822–3829.

[25] Feller A., Barth J., Guzman A., Zuazo I., Lercher J.A., *J. Catal.*, 220 (2003) 192–206.

[26] Sievers C., Zuazo I., Guzman A., Olindo R., Syska H., Lercher J.A., *J. Catal.*, 246 (2007) 315–324.

[27] Querini C.A., *Cat. Today*, 62 (2000) 135–143.

[28] Klingmann R., Josl R., Traa Y., Gläser R., Weitkamp J., *Appl. Catal. A: General*, 281 (2005) 215–223.

[29] Ginosar D.M., Thompson D.N., Burch K.C., *Appl. Catal. A: General*, 262 (2004) 223–231.

[30] Thompson D.N., Ginosar D.M., Coates K., *Fuel Chem. Prepr. ACS*, 46 (2001) 422–427.

[31] (a) Imai T., Kocal J.A., Vora B.V., *Sci. Techn. Catal.* (1994) 339–341; (b) Kocal J.A., Vora B.V., Imai T., *Appl. Catal. A: General*, 221 (2001) 295–301.

[32] Knifton J.F., Anantaneni P.R., Dai P.E., Stockton M.E., *Catal. Lett.*, 75 (2001) 113–117.

[33] Da Z., Magnoux P., Guisnet M., *Catal. Lett.*, 61 (1999) 203–206.

[34] Da Z., Magnoux P., Guisnet M., *Appl. Catal. A: General*, 219 (2001) 45–52.

[35] Han M., Cui Z., Xu C., Chen W., Jin Y., *Appl. Catal. A: General*, 238 (2003) 99–107.

Part V: Case Studies

Chapter 13

DEACTIVATION AND REGENERATION OF FCC CATALYSTS

H.S. Cerqueira and F. Ramôa Ribeiro

13.1. Introduction

Fluid catalytic cracking (FCC), the most important conversion refining process, is used to convert heavy feedstocks: gasoils from vacuum distillation towers and/or residues from atmospheric distillation towers into lighter, more valuable products, the most desired being naphtha (i.e. the major constituent of the gasoline pool), and more recently propylene. The operating conditions are relatively severe, in particular with reaction temperatures between 750 and 850 K depending on the position in the reactor. More than 400 FCC units are presently operated worldwide. Cracking units are very large, some of them processing more than 6 Mt/y.

A multicomponent catalyst with an acid FAU (Y) zeolite as the main active phase is responsible for the incredibly large number of reactions involved in the FCC process. A large diversity of catalysts is available on the market. The major global suppliers of FCC catalysts worldwide are Albemarle Corporation (formerly Akzo Nobel Catalysts), BASF Catalysts (formerly Engelhard) and W.R. Grace and Company.

An important particularity of the FCC process is that a significant portion of the feedstock: ~6 wt% from a typical vacuum gasoil + residue feedstock [1] is converted into coke, i.e. a mixture of compounds "dragged" or retained on/in the catalyst after stripping. This coke rapidly deactivates the acid sites of the catalyst, resulting in a significant activity loss [2]. It is for this reason that the process was conceived to organize a continuous circulation of the catalyst from the riser reactor to the regenerator in which coke is removed from the catalyst by combustion at high temperatures (typically 950–1030 K), and vice versa. One advantage of this continuous catalyst circulation is that FCC units operate in heat balance, i.e. the amount of heat released during the burning of the coke deposited on the spent catalyst is utilized to vaporize and heat the feed (steam and additional quench streams) to the reaction temperature, to provide heat to the endothermic cracking reactions, heat the regenerator flue gas, and compensate for heat losses.

In addition to coke formation, there are many other causes of deactivation of FCC catalysts originating either from feed contaminants (e.g. basic nitrogen containing compounds, metal derivatives) or from mechanical or structural degradation of the catalysts under the very severe conditions of regeneration: high temperature in the presence of steam.

The present chapter addresses the main aspects of this essential refining process, the focus being placed on the phenomena involved in catalyst deactivation and regeneration.

13.2. Main Characteristics of the FCC Process

13.2.1. *Feedstock, products and reactions*

13.2.1.1. *Feedstock*

The feedstock of FCC units is typically vacuum gas oil (VGO) but can include many other heavy streams, such as straight-run gas oil, coker gas oil, hydrocracked gas oil, atmospheric distillation bottoms and vacuum bottoms. All these feeds consist essentially of a mixture of high-boiling-point hydrocarbons: paraffinic, naphthenic, aromatic and polyaromatic hydrocarbons and combinations thereof, but no olefinic hydrocarbons. Small amounts of other organic compounds containing heteroatoms such as sulphur, nitrogen and oxygen as well as metals such as copper, iron, nickel and vanadium often in the form of porphyrins are also present. Large differences can be found in the FCC feedstock characteristics, generally dependent on the crude oil source. Thus, vacuum distillation bottoms are heavier than gas oils, containing metals such as Ni and V in amounts close to 5 ppm (against less than 1 ppm for gas oils) as well as more highly polyaromatic compounds like Conradson carbon, that can represent up to 5 wt% of the feed (against less than 1 wt% for gas oils) [3].

Discoveries in the last decade of heavy oil deposits have incited many FCC units to begin processing feedstocks with a higher tendency to form coke, leading to the development of residue fluid catalytic cracking (RFCC) units, which are designed to convert 100% of residue from the atmospheric distillation tower. Moreover, high crude oil prices press refiners to find alternative feedstock components originating from biomass, in particular vegetable oils, and lignocellulose-based bio-oils. These are widely different from gasoils in terms of viscosity and stability and moreover are rich in oxygenates and present several acid compounds. Typical bio-oil compounds are lighter than gas oil, leading to an increase in FCC conversion when used as a co-feed [4]. Depending on the bio-oil characteristics, an aromaticity increase of gasoline may also be observed.

13.2.1.2. *Products*

The FCC unit leads to the following products: fuel gas, liquefied petroleum gas (LPG), cracked naphtha, light cycle oil (LCO), heavy cycle oil (HCO) and coke. The typical fuel gas yield is between 3 and 5 wt%; it consists of a blend of hydrogen, methane, ethylene and ethane. LPG (yield ~12–22 wt%) corresponds to the C_3-C_4 cut; in some refineries, this cut is sent to a special distillation tower in order to recover propylene. The yield of cracked naphtha is the desired cut, representing 45–60 wt% of the products and has a boiling point (b.p.) in the range of 310 to 490 K, which corresponds to compounds presenting between 5 and 10 carbon atoms. LCO (yield ~8–20 wt%, b.p. ~490–620 K) contains molecules with 10 to 22 carbon atoms and after hydrotreating leads to diesel and jet fuel cuts. HCO that typically presents a b.p. higher than 620 K (yield ~5–10 wt%) can be used as fuel oil. The last product, coke (yield ~5–8 wt%) which is not desorbed from the catalyst is burned in the regenerator.

13.2.1.3. *FCC chemistry*

Under the FCC conditions, a very significant number of reactions can be catalysed on the acid sites, which explains the large diversity of the products. These reaction steps can be classified into four main categories:

(1) Cracking reactions (scission of C–C bonds) which are endothermic, hence thermodynamically favoured at high temperatures. All the hydrocarbon components of the feed (except polyaromatics) — alkanes, cycloalkanes (naphthenes), alkyl aromatics — can be cracked and moreover the resulting products undergo secondary cracking.
(2) The reverse reactions, i.e. condensation, including cyclization (intramolecular condensation).

(3) Hydrogen transfer which plays an essential role in the distribution FCC product (e.g. Eq. 13.1).

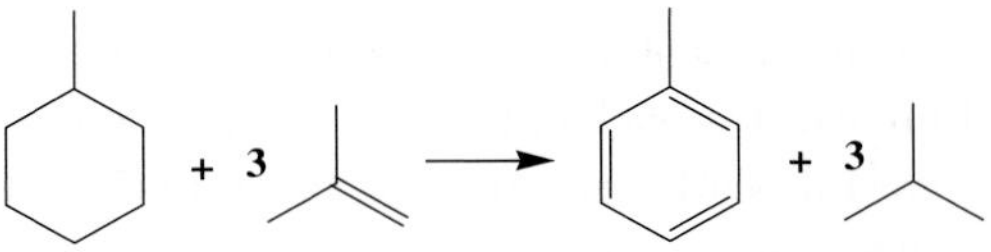

(13.1)

(4) Rearrangement of the reactant and of the product molecules.

It should moreover be noted that a radical mechanism may play a role in the formation of certain products (e.g. methane) and that metals (especially Ni deposited on the catalyst) catalyse reactions of dehydrogenation, hydrogenolysis, etc.

The heterogeneously catalysed cracking can be modelled as a pseudohomogeneous process by a set of partial differential equations. The traditional lump approach [5, 6] divides the feed into pseudocomponents of interest and has been used by several other authors [7, 8]. A typical way of lumping is, for instance, fuel gas, LPG, gasoline, LCO, HCO and coke. Alternatively, the feed can be divided into different groups, depending on the functionality and/or reactivity of its components.

Catalyst deactivation can be included in the mathematical model by means of a deactivation function. Several equations are available in the literature; the most common are exponential and power law types, relating catalyst decay to the coke content of the catalyst [9]. For a better description of the phenomenon, coke can be divided into two different types: adsorbed hydrocarbons that can be totally removed by stripping and hydrocarbons that cannot be removed by stripping [7].

13.2.2. *FCC catalysts, formulations in permanent evolution*

A good FCC catalyst needs to have at least two main features: good selectivity towards the desired products (i.e. gasoline with a high octane number and propylene) and good stability with respect to high temperatures, steam and attrition. The changes in feedstock (heavier with high metal content), the increasing demand for propylene as well as the increase in environmental constraints have led to significant changes in catalyst formulation.

FCC catalysts are manufactured under the form of microspheres of 60–100 μm diameter. They are constituted by a mixture of a FAU (Y) zeolite, an inert matrix, an active matrix and a binder. Other specific functional ingredients (filler) may also be present or, more commonly, the catalyst can be used in a blend with one or more additives. Zeolite content can range from about 15 to 50 wt% of the total weight. The matrix contains amorphous alumina which has large pores accessible by bulky molecules and presents some catalytic activity. The resulting small product molecules can accede to the zeolite active sites and undergo various transformations. In addition to zeolite and matrix activity, many of the catalyst's physical and chemical properties contribute to increased conversion and selectivity differences. These include zeolite type, pore size distribution, relative matrix to total surface area, and chemical composition. The binder and filler components provide the needed physical strength and integrity of the catalyst. The binder is usually silica or silica-alumina and the filler is usually clay (kaolin).

The importance of additives in the FCC process has been increasing in the last decades [10]. The most common are the following: MFI (ZSM-5) zeolite used for increasing propylene production and octane gasoline, combustion promoters based on noble metals, and different special matrices developed to reduce the deleterious effect of Ni and to reduce the sulphur content of gasoline, the emissions of SOx and NOx from the regenerator or to enhance catalyst resistance to attrition.

Due to the cyclic nature of the FCC process, the catalyst particles may break, producing fines that will result in particulate emissions. In order to cope with the loss of fines because of catalyst attrition, hence to maintain the catalyst activity, addition of fresh catalyst is frequently needed. For FCC units processing feedstocks with a high level of metals, it is also common to replace a portion of the inventory with fresh catalyst in order to keep contaminant metals at an acceptable level. This necessity to regularly add fresh catalyst makes FCC the most important market for catalysts [11]. Another consequence is that one of the FCC catalyst is a heterogeneous mixture, constituted by particles with different ages, from very young (fresh, hence with high activity and very low metal concentration) to old (aged, hence with low activity and high metal concentration). This mixture is called equilibrium catalyst (e-cat).

As the operational variables play an important role, the performances of the commercial FCC catalysts have to be evaluated under comparable conditions — feed composition, flow rate, temperature, etc. — with, in addition, characterization of the e-cat samples through several complementary techniques.

13.2.3. *FCC process*

As briefly presented in the introduction, FCC is a continuous regenerative process. The most common unit configurations of FCC units are the "stacked", where the reactor and the catalyst regenerator are contained in a single vessel, with the reactor being located above the regenerator, and the "side-by-side" where the reactor and regenerator are in two separate vessels (Fig. 13.1). The three main parts of the FCC process — reactor, devices used for the product-catalyst separation and regenerator — are described hereafter.

13.2.3.1. *Reactor*

The feed preheated at 450–550 K is injected at the bottom of the riser reactor and mixed with the freshly regenerated hot catalyst at temperatures in the range of 900–1,000 K. Vaporization and cracking of the feed cause

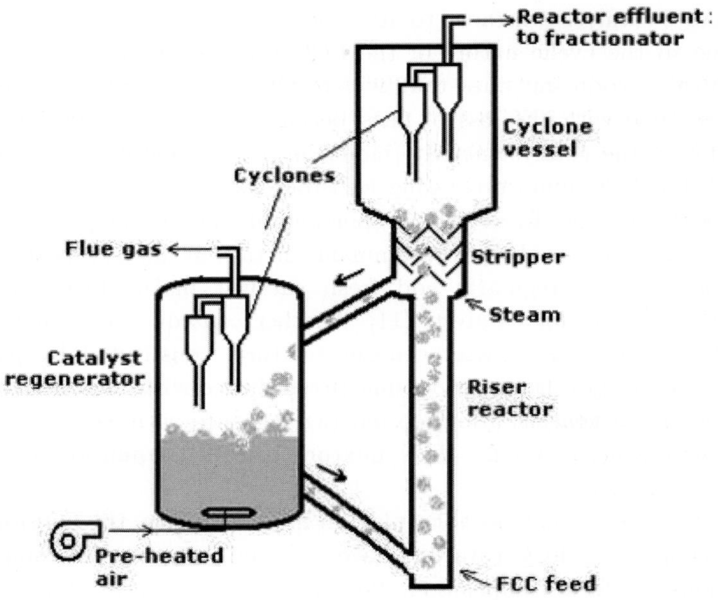

Fig. 13.1. Scheme of a side-by-side FCC unit. Adapted from [13].

a significant dilatation, which favours a rapid transport of the catalyst-oil mixture upward in the riser. The catalyst-to-oil weight ratio is 5 to 8, and owing to the endothermicity of cracking reactions, there is a temperature decrease from $\sim$850 K at the bottom to $\sim$790 K at the outlet.

A feed injection system with multiple nozzles and dispersion steam is critical to ensure adequate feed atomization, easing its vaporization. In the riser, the solids concentration is always higher near the wall than in the centre (core-annulus effect) which can result in a broad range of local catalyst-to-oil (CTO) ratios. This can also be affected by catalyst back-mixing. In order to minimize these two effects and to have a more even radial catalyst distribution, small internal baffles can be used to improve riser hydrodynamics and minimize the core-annulus effect [12].

The main process variables that affect FCC performance are reactor temperature and residence time. An increase in residence time (longer or fatter riser) increases conversion. A high reactor temperature increases unit conversion, primarily through a higher rate of the endothermic cracking reactions. A higher gasoline octane and LPG olefinicity could also happen.

It should be noted that most commercial units are already operated close to an optimum conversion level for a given feed rate, feed quality, set of processing objectives and catalyst, being limited by one or more unit constraints (e.g. wet gas compressor capacity, fractionation capacity, air blower capacity, reactor temperature, regenerator temperature, catalyst circulation). However, the operator can always replace the catalyst, targeting to remove operating constraints, shifting operation to a more profitable steady-state condition.

13.2.3.2. *Separation devices*

Old FCC units used to have the reactor outlet discharged directly into a separator vessel, leading to overcracking due to substantial post-riser residence time. Modern designs include improved riser termination called close-coupled cyclones, which prevent product vapours from entering the separation vessel.

The catalyst falls into the stripper system in which the hydrocarbons retained on the catalysts are desorbed by steam treatment in a dense fluidized bed. This desorption is a key step. Indeed, it determines the coke content of the catalyst entering the regenerator, which plays an essential role in the thermal balance of FCC units. Thus, it is not surprising that various internal devices from simple baffles to disc, doughnut and then structured packing were developed to improve the contact between steam and catalyst, thus the stripper efficiency.

The close-coupled cyclones can play a role in pre-stripping a part of the dragged hydrocarbons. It is estimated that between 5 to 30 wt% of the material adsorbed on the catalyst surface can be recovered in the stripper [14]. Moreover, strippable hydrocarbons present a average H/C ratio higher than the coke compounds, resulting in an excess release of heat in the regenerator that negatively impairs unit circulation.

The stripping is basically a mass-transfer process. However, because of the high temperature (~770 K) and the relatively long residence time (90–200 s), transformations of the desorbed products and of the residual coke (increase in polyaromaticity) may also occur [15].

13.2.3.3. *Regeneration*

The catalyst entering the regenerator typically contains between 0.8 and ~1.5 wt% of coke with an atomic H/C ratio ≤0.4. The regeneration of the spent catalyst is achieved by burning coke in the presence of air. In order to ensure a proper regeneration of the FCC catalyst, it is common to operate the regenerator at high temperatures (950–1030 K). Uneven temperature distribution is highly undesirable, since hot spots can destroy the catalyst structure or even damage the regenerator vessel. This can be avoided by the addition of internal devices designed to better distribute the spent catalyst coming from the stripper. The regenerated catalyst going back to the riser inlet usually has a coke content in the range of 0.1 to 0.2 wt%.

The heats of partial oxidation of carbon atoms into CO, as well as of CO oxidation into CO_2, are indicated in Eqs. 13.2 and 13.3, respectively:

$$C + \frac{1}{2}O_2 = CO \quad \Delta H = -110\,\text{KJ mol}^{-1} \tag{13.2}$$

$$CO + \frac{1}{2}O_2 = CO_2 \quad \Delta H = -283\,\text{KJ mol}^{-1} \tag{13.3}$$

Comparison of the ΔH values shows that a significant amount of heat can be lost when the coke combustion is not complete.

Depending on the FCC design, the coke combustion may be complete in the regenerator or not. In the former case, the regenerator operates with excess oxygen in order to favour CO_2 formation, whereas in the latter case, the air flow is controlled so as to provide the desired CO/CO_2 ratio. In this latter case, the flue gas is then routed to a CO boiler, where CO is burned as fuel to provide steam, while complying with environmental regulations.

In the regenerator, the catalyst bed is divided into two main regions of different gas-solid densities, named the dense and diluted phases. At total combustion operation, it is important to favour the burning of the coke mostly in the regenerator dense phase; otherwise, unconverted CO escapes from this region, reacting with the excess oxygen in the diluted phase where the combustion gases are separated from the catalyst. This leads to undesirable hot spots. Similarly, in the partial combustion operation, non-reacted oxygen that escapes from the dense phase to react with CO in the diluted phase must be prevented. This undesirable phenomenon, known as "afterburning", is quantified by the temperature difference between the dense and diluted phases. CO combustion promoters (e.g. supported noble metal) are used worldwide as additives in FCC operations, to minimize the "afterburning" [16].

Some residue fluid catalytic cracking (RFCC) units have a two-stage regenerator, an option when full combustion would result in excessive regenerator temperatures. The first stage operates in partial combustion, the second stage in full combustion. The excess coke produced in RFCC units results in a surplus of energy during the burning in the regenerator. In order to maximize the profitability of those residue FCC units, part of the heat produced by coke combustion is recovered by means of catalyst coolers that control the regenerator dense-phase temperature and produce steam. Catalyst coolers are relatively easy to add to all styles of regenerators.

13.3. Catalyst Deactivation and Regeneration

As previously discussed, it is possible to regenerate the FCC catalyst by coke combustion with air and to recover the larger part of the activity by this treatment. However, this reversibility is only partial due to the simultaneous presence of irreversible deactivation that forces withdrawal of e-cat and addition of fresh catalyst. Both modes will be presented and discussed.

13.3.1. *Reversible deactivation*

13.3.1.1. *Coke formation*

The deactivation of zeolite-based catalysts by coke is caused either by poisoning of acid sites or by pore blockage [2, 18]. In the former case, one coke molecule blocks one active site, and may also affect the reaction selectivity [17]. Nevertheless, the deactivating effect is much more pronounced in the case of pore blockage, with one coke molecule blocking the access of reactants to, on average, more than one active site [18].

Diffusive mass transport plays a key role in catalytic cracking under commercial conditions [19]. An important parameter related to mass transfer limitations is the accessibility of reactant molecules to the active sites located inside the catalyst particles. The matrix can play a beneficial role by trapping voluminous molecules containing heteroatoms and contaminant metals, indirectly reducing the amount of coke formed [20].

An estimation of the accessibility could be made either with selected model molecules and catalyst components or with real FCC feedstock and commercial catalysts. More important than the catalyst accessibility itself is the accessibility to the acid sites. Clearly, coke deposits can severely reduce the accessibility to the acid sites, particularly because the coke molecules tend to adsorb onto those sites [21]. The measurement of accessibility to the acid sites is a non-trivial problem, especially where coked catalysts are concerned.

There are five main types of coke identified in catalytic cracking:

- Catalytic Coke: from condensation and dehydrogenation reactions.
- Catalyst-to-Oil Coke: the higher the catalyst-to-oil ratio, the higher the catalyst circulation and thus the higher the amount of hydrocarbons entrained in the small pores that may not be removed by stripping.
- Thermal Coke: formed through a free radical mechanism that is favoured at high reaction temperatures and also yields hydrogen; at typical FCC conditions, it is less important than catalytic coke due to the low extent of thermal cracking.
- Additive Coke (or Conradson Coke): resulting from the simple deposition on the catalyst of very heavy molecules of the feed. Its amount correlates directly with the Conradson carbon residue (measured by feed pyrolysis up to 1073 K).
- Contaminated Coke: from dehydrogenation reactions catalysed by metals such as Ni, Fe and V.

Coke formation has a strong impact on the performance of the catalyst, reducing the conversion, sometimes changing product selectivity, besides playing an essential role in the heat balance of the FCC unit. Nevertheless, the different types of coke do not have the same influence on the cracking activity, e.g. depending on the feedstock characteristics; catalytic coke may have a greater influence on catalyst activity than additive coke [18]. The time scale in which the different types of coke are deposited on the catalyst surface may also vary, determining the change in the deactivation mode from active site deactivation to pore blockage [22].

Another interesting aspect observed by several authors is the presence of significant (residual) activity on the coked FCC catalysts [23]. Clearly, FCC catalyst deactivation by coking is a complex phenomenon where the different variables in the process (feed composition, catalyst composition and process conditions) are all inter-related.

13.3.1.2. *Effect of heteroatoms: coking and poisoning*

As mentioned before, FCC feedstocks are composed mainly by naphthenes, aromatics and alkanes. However, apart from hydrocarbons, FCC feeds also contain non-negligible amounts of oxygen (0–2%), sulphur (0–7.5%) and nitrogen (0–0.4%). These heteroatom-containing molecules exhibit quite distinct chemical properties.

The damaging effect of nitrogen compounds has been known for several decades [24] and is still considered a relevant topic. The FCC feed is mainly composed of vacuum gas oil (VGO) containing approximately 25–30 wt% of the nitrogen existing in crude oil. Nevertheless, in the last decade, increasing amounts of vacuum residue that are richer in nitrogen (70–75 wt% of the crude oil nitrogen) have been added to FCC feedstocks, increasing the nitrogen content of the charge. For the lighter fractions of crude oil, nitrogen is mainly in the form of basic compounds, while in the heavier fractions non-basic compounds are predominant.

Progress has been made in the identification and characterization of nitrogen compounds present in different oil fractions and feeds. In most cases, the nitrogen present in crude oil occurs in either high-molecular-weight molecules, containing other heteroatoms such as S and O, or small- and medium-molecular-weight with well-defined structures that include alkyl derivatives of pyridine, quinoline, isoquinoline, acridine and phenanthridine (basic) or derivatives of pyrrol, indole and carbazole (non-basic).

Nitrogen bases are thought to deactivate FCC catalysts by interacting with the acid sites responsible for the cracking reaction, hence decreasing activity [25]. The acid matrix present in current catalytic cracking catalysts seems to partially prevent nitrogen poisoning. Lewis acid sites present in the active matrix can adsorb, even if weakly, the nitrogen-containing molecules, thus preventing the interaction with the zeolite active sites. As a consequence, catalysts containing special matrices with high acidity are less affected by basic molecules present in the feedstock [26].

Several approaches are possible to minimize the harmful effects of nitrogen compounds prior to catalytic cracking:

(1) Hydrotreatment: a well-known process that can be applied to decrease the nitrogen content in feedstocks. This catalytic process requires high hydrogen pressures and temperatures.
(2) Adsorption: use of acidic solid adsorbents to capture the basic nitrogen compounds.
(3) Liquid/liquid extraction: use of an immiscible solvent to extract nitrogen compounds.
(4) Neutralization: use of acid additives to neutralize basic nitrogen compounds.

At iso-conversion and in the presence of nitrogen-containing compounds, fuel gas and coke yields are increased, whereas gasoline yield decreases [27]. In the commercial FCC unit, those changes can be explained as follows: if the refiner wants to change from a given feedstock to another with higher nitrogen content maintaining the conversion, in the latter case the FCC catalyst will have a lower acid site density due to site poisoning, thus requiring a higher catalyst circulation. However, the higher the catalyst circulation is, the higher the catalyst-to-oil ratio, favouring fuel gas and coke at the expense of gasoline.

Another common heteroatom is sulphur. Environmental regulations state a limit on the amount of sulphur allowed in the gasoline pool, 10 ppm being imposed by the European Parliament and Council in 2009. It is well known that from all their constituents (alkylate, reformate, isomerate, hydrocracking gasoline, catalytic cracking gasoline), the only contributor to the final amount of sulphur is FCC gasoline. Indeed, several processes are commercialized worldwide to hydrotreat FCC gasoline in order to obtain a low sulphur product. Oxygenate compounds can also be removed by hydrotreating.

13.3.2. *Irreversible deactivation*

The water which is added or formed at different stages of the FCC process plays a major role in catalyst deactivation. At the riser inlet, the steam added (0.5–5 wt%) is crucial in order to facilitate the atomization of the feed. This steam also causes changes in the conversion and product distribution, namely reducing coke formation. At the stripper, more steam is added to the process. Lastly, in the regenerator, water is produced by the combustion of coke, hence the catalysts are submitted in the presence of steam to very high temperatures (close to 1,000 K).

The severity of the regeneration step is responsible for Y zeolite dealumination, reducing its number of active acid sites and hence the activity. This dealumination has however positive effects in making more thermally and hydrothermally stable the zeolite catalyst component. This is one of the reasons why before use in the FCC process, Y zeolites are usually steamed in optimized conditions, which explains their denomination as USY (ultrastable Y) zeolites. The resistance to the combined effect of steam and temperature is called hydrothermal stability. Besides dealuminating the zeolite, with effects on the acid site, number and strength, steaming also creates mesopores that largely contribute to reducing mass-transfer limitations, increasing catalyst activity in the cracking of VGO.

Another possibility to increase the thermal stability of H-Y zeolites is through the introduction of rare-earth metals (La, Ce, Pr) [28]. These cations also seem to provide an activity increase for moderate exchange levels. Several commercial FCC catalysts contained H-USY zeolites exchanged with a mixture of rare-earth components (H-REUSY). Note however that the exchange of the Y zeolite with rare-earth elements causes an increase in the selectivity to hydrogen transfer, which favours coke formation.

As underscored in Section 13.2.1.1, the FCC feedstock may present different amounts of contaminant metals — Ni, V, Na, Fe, etc. — which can affect the reaction selectivity and accelerate the zeolite degradations during the regeneration step.

Nickel mainly affects the reaction process, promoting dehydrogenation steps (with namely additional H_2 formation in the fuel gas) and increasing the selectivity to coke [29]. In the FCC catalyst, the presence of extraframework alumina species decreases nickel mobility and reducibility. The passivation of nickel is generally achieved by antimony addition to the feed, favouring a Ni-Sb alloy which is relatively inactive in dehydrogenation reactions. Another option is by means of special alumina matrices that favour the formation of nickel aluminates as follows [3]:

$$\mathrm{NiO} + x\mathrm{Al_2O_3} \rightarrow \mathrm{NiAl_2O_4} \cdot (x-1)\mathrm{Al_2O_3} \tag{13.4}$$

Similarly to nickel, vanadium is also responsible for an enhancement of the USHY zeolite coke formation [30, 31]; but in the commercial FCC catalysts, in the presence of rare-earth elements, the amount of coke formed due to those metals is reduced [32]. Contrarily to nickel, both vanadium and sodium compounds present mobility over the catalyst surface [33]. Sodium preferentially interacts with the alumina phase and tends to migrate to the fresh particles due to the higher availability of acid sites for exchange, thus reducing catalyst activity. Both vanadium and sodium are also responsible for the permanent damage of the zeolite structure in the presence of steam at high temperatures. The existence of these metals is the main cause for the fresh catalyst addition needed to maintain the activity of the inventory. To limit the damage caused by vanadium, a possibility is to choose a catalyst that presents a more stable Y zeolite, i.e. with low Na and Al contents. Specific vanadium traps, comprised of solid bases such as MgO, CaO, SrO, $BaTiO_3$ or La_2O_3, are able to react with H_3VO_4 under regenerator conditions to form stable vanadates [3].

Iron can affect the morphology of the FCC catalyst particles due to a phenomenon often referred to as nodulation. Large iron species are preferentially located on the external surface of the FCC catalyst particles [34]. Above a critical iron content, the accessibility of the particles can be severely reduced. The presence of paramagnetic iron in the FCC catalyst can be exploited to selectively discard the relatively older e-cat with high iron content, therefore producing a high-activity/low-metal catalyst to recycle, that will be responsible for lower hydrogen, dry gas and coke yields and higher wet gas and octane productions [35].

13.4. Deactivation of Additives

The knowledge of how these additives deactivate is of extreme importance for their performance in the commercial FCC unit, influencing its development and evaluation as well. Combustion promoters are based on noble metals and its deactivation depends primarily on the metal-support interaction. Under the FCC unit regenerator conditions, sintering of the active metal may occur, reducing the available metallic surface. The combination of cyclic deactivation and coke combustion assays influences the ranking of combustion-promoting additives [16, 36].

ZSM-5 zeolite has a lower tendency to form coke, compared to Y zeolite, due to its narrow pores that limit the formation of bulky coke intermediates. The main cause of ZSM-5 additive deactivation is dealumination due to the presence of steam at high temperatures, which leads to a partial destruction of its framework structure. Several studies reported changes in the hydrothermal stability of ZSM-5 zeolites after impregnation with phosphorus [37, 38]. Even so, before the steaming treatment, the impregnation with phosphorus was said to produce several counterproductive effects: (i) reversible decrease in activity due to the interaction of P species with the protonic sites; (ii) external surface blockage; (iii) decrease in the microporous volume; and even (iv) dealumination. Despite these facts, the phosphorus-impregnated samples seemed to retain their acidity and activity during the steaming treatment to a higher level than the untreated zeolite. This means that the P species formed in the treatment reinforce the zeolite structure and prevent dealumination [39]. Optimal phosphorus contents (highest activity) depend on the zeolite framework Si/Al ratio.

13.5. Final Remarks

FCC has been a refinery workhorse for more than half a century. It is a complex process that converts a heavy feed with variable composition and contaminants into a very large number of product molecules through several interconnected reactions, in which deactivation has various origins. Coke formation is the main deactivation cause which, being reversible, determines the key characteristics of the process (cyclic and autothermal) and of the catalyst system (multifunctional). The severity of the process (high temperatures, presence of water), combined with feed contaminants, is responsible for the irreversible deactivation. This deactivation can be mitigated by the use of additives or compensated for by frequent replacements of a fraction of the e-cat inventory by fresh catalyst.

Although catalytic cracking is considered to be a mature technology, improvements in both process design and catalyst formulations have never stopped since its original conception. One can be sure that this field will continue to receive attention in the coming decades.

References

[1] Kouwenhoven, H.W., de Kroes N., *Stud. Surf. Sci. Catal.*, 137 (2001) 673–706.

[2] Forzatti P., Lietti L., *Catal. Today*, 52 (1999) 165–181.

[3] Habib E.T., Zhao X., Cheng W.C., Boock L.T., Gilson, J.-P., in *Zeolites for Cleaner Technologies*, Guisnet M., Gilson J.-P. (Eds.), Imperial College Press, London (2002) Chapter 5, 105–130.

[4] Graça I., Ramôa Ribeiro F., Cerqueira H.S., Lam Y.L., Almeida M.B.B., *Appl. Catal. B*, 90 (2009) 556–563.
[5] Weekman Jr. V.W., *Ind. Eng. Chem. Process Des. Dev.*, 7 (1968) 90–95.
[6] Jacob S., Gross B., Voltz S.E., Weekman Jr. V.W, *AIChE J.*, 22 (1976) 701–713.
[7] Vieira R.C., Pinto J.C., Biscaia Jr. E.C., Baptista C.M.L.A., Cerqueira H.S., *Ind. Eng. Chem. Res.*, 43 (2004) 6024–6034.
[8] Fernandes J.L., Pinheiro C.I.C., Oliveira N.M.C., Inverno J., Ramôa Ribeiro F., *Ind. Eng. Chem. Res.*, 47 (2008) 850–866.
[9] Froment G.F., Bischoff K.B., *Chem. Eng. Sci.*, 17 (1962) 105–114.
[10] Cheng W.-C., Kim G., Peters A.W., Zhao X., Rajagopalan K., Ziebarth M.S., Pereira C.J., *Catal. Rev. Sci. Eng.*, 40 (1998) 39–79.
[11] Biswas J., Maxwell I.E., *Appl. Catal.*, 63 (1990) 197–258.
[12] Chen Y.-M., *Powder Technology*, 163 (2006) 2–8.
[13] Guisnet M., Cerqueira H.S., Figueiredo J.L., Ramôa Ribeiro F., in *Desactivação e Regeneração de Catalisadores*, Guisnet M., Cerqueira H.S., Figueiredo J.L., Ramôa Ribeiro F. (Eds.), Fundação Calouste Gulbenkian, Lisboa (2008).
[14] Baptista C.M.L.A., Cerqueira H.S., *Stud. Surf. Sci. Catal.*, 149 (2004) 287–295.
[15] Magnoux P., Cerqueira H.S., Guisnet M., *Appl. Catal. A*, 235 (2002) 93–99.
[16] Carvalho M.C.N.A., Morgado Jr. E., Cerqueira H.S., Resende N.S., Schmal M., *Ind. Eng. Chem. Res.*, 43 (2004) 3133–3136.
[17] Nam I.-S., Froment G.F., *J. Catal.*, 108 (1987) 271–282.
[18] Guisnet M., Magnoux P., Martin D., *Stud. Surf. Sci. Catal.*, 111 (1997) 1–19.
[19] de la Puente G., Ávila A., Chiovetta G., Martignoni W.P., Cerqueira H.S., Sedran U., *Ind. Eng. Chem. Res.*, 44 (2005) 3879–3886.
[20] Scherzer J., *Appl. Catal.*, 75 (1991) 1–32.
[21] Cerqueira H.S., Ayrault P., Datka J., Guisnet M., *Microporous Mesoporous Mater.*, 38 (2000) 197–205.
[22] den Hollander M., Makkee M., Moulijn J.A., *Appl. Catal. A*, 187 (1999) 3–12.
[23] Corma A., Melo F.V., Sauvanaud L., *Appl. Catal. A*, 287 (2005) 34–46.
[24] Ho T.C., Katritzky A.R., Cato S.J., *Ind. Eng. Chem. Res.*, 31 (1992) 1589–1597.
[25] Hughes R., Hutchings G.J., Koon C.L., McGhee B., Snape C.E., Yu D., *Appl. Catal. A*, 144 (1996) 269–279.
[26] Corma A., Mocholi F.A., *Appl. Catal. A*, 84 (1992) 31–46.
[27] Caeiro G., Costa A.F., Cerqueira H.S., Magnoux P., Lopes J.M., Matias P., Ramôa Ribeiro F., *Appl. Catal. A*, 320 (2007) 8–15.
[28] Maugé F., Gallezot P., Courcelle J.C., Engelhard P., Grosmangin J., *Zeolites*, 6 (1986) 261–266.
[29] Xu M., Liu X., Madon R.J., *J. Catal.*, 207 (2002) 237–246.
[30] Kugler E.L., Leta P.D., *J. Catal.*, 109 (1988) 387–395.

[31] Escobar A.S., Pereira M.M., Pimenta R.D.M., Lau L.Y., Cerqueira H.S., *Appl. Catal. A*, 286 (2005) 196–201.
[32] Oliveira H.M.T., Herbst M.H., Cerqueira H.S., Pereira M.M., *Appl. Catal. A*, 292 (2005) 82–89.
[33] Wormsbecher R.F., Cheng W.-C., Harding R.H., *ACS Div. Pet. Chem.*, 40 (1995) 482.
[34] Bayraktar O., Kugler E.L., *Catal. Lett.*, 90 (2003) 155–160.
[35] Goolsby T.L., Moore H.F., *Sep. Sci. Technol.*, 32 (1997) 655–668.
[36] Liu L., Rainer D., Gonzalez J.A., *Appl. Catal. B*, 72 (2007) 212–217.
[37] Kaeding W.W., Butter S.A., *J. Catal.*, 61 (1980) 155–164.
[38] Caro J., Bülow M., Derewinski D., Haber J., Hunger M., Kärger J., Pfeifer H., Storek W., Zibrowius B., *J. Catal.*, 124 (1990) 367–375.
[39] Caeiro G., Magnoux P., Lopes J.M., Ramôa Ribeiro F., Menezes S.M.C., Costa A.F., Cerqueira H.S., *Appl. Catal. A*, 314 (2006) 160–171.

Chapter 14

HYDROCRACKING

C. Henriques and M. Guisnet

14.1. Introduction

Hydrocracking is one of the most important oil conversion processes. The first modern hydrocracker was commercialized in 1958, i.e. about 15 years after the first FCC unit. Presently, the worldwide hydrocracking capacity is ~300 millions tons of feed per year [1], ~4 times lower than that of catalytic cracking [2]. A steady growth in the hydrocracking capacity (3–5% per year) can be observed due to: (i) the continuous increased demand for middle distillates and (ii) the increase in severity of the environmental regulations for fuels [1].

Like FCC, hydrocracking aims to convert heavy high-boiling-point feedstocks into fuels. However, large differences exist between the processes:

- FCC units are usually operated to maximize gasoline production: an acid catalyst is operated at high temperatures (750–850 K) and low pressures to convert feedstocks typically constituted by vacuum gas oil or VGO (but which include often heavier cuts, such as atmospheric distillation bottoms and vacuum bottoms). An important particularity of the process is the fast catalyst deactivation by coking which imposes a continuous circulation of the catalyst from the riser reactor to the regenerator and vice versa (Chapter 13).
- Hydrocracking units aim to maximize middle distillate production and use bifunctional catalysts, i.e. associating an acidic function (cracking) with a hydrogenating one, operated in a fixed-bed reactor at relatively low temperatures (570–720 K) and high hydrogen pressures (80–200 bar). The typical feedstocks are either VGO (outside North America) or lighter cuts

(USA) but heavier cuts are sometimes processed. The major advantages of hydrocracking are a remarkable flexibility in the product distribution and their good quality both from performance and environment points of view, e.g. gasoil with a cetane number and sulphur content of 55 and 40 wt ppm respectively against 27 and 10,000 wt ppm from FCC [1]. Another advantage is the slow catalyst deactivation by coking. However, hydrocracking which requires costly pressure equipment and significant consumption of high-pressure hydrogen is a more expensive process than FCC.

After a brief description of the main characteristics of hydrocracking, the main causes of catalyst deactivation (coking, poisoning) and the means of controlling and curing this deactivation will be discussed.

14.2. Main Characteristics of the Hydrocracking Process

14.2.1. *Process schemes*

Two configurations are mostly used: two-stage and single-stage processes [1]. The first one comprises two fixed-bed reactors with intermediate product separation. In the first reactor, catalysts are generally Ni-Mo or Ni-W/alumina or silica-alumina. On these weakly acidic catalysts, hydrocracking reactions occur to a limited extent (10–50%) while the organic sulphur, nitrogen and oxygen components of the feed are converted into hydrocarbons plus hydrogen sulphide, ammonia and water respectively and a large part of the aromatic hydrocarbons are hydrogenated. The effluents of this reactor are separated into gas and liquid streams; the gas stream which contains H_2S and NH_3 is sent to a gas scrubber, the liquid stream to a fractionator, the bottom distillation cut (hydrotreated but non-converted oil) being sent to the second reactor to be hydrocracked. The amorphous silica-alumina originally used to support the hydrogenating components of the hydrocracking catalyst was replaced by zeolites which are significantly more acidic and more active, especially in the presence of NH_3. Moreover, the possibility to operate this second stage with very low H_2S partial pressures makes possible the use of noble metals as hydrogenating components.

The simplest single-stage process (the two consecutive stages are in series without intermediate gas and liquid separation) issues from the introduction of acid zeolites as a support for the hydrogenating component. There are various configurations: with one reactor or several reactors in series and with full or partial conversion of the feedstock. Milder operating conditions (lower hydrogen pressure and hydrogen/oil ratios) can also be chosen. In this less costly mild hydrocracking, the conversion level is lower and the products are of lower quality [1].

14.2.2. *Feedstocks, products catalysts and reactions*

14.2.2.1. *Feedstocks and products*

The common feedstock of hydrocracking units (outside North America) is straight vacuum gas oil (VGO) with a typical 620–770 K boiling point range. Various types of hydrocarbons are present, corresponding to three main types: paraffinic, naphthenic and aromatic, often associated (e.g. alkyl polyaromatic) with a total number of C atoms between 15 and 45–50. In addition, VGO contains organic molecules with hetero-elements: S, N, O, metals. However, as indicated in the introduction, feedstocks differing in their heaviness can be used, this characteristic playing a major role in hydrocracking units. Indeed, the amount of catalyst poisons (metals, polyaromatic coke precursors, organic nitrogen-containing compounds, etc.) generally increases with this feature. Metals cause the irreversible deactivation of the hydrotreating catalysts, whereas nitrogen compounds reduce the hydrocracking activity and coke affects both the hydrotreating and hydrocracking activities. It should also be underscored that the amount of the hydrotreatment catalyst necessary to convert the organic nitrogen compounds into the less poisoning ammonia is determined by both their amount (from ~700 to 3000 wt ppm) and their nature: basic compounds are more reactive than non-basic ones [3].

14.2.2.2. *Hydrocracking catalysts*

Catalysts used for hydrotreating the feed are generally Ni-Mo and Ni-W systems supported on alumina or silica-alumina. The hydrocracking catalysts are bifunctional, associating acidic and hydrogenating components such as those indicated in Table 14.1. The choice of these components should depend on the catalyst environment, hence on the process configuration, e.g. a more favourable situation in the two-stage configuration with no NH_3 in the hydrocracking stage and moderate to little H_2S compared to the single-stage configuration in which the hydrocracking catalyst should operate at high temperatures in the presence of organic nitrogen compounds and relatively high partial pressures of NH_3 and H_2S.

Table 14.1. Potential acid and hydrogenating components of hydrocracking catalysts. Adapted from [4].

Acidic Function		Hydrogenation Function	
	Al_2O_3	Ni/Mo	Increasing
Increasing	Al_2O_3/halogen	Ni/W	Hydrogenating
Acidity	SiO_2/Al_2O_3	(Pd, Pt)*	Activity
	Zeolites		

*used only under low H_2S conditions.

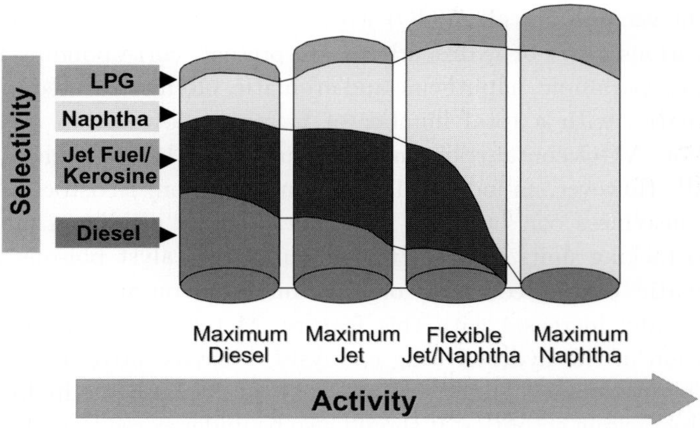

Fig. 14.1. Optimization of the hydrocracking product distribution [5].

14.2.2.3. *Reaction scheme*

The desired product distribution depends on the local requirements, e.g. mainly diesel fuel in Western Europe (especially in France, Belgium and Luxembourg), mainly gasoline in the US, etc. Whatever the fuel to be maximized, successive cracking reactions are necessary. Obviously, the lower the number of C atoms of the fuel, the higher the number of reaction steps, hence the higher the hydrocracking activity should be (Fig. 14.1).

Main reactions occurring after the hydrotreating stage are the following: (i) hydrogenation of mono- and polyaromatics, (ii) hydrodealkylation of alkylnaphthene or aromatics, (iii) hydrodecyclization of naphthenes, (iv) hydroisomerization and hydrocracking of alkanes, (v) coke formation. All these reactions (except i) occur through bifunctional mechanisms involving hydrogenation and dehydrogenation steps on the noble metal or mixed sulphide components, isomerization or cracking steps on the acid sites and diffusion steps on the olefinic intermediates. As a consequence, the activity, selectivity and even stability of the catalyst should depend on the characteristics of the acid and hydrogenating sites, in particular on the balance between the acid and hydrogenating functions [6a–c, 7a–c, e] and on the features of the path of diffusion of the olefinic intermediates, i.e. those of the micropores in the case of an acid zeolite component [7c, d, 8a, b, 9].

The effect of the characteristics of the sites was quantitatively shown on model reactions, in particular on the hydroisomerization and hydrocracking of pure n-alkanes over a series of Pt-HFAU zeolite catalysts differing in the concentration of Pt sites and/or acid sites [7c, e]. The change in activity, selectivity and stability was established as a function of the ratio between the concentration of accessible Pt and acid sites ($C_{\mathrm{Pt}}/C_{\mathrm{A}}$) (Table 14.2). Note that under the operating conditions (feed without S and N poisons), deactivation was only due to coke formation.

Table 14.2 distinguishes between three domains of $C_{\mathrm{Pt}}/C_{\mathrm{A}}$ with differences in the catalytic characteristics:

(i) At low $C_{\mathrm{Pt}}/C_{\mathrm{A}}$ values (<0.03), there are not enough Pt sites for all the acid sites to be fed with intermediate decenes, i.e. reactant dehydrogenation is the limiting step of n-decane transformation. The number of acid sites that intermediate molecules can encounter during their diffusion from one Pt site to another is high, with consequently a high probability of successive transformations before hydrogenation (Fig. 14.2a), hence an apparent direct formation of dibranched isomers and cracking products. Condensation reactions can also occur with

Table 14.2. n-Decane transformation over a series of Pt-HFAU catalysts: Influence of the balance between the metal and acid functions (characterized by the ratio of the corresponding active sites: $C_{\mathrm{Pt}}/C_{\mathrm{A}}$) on activity (per acid site), stability and reaction scheme. $n\mathrm{C}_{10}$, M, B, C correspond respectively to n-decane, monobranched and dibranched isomers, and cracking products. Adapted from [7e].

$C_{\mathrm{Pt}}/C_{\mathrm{A}}$	>0.10	>0.03 <0.10	<0.03
Activity	Maximal	Maximal	Low
Stability	Perfect	Average	Low
Reaction scheme	$n\mathrm{C}_{10} \leftrightarrows \mathrm{M} \leftrightarrows \mathrm{B} \rightarrow \mathrm{C}$	$n\mathrm{C}_{10} \leftrightarrows (\mathrm{M, B}) \rightarrow \mathrm{C}$	$n\mathrm{C}_{10}$ ⇉ M; $n\mathrm{C}_{10}$ ⇉ B; $n\mathrm{C}_{10}$ → C

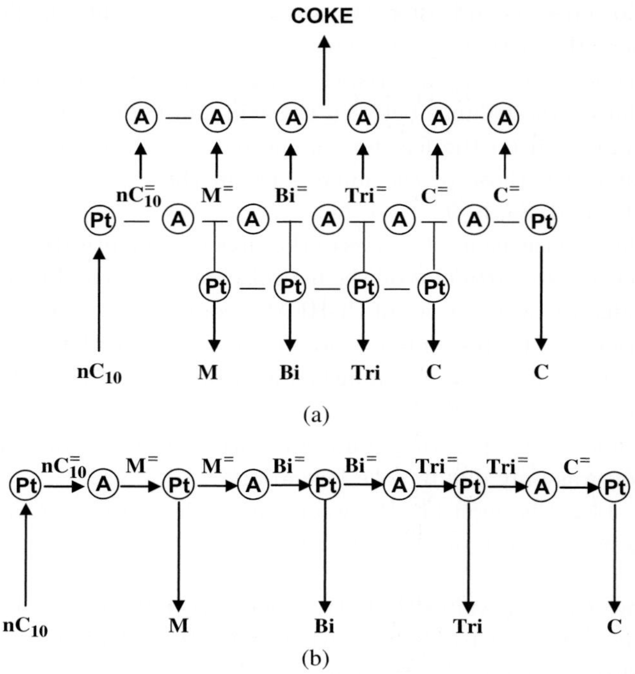

Fig. 14.2. Scheme of n-decane — nC_{10} — transformation over bifunctional Pt/acid zeolite catalysts: (a) with a low value of the hydrogenating/acid balance ($n\text{Pt}/n\text{A} < 0.03$); (b) with a high value of this balance ($n\text{Pt}/n\text{A} > 0.17$); $nC_{10}^{=}$: *n*-decenes; $M^{=}, Bi^{=}, Tri^{=}$: mono-, bi- and tribranched decenes; $C^{=}$: cracking products [10].

formation of coke molecules, hence a rapid deactivation of the catalyst (Table 14.2, column 3).

(ii) For $C_{\text{Pt}}/C_{\text{A}} > 0.03$, there is no more increase in the activity, which means that the acid transformation of decene intermediates is the limiting step. However, up to $C_{\text{Pt}}/C_{\text{A}} = 0.17$, dibranched isomers (but not cracking products) continue to appear as primary products and catalyst deactivation can be observed (Table 14.2, column 2).

(iii) For $C_{\text{Pt}}/C_{\text{A}} > 0.17$, *n*-decane transforms through a successive scheme, the activity is maximal and no deactivation can be observed (Table 14.2, column 1). This means that the number of acid sites encountered by the decene intermediates is such that only one transformation (branching isomerization or scission) can occur before hydrogenation (Fig. 14.2b). Therefore, the activity per acid site is maximal and the reaction scheme is the same as the scheme of alkene intermediate transformation, hence these bifunctional catalysts can be considered as ideal.

From Fig. 14.2b, it can be suggested that with a successive scheme, ideal bifunctional catalysis is the best way to maximize the desired production of middle distillates (low formation of undesired gasoline and LPG) or that of gasoline (low LPG formation) (Scheme 1) with a limited deactivation by coke [10].

$$\text{VGO} \rightarrow \text{Middle Distillates} \rightarrow \text{Gasoline} \rightarrow \text{LPG}$$

SCHEME 1

Although many other types of acid zeolite supports have been tested, commercial hydrocracking catalysts are generally based on FAU zeolites. This can be related to the large use of these zeolites in catalytic processes (e.g. FCC) and to the three-dimensional micropore system with large openings well-adapted to the transformation of bulky molecules. With zeolites having smaller pore openings and/or a one-dimensional pore system, hydrocracking is more strongly diffusion limited with as a consequence a lower selectivity to intermediates in Scheme 1. This explains why under commercial conditions, lower selectivity to the desired middle distillates and significant increase in undesirable gas products can be observed, e.g. from 5 wt% with a dealuminated HFAU support to 25 and 35 wt% with HBEA and HMTW respectively [1]. It should be noted that this selectivity change could be expected from the isomerization/cracking values obtained in *n*-heptane hydrocracking over Pt/H-zeolite catalysts, e.g. 20 with FAU against 2.5 with HBEA [11]. Even with FAU supports, there are problems of access to the inner acid sites of the heavier reactant molecules. It is why adding of amorphous silica-alumina on which the heavier molecules can be transformed is sometimes practised [1]. Internal mesopores created by dealumination or desilication can reduce diffusion limitations during hydrocracking, and hence can increase the selectivity to middle distillates to the detriment of gas formation and favour the access of heavy molecules to the inner active sites. It should be shown in Section 14.3 that these mesopores increase the catalyst stability by limiting pore blockage by coke.

14.3. Catalyst Deactivation

Hydrocracking catalysts operate in the presence of various potential poisons of their acid and/or of their hydrogenating active sites, namely nitrogen- and sulphur-containing compounds and polyaromatics. Nitrogen-containing molecules deactivate the acid sites, sulphur molecules deactivate essentially the noble metal component of certain hydrocracking catalysts but could also poison the acid sites [11, 12]. This poisoning of the acid sites was ascribed to H_2S adsorption on the acid hydroxyl groups [11], to an additional formation of coke [12a] or to pore blockage by dimethyl polysulphides, resulting from the transformation of dimethyl disulfide (DMDS) used as a H_2S precursor [12b]. Polyaromatic molecules, which are well-known as coke maker molecules in acid catalysis (Chapter 7), also inhibit by competitive adsorption the acid transformation of the olefinic intermediates of bifunctional reactions [13]. Lastly, with noble metal/acid zeolite catalysts, metal sintering is an additional cause of deactivation.

14.3.1. *Deactivation by poisoning*

The poisoning effect of NH_3 and H_2S on an "ideal" bifunctional catalyst (noble metal/acid zeolite) was shown in Fig. 6.2. While NH_3 caused an activity decrease in agreement with an acid-controlling reaction, the poisoning effect of H_2S only appeared when the added amount was enough for the metal reactions to become rate-controlling. A large poisoning effect of NH_3 was also shown under industrial conditions. Thus an increase of the operating temperature of $\sim$100 K was shown to be necessary to compensate for the loss in the hydrocracking activity of a NiMo/HFAU zeolite catalyst when co-feeding 2000 wt ppm NH_3. However, as suggested by Scheme 1, the selectivity to middle distillates was significantly increased [14, 15]. This large effect of NH_3 on catalyst performances provides an opportunity to tune the product distribution [1].

Highly dispersed noble metals, supported over acid zeolites, were shown to be sulphur-tolerant, which was related to the electron deficiency of small metal clusters located near the inner acid sites. Alternative explanations — partial sulphuration of metal clusters, participation in hydrogenation reactions of the protonic sites close to Pt particles — as well as different ways to increase this sulphur tolerance — high metal dispersion, alloying with another metal, promoters, etc. — have also been proposed (Chapter 6). This sulphur tolerance of noble metals allows the use of noble metal/acid zeolite (FAU type) in the second stage of the hydrocracking unit.

14.3.2. *Deactivation by coking*

Polyaromatic molecules which are not hydrogenated during the hydrotreating stage are the main factor responsible for coke formation, primarily because of the long contact time of these bulky and relatively basic molecules with the acid sites. The first way to limit this coke formation is to operate hydrocracking under high hydrogen pressure, which allows the fast hydrogenation of these coke maker molecules and, in addition, limits for thermodynamic equilibrium reasons the concentration of dehydrogenated species at the proximity of acid sites, and hence the condensation steps leading to coke.

Although coking is the main cause of deactivation of hydrocracking catalysts, there are no detailed reports on the coke composition and mode of formation. The operating conditions however suggest a highly polyaromatic composition and a bifunctional mode of formation. Like with reforming catalysts, acid sites would catalyse the condensation and rearrangement steps of unsaturated species, and the metal sites the dehydrogenation steps [16]. By analogy to the nature of carbonaceous compounds formed during toluene hydrogenation at low temperatures (383 K) over bifunctional Pt-HFAU zeolites, it seems likely that condensation reactions between aromatics and olefinic naphthene intermediates, which will easily occur through acid catalysis, would be more favourable than condensation of aromatic molecules. Furthermore, while the smaller polyaromatic coke components could be located within the FAU zeolite supercages, the heavier coke molecules can result either from the growth of polyaromatic molecules located near the mesopore surface with molecules overflowing onto this surface or from their desorption followed by thermal or catalytic condensation during their slow migration along the fixed-bed reactor (Chapter 7). In agreement with a significant role in coke formation on the inner acid sites (hence in constrained location), zeolite catalysts were found to be more stable in commercial operations than silica-alumina catalysts [1].

More attention was paid to the hydrocracking catalyst features which can limit deactivation by coking. The major role that the balance between hydrogenating and acidic functions played on the product distribution and on the deactivation by coking was shown in Section 14.2.2.3. It should be remarked that the ideal balance should be specified by both the relative concentration of hydrogenating and acid sites and their turnover frequency (TOF) which depends respectively on the nature of the hydrogenating component and on the acid strength (and also on the sensitivity to poisoning). In addition to chemical steps, diffusion steps of reaction intermediates between hydrogenating sites are involved in bifunctional schemes, which explains the positive effect of the intimacy of hydrogenating and acid sites on the catalyst activity and selectivity and on coke formation. Experiments carried out with Ni/W FAU and NiMo-alumina/FAU systems seem to indicate that coke formation is reduced with increased intimacy of the two functions at the submicron level [17, 18]. Note that while clusters of noble metals can be located within the FAU zeolite supercages, this seems impossible for mixed sulphides which are located on the binder and/or within zeolite mesopores, hence at a relatively long distance from the acid sites.

One of the great advantages of zeolite acid components is the possibility to easily tune their physical and chemical features during the synthesis or through post-synthesis treatments. Thus, the zeolite can be synthesized with different crystal sizes, mesopores can be created by dealumination or desilication, the density of protonic acid sites can be varied by changing the Si/Al ratio, their acid strength can be increased by creating extraframework Al (EFAL) species or decreased by isomorphous substitution (Ga or Fe substitution of framework Al), etc. (Chapter 2). When considering Fig. 14.2, reducing the density and/or the strength of the acid sites while maintaining a constant hydrogenation activity should limit both the formation of gases at the benefit of the desired middle distillates and coke formation [1]. However, this reduction of the acid function causes a decrease in the hydrocracking activity which has to be compensated for by an increase in the operating temperature. Furthermore, the creation of internal mesopores [19, 20] or the use of small (nano-sized) zeolite crystals [21, 22], which reduce mass transport limitations, have also a positive effect on the selectivity to middle distillates. Moreover for heavy feeds, an increase in activity can be expected due to an easier transformation of the bulkiest molecules. Such an effect was demonstrated by comparing the hydrocracking of tetralin and atmospheric residue over NiMo HFAU catalysts with and without mesopores [19]. These mesopores can also accommodate the support of NiMo sulfides, e.g. extraframework Al species created during zeolite dealumination by steaming or titanium dioxide, with high hydrogenating activity and relatively close proximity of the acid and hydrogenating sites, both features being favourable for selectivity to middle distillates and catalyst stability [20, 23].

14.4. Regeneration and Recycling

Since the mid-1970s, the regeneration of the processing catalysts including hydrocracking is generally carried out *ex situ*. Indeed, this *ex situ* regeneration is preferable to the *in situ* mode for many reasons: gain of time, well-resolved issues of corrosion, safety and environment, better defined conditions (especially more precise and homogeneous temperature control during coke combustion) and hence better activity recovery [24]. This preference for *ex situ* regeneration explains why there are very few papers dealing with the regeneration of hydrocracking catalysts. Companies that specialize in *ex situ* regeneration (the main ones being Eurecat, Porocel and Tricat) obviously considered most of the relevant information as proprietary. However the general steps involved in the complete life cycle of hydroprocessing catalysts, including the regeneration steps, were described in a recent paper from Eurecat (Fig. 14.3) [24]. The following text is essentially based on this very informative paper.

The deactivation of hydrocracking catalysts is due to three causes: coking, poisoning and sintering, the main one being by far coke formation. The loss in activity is relatively low, which explains the use of fixed-bed reactors (Chapter 10) and can be simply compensated for a relatively long

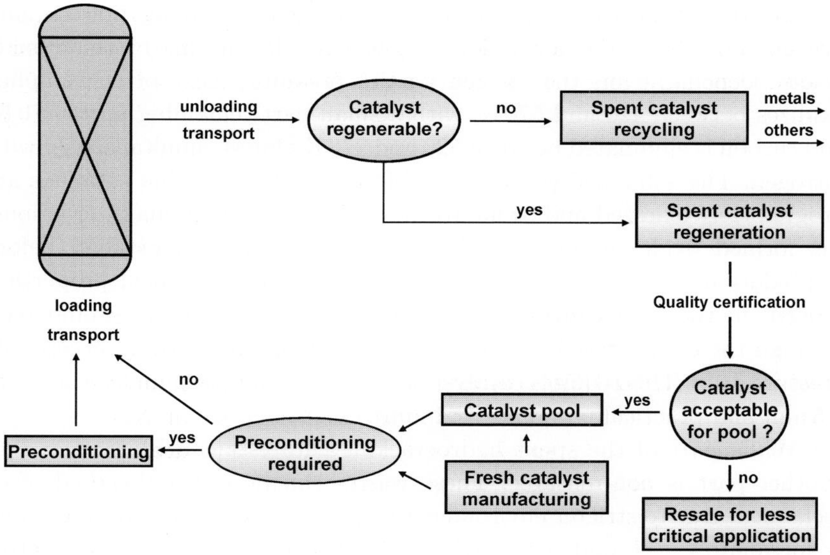

Fig. 14.3. Hydroprocessing catalyst life cycle [24].

time (1–2 years in the case of VGO processing [24]) by increasing the operating temperature. The end of the cycle is usually determined by the maximum temperature at which the product specifications can no more be satisfied. However it can also be due to technical problems such as compressor failure, hydrogen shortage, etc. or to a scheduled unit shutdown. In the first case, the catalysts that were pushed to the extreme limit in time-on-stream and temperature will contain a very large amount of coke much greater than in the second and moreover this coke will be more highly polyaromatic [25].

At the end of the cycle, careful analyses are required to specify the properties of spent catalysts so as to check their suitability for reuse after regeneration and to prevent safety problems (due for instance to entrapped volatile gases, including hydrogen) during catalyst withdrawal from the reactor [24, 25]. As the degree of contamination of the catalyst by the deactivating species varies along the reactor, the sampling for analysis should be organized so as to lead to information on the contaminant distribution. A good sampling of the unloaded catalyst containers is also essential to allow a proper segregation between good and contaminated materials.

Whatever its composition, coke can be eliminated by combustion under an oxidizing atmosphere at a temperature of 723–823 K. In addition, regeneration removes the sulphur- and nitrogen-containing poisons and converts the sulphide phase (which can be partially sintered) into an oxide phase. Depending on the oxygen partial pressure, most of the sulphur oxidizes between 420 and 570 K, with a small part remaining up to 770 K, and carbon is eliminated between 520 and 770 K almost simultaneously with nitrogen. The sulphided phases of CoMo and NiMo/alumina catalysts are progressively oxidized and owing to the strong metal-alumina interactions, the formed oxide phase is similar to that of the fresh catalyst before sulphidation [26]. There are also specifications of mechanical properties. Indeed, in the hydrocracking unit, the gas and liquid flows should pass through the catalyst bed without any channelling and with an acceptable pressure drop. This requires catalyst particles with a minimum length above 2.5 mm and practically no fines ($<$1 mm).

While part of the spent hydrocracking catalysts can be regenerated, another part is non-reusable. These wastes can be either disposed of or reclaimed. Due to stricter environmental regulations, disposal by landfilling is fortunately less and less used. The few companies which are active in recycling use two types of reclamation processes: hydrometallurgy and pyrometallurgy, this last route having the advantage of allowing a complete recycling of the catalyst components: metals for manufacturing special alloys, the inert slag for manufacturing rock wool [24].

14.5. Conclusion

The main purpose of hydrocracking is to transform heavy oil cuts such as vacuum gas oil into transportation fuels, especially into kerosene and gasoil of high quality. The hydrogenating-acid bifunctional catalysts used in this process suffer from deactivation, essentially due to coking; the other causes, i.e. the poisoning by nitrogen- and sulphur-containing compounds, sintering of the sulphide or noble metal hydrogenating phases play a minor role.

The primary way to limit deactivation by coking is to operate at high hydrogen pressures so as to hydrogenate the polyaromatic components of the feed which are well-known as coke maker molecules. The so-called ideal bifunctional catalyst is well-adapted for this purpose and in addition can orient the selectivity to the desired middle distillate products. This catalyst must have not only a high hydrogenating activity but also a relatively high cracking activity, which explains the choice of zeolites as acid components. However, the balance between the hydrogenating and cracking functions must be in favour of the first one, which means that the cracking of olefinic intermediates over the acid sites must be the controlling step of hydrocracking. Moreover, both the number of acid sites encountered by the intermediates during their diffusion from one hydrogenating site to another and their contact time must be very limited so as to avoid condensation reactions leading to coke molecules. This imposes a close intimacy between the hydrogenating and acid sites as well as the absence of diffusion limitations, hence the choice of a zeolite with large pores such as the FAU zeolite.

There is a significant economic incentive to use heavier hydrocracking feedstocks [3]. The first problem with these feedstocks originates from their relatively high concentration of polyaromatics with as a consequence a much faster catalyst deactivation by coking. One possible solution is to substitute the fixed-bed reactor for a reactor allowing a continuous withdrawal of deactivated catalyst such as the ebullated-bed reactor, proposed in several processes. The second problem to be solved is the low activity of zeolite-based hydrocracking catalysts due to a restricted access to the inner acid sites of bulky molecules. Various solutions were tested: development of zeolites with larger micropores, of nano-sized or of delaminated zeolites, creation of inner mesopores, etc. However, despite some improvements, significant advances still remain to be accomplished.

References

[1] van Veen J.A.R., Minderhoud J.K., Huve L.G., Stork W.H.J., in *Handbook of Heterogeneous Catalysis*, Ertl G., Knözinger H., Weitkamp J., Schuit G.C.A. (Eds.), Wiley-VCH Verlag, Weinham (2008) Section 13.6, 2778–2808.

[2] Cheng W.-C., Habib E.T. Jr., Rajagopalan K., Roberie T.G., Wormsbecher R.F., Ziebarth M.S., in *Handbook of Heterogeneous Catalysis*, Ertl G., Knözinger H., Weitkamp J., Schuit G.C.A. (Eds.), Wiley-VCH Verlag, Weinham (2008) Section 13.5, 2741–2778.

[3] Ferdous D., Dalai A.K., Adjaye J., *Energy & Fuels*, 17 (2003) 164–171.

[4] van Veen J.A.R., *Zeolites for Cleaner Technologies*, Guisnet M., Gilson J.-P. (Eds.), Imperial College Press, London (2002) Chapter 6, 131–152.

[5] Jensen R.H., *Zeolites for Cleaner Technologies*, Guisnet M., Gilson J.-P. (Eds.), Imperial College Press, London (2002) Chapter 4, 75–103.

[6] (a) Jacobs P.A., Uytterhoeven J.B., Steijns M., Froment G., Weitkamp J., in *Proceedings of the 5th International Zeolite Conference*, Naples 1980, Rees L.C.V. (Ed.), Heyden, London (1980) 607; (b) Weitkamp J., *Ind. Eng. Chem. Prod. Res. Dev.*, 21 (1982) 550–558; (c) Weitkamp J., Ernst S., in *Guidelines for Mastering Properties of Molecular Sieves*, Barthomeuf D., Derouane E.G., Höeldrich W. (Eds.), NATO ASI Ser. B, 221, Plenum, New York (1990) 343.

[7] (a) Guisnet M., Pérot G., in *Zeolites: Science and Technology*, Ramôa Ribeiro F., Rodrigues A., Rollmann L.D., Naccache C. (Eds.), NATO ASI Ser. B, 80, Martinus Nijhoff, The Hague/Boston/Lancaster (1984) 397; (b) Alvarez F., Giannetto G., Pérot G., Guisnet M., *Appl. Catal.*, 34 (1987) 353–365; (c) Guisnet M., Alvarez F., Giannetto G., Pérot G., *Catal. Today*, 1 (1987) 415–433; (d) Giannetto G., Alvarez F., Ramôa Ribeiro F., Pérot G., Guisnet M., in *Guidelines for Mastering Properties of Molecular Sieves*, Barthomeuf D., Derouane E.G., Höeldrich W. (Eds.), NATO ASI Ser. B, 221, Plenum, New York (1990) 355; (e) Alvarez F., Ramôa Ribeiro F., Perot G., Thomazeau C., Guisnet M., *J. Catal.*, 162 (1996) 179–189.

[8] (a) Martens J.A., Tielen M., Jacobs P.A., *Stud. Surf. Sci. Catal.*, 46 (1989) 49–60; (b) Martens J.A., Jacobs P.A., in *Zeolite Microporous Solids: Synthesis, Structure and Reactivity*, Derouane E.G., Lemos F., Naccache C., Ramôa Ribeiro F. (Eds.), NATO ASI Ser. C., 352, Kluwer Academic, Dordrecht (1992) 511.

[9] Taylor R.J., Petty R.H., *Appl. Catal. A: General*, 119 (1994) 121–138.

[10] Guisnet M., *Polish J. Chem.*, 77 (2003) 637–656.

[11] Dauns H., Ernst S., Weitkamp J., *Proc. 7th Intern. Zeolite Conf.*, Murakami Y. *et al.* (Eds.), Kodansha, Elsevier Publishers (1986) 787–794.

[12] (a) Guisnet M., Lemberton J.L., Thomazeau C., Mignard S., *Proc. 9th Intern. Zeolite Conf.*, van Balmoos R. *et al.* (Eds.), Butterworth-Heinemann, Stonehan, MA (1993) 413–420; (b) Thomazeau C., Canaff C., Lemberton J.L., Guisnet M., *Appl. Catal. A: General*, 103 (1993) 163–171.
[13] Guisnet M., Fouché V., *Appl. Catal.*, 71 (1991) 307–317.
[14] Nat P.J., *Erdöl & Kohle Erdgas Petrochem*, 42 (1989) 447–451.
[15] Dufresne P., Quesada A., Mignard S., *Stud. Surf. Sci. Catal.*, 53 (1989) 301–308.
[16] Guisnet M., Magnoux P., *Appl. Catal. A: General*, 212 (2001) 83–96.
[17] Lemberton J.L., Touzeyidio M., Guisnet M., *Appl. Catal. A: General*, 79 (1991) 115–126.
[18] Sato H., Iwata Y., Miki Y., Shimada H., *J. Catal.*, 186 (1999) 45–56.
[19] Sato K., Nishimura Y., Honna K., Matsubayashi N., Shimada H., *J. Catal.*, 200 (2001) 288–297.
[20] Shimada H., Sato K., Honna K., Enomoto T., Ohshio N., *Catal. Today*, 141 (2009) 43–51.
[21] Camblor M.A., Corma A., Martinez A., Martinez-Soria V., *J. Catal.*, 179 (1998) 537–547.
[22] Landau M.V., Vradman L., Valtchev V., Lezervant J., Liubich E., Talianker M., *Ind. Eng. Chem. Res.*, 42 (2003) 2773–2782.
[23] Ohshio N., Enomoto T., Honna K., Ueki H., Hashimoto Y., Aizono H., Yoshimoto M., Shimada H., *Fuel*, 83 (2004) 1898–1898.
[24] Dufresne P., *Appl. Catal. A: General*, 322 (2007) 67–75.
[25] Furimsky E., *Catal. Today*, 30 (1996) 223–286.
[26] Yoshimura Y., *Ind. Eng. Chem. Res.*, 30 (1991) 1092–1099.

Chapter 15

METHANOL TO OLEFINS: COKE FORMATION AND DEACTIVATION

D. Chen, K. Moljord and A. Holmen

15.1. Introduction

The catalytic conversion of methanol to lower olefins (MTO) is a promising way of converting natural gas and coal to chemicals via methanol [1, 2]. The MTO reaction has been studied over different types of zeolites or molecular sieves (ZSM-5, SAPO-34, MOR, etc.) and in different reaction conditions. Several reviews have dealt with the catalysts, reaction mechanisms and processes of converting methanol to hydrocarbons including olefins [3–20]. The catalytic processes using zeolites often include side reactions leading to the formation of carbonaceous material with catalyst deactivation as a result, defined as "coke". More exact knowledge about the mechanism and kinetics of coke formation is a basis for improving these catalytic processes. Coke deposition is known to be the major cause of deactivation in the MTO reaction over zeolites, and both catalyst activity and selectivity are influenced by coke deposition [21, 22].

The present contribution starts with the historical development of the methanol-to-olefin process, followed by a summary of coke formation and deactivation in MTO. The focus is on reaction mechanisms involved in coke formation and catalyst deactivation in the MTO process. Various topics — reaction pathways leading to coke formation, location of coke, effects of zeolite crystal size, cage structure and composition, acidic site density and strength, as well as effect of coke on catalyst activity and selectivity, mostly on the SAPO type of catalysts — are included.

15.2. Methanol to Olefins

In 2008, the world proven reserves of natural gas were about 182 trillion cubic meters (tcm), and the annual production was 3 tcm, an increase of more than 47% from the volume produced in 1983 [23]. There are five major options to bring the energy potential of natural gas to the market: pipeline, liquefied natural gas (LNG), compressed natural gas (CNG), gas to wire (GTW) by generating power and finally gas to liquids (GTL) by converting the gas to high-quality liquid fuels and chemicals [19]. Conversion of gas to liquids represents an alternative for cost-effective transport over long distances. In addition, the hydrogen-rich character of natural gas makes it very suitable for conversion to clean transportation fuels and power. Indirect conversion (via synthesis gas) of methane to high-value products has been studied for the last 50 years, including methanol and mixed alcohols synthesis, Fischer–Tropsch synthesis to transportation fuels, and methanol to gasoline (MTG) and distillates (MOGD) and olefins (MTO) [19]. Natural gas can also be directly converted to chemicals and fuels via partial oxidation to methanol, oxidative coupling of methane, reductive coupling of methane and aromatization on zeolites, but these routes are still not able to compete with indirect routes via synthesis gas in terms of selectivity and economy [24].

Since the increase in crude oil prices in 1973, there has been a considerable interest in new technologies for the production of chemicals and liquid transportation fuels from fossil sources other than oil. Initially, a large part of the research effort was directed at improving the existing industrial pathway of the Fischer–Tropsch synthesis based on CO hydrogenation. However, the observation that methanol could be converted to high-octane gasoline over a zeolite catalyst led to the development of a new process which was commercialized in New Zealand in 1985 [25, 26]. The MTG plant was closed later, due to the small price differential between gasoline and methanol. High-quality distillates can also be produced, either by Fischer–Tropsch synthesis or by oligomerization of light olefins (which can be produced from methanol) over ZSM-5 catalyst. Mobil's technologies via methanol over ZSM-5 are summarized in Fig. 15.1.

Due to increasing demand for light olefins (especially for propene), methanol-to-olefins (MTO) has experienced renewed attention. Steam cracking is the main industrial process for the production of light olefins, but on-purpose technologies including propane dehydrogenation (PDH), ethene/butene metathesis, methanol-to-olefins (MTO), methanol-to-propene (MTP) and high-severity FCC represent solutions for

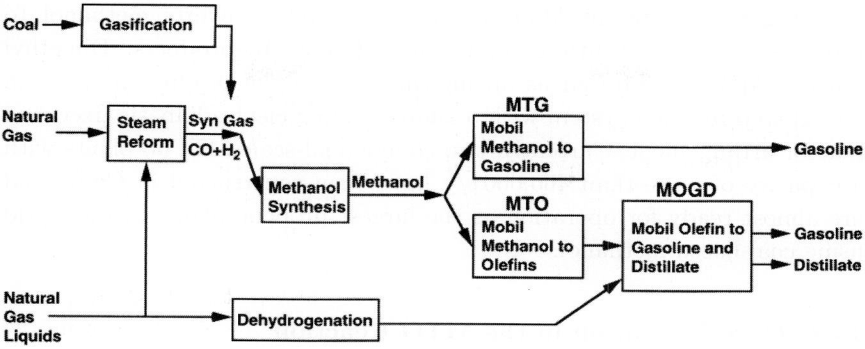

Fig. 15.1. Synthetic fuels production using Mobil zeolite technology [26].

increasing propene capacity which is decoupled from ethene production. Commercialization of mega-methanol plant technologies with production capacities of 5,000 t/d or higher will probably improve the economy of the MTO process further.

MTO originates from Mobil's MTG process on the MFI structure of H-ZSM-5 (i.e. a three-dimensional 10-ring structure with 5.1*5.5 and 5.3*5.6 Å window dimensions). Mobil's MTO process operated under partial methanol conversion to suppress the heavier product formation. The main challenge of this process was to increase the olefin selectivity. The chemistry, modification of the catalysts and development of new ones to increase the olefin selectivity have been reviewed by Chang [3]. In November 1995, UOP and Norsk Hydro announced an indirect conversion of natural gas to olefins via methanol, called the UOP/HYDRO MTO process. The process can be used as the second step of a two-step GTO (gas-to-olefins) process. The key to the UOP/HYDRO process was the discovery of the CHA structure of the molecular sieve catalyst SAPO-34 (silicoaluminophosphate) by Union Carbide in the 1980s. SAPO-34 has a three-dimensional ellipsoidal cage (11 Å long and 6.5 Å wide) structure with 3.8*3.8 Å window dimensions. Because of these small pore openings, SAPO-34 has a higher selectivity to ethene and propene, primarily a much higher selectivity to ethene, than ZSM-5 [1, 2, 27, 28]. Therefore, the UOP/HYDRO process can combine 100% methanol conversion with 90% or higher selectivity to C_2–C_5 olefins in the hydrocarbon fraction. Ethene, propene and butene are the primary products. The performance has been extensively studied in a pilot plant (0.5 tons/day of methanol) located at the Norsk Hydro Research Center in Norway.

Lurgi's more recent MTP process selectively converts methanol to propene on ZSM-5, with gasoline and LPG as by-products. Dimethyl ether (DME) is produced as an intermediate to the product propene in a dual-step reaction system with a simple and efficient adiabatic fixed-bed reactor arrangement. Currently, two commercial-scale MTP® plants with a capacity of more than 460,000 t/y have been constructed in China and are almost ready for operation as the largest propene plants in the world using coal-based methanol.

15.3. Coke Formation in the MTO Reaction

15.3.1. *Techniques for studying coke deposition*

Measurement of coke deposition and deactivation is essential for a better understanding of the mechanism of deactivation by coke deposition [16, 29, 30]. Deactivation studies on zeolites in fixed-bed reactors have been reported in the literature [31–34]. The drawback of the method is that it is time-consuming and that it has relatively large uncertainty due to temperature and concentration gradients in the fixed-bed reactor. Grønvold *et al.* [35] have studied coke formation and deactivation during MTO over SAPO-34 and H-ZSM-5 in a conventional microbalance reactor. As discussed previously [14], it is very difficult to obtain kinetic data for the coking reaction and the resulting deactivation during MTO over SAPO-34 using a conventional microbalance, due to a large fraction of gas bypass with a not well-defined flow in the reactor as a result. This is particularly problematic for a fast reaction such as MTO, where important concentration and temperature gradients will occur in the basket of the microbalance reactor. The tapered element oscillating microbalance (TEOM) reactor with its fixed-bed characteristics is a much more suitable technique for studying fast coke formation and deactivation processes, because differential conversions can be ascertained [14, 36–38].

In recent years different advanced spectroscopic techniques have been used to investigate coke formation in MTO, such as Raman [39–40], IR, C-13 NMR and UV-Vis spectroscopic methods [41]. FTIR, UV-Vis, ESR and NMR spectroscopy are suitable methods for *in situ* investigations of zeolites and reactions catalysed by these materials. During the past decade, an increasing number of research groups have been dealing with the development and application of new techniques allowing *in situ* studies under batch and continuous-flow conditions. Hunger [42] has recently reviewed *in situ* spectroscopic studies in zeolite catalysis. Very recently, a method using simultaneous synchrotron powder X-ray diffraction (PXRD) and Raman spectroscopy with online analysis of products by mass spectrometry has been developed to study MTO on SAPO-34 under real working conditions. These technologies provide information on the nature of surface species and coke molecules, the location of coke and also the kinetics [40].

15.3.2. *Coke formation on SAPO-34*

SAPO-34 has a high catalytic activity for the MTO reaction, but the conversion unfortunately includes a fast coke formation responsible for deactivation [21, 22, 33, 39, 43–57]. The rate of deactivation is so high that it is difficult to decouple from the kinetics of the main reactions. Also the high heat of reaction (−100 to −300 cal/g depending on the selectivity to olefins [3]) makes the kinetics of this reaction difficult to measure in conventional laboratory reactors.

Coke deposition during MTO on SAPO-34 was studied in detail in a TEOM reactor as a function of temperature, space velocity and partial pressure of methanol [21, 22, 53, 55]. The effect of space velocity on coke formation and deactivation was investigated at 698 K in a WHSV range of 57–384 g/(g of catalyst × h) with a methanol partial pressure of 7.2 kPa. Because the catalyst underwent rapid deactivation, the MTO reaction was investigated by using 3-min interrupted pulses with GC analysis carried out after 2 min for each pulse (the integrated pulse method). At high methanol pressure and high space velocity, SAPO-34 can deactivate completely in less than one minute. Small-pulse experiments were used to study coke deposition in industrially relevant conditions.

Coke deposition is often studied as a function of time-on-stream [32–33]. Marchi and Froment [52] studied deactivation during MTO over SAPO-34 as a function of the cumulative amount of methanol fed to the catalyst. The interpretation of the measured coke content as a function of time-on-stream and as a function of the cumulative amount of methanol fed to the catalyst could lead to contradictory results [21]. It was found that the cumulative amount of methanol fed to the catalyst is a better parameter for describing the coke deposition in different conditions, something which has recently been supported by Qi *et al.* [58]. Figure 15.2 presents coke deposition at different space velocities. A lower space velocity resulted in a higher coking rate, as a result of a high average concentration of olefins and low average concentration of oxygenates (MeOH/DME mixtures) in

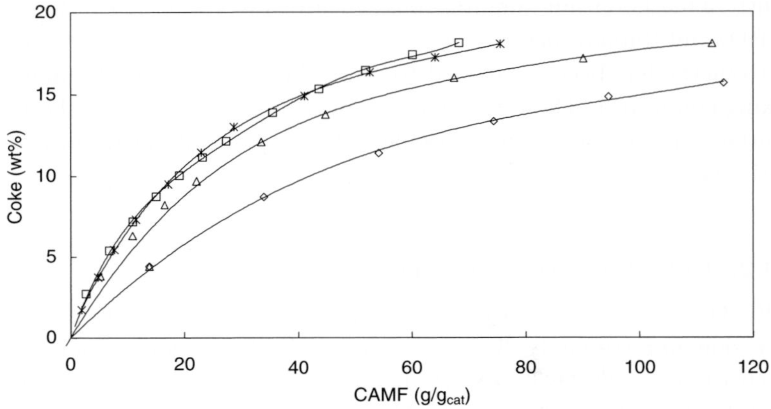

Fig. 15.2. Coke formation during MTO over SAPO-34 (S) versus cumulative amount of methanol fed to the catalysts (g/g_{cat}) at 425°C and a methanol partial pressure of 7.2 kPa for different pace velocities WHSV ($g/(g_{cat}\bullet h)$): □: 57, *: 82, Δ: 113, ◇: 385 [21].

the reactor. It indicates that the rate of coke deposition is related to the concentration of olefins, which is in good agreement with the results reported for methanol conversion to gasoline (MTG) over HZSM-5 [59]. Aguayo [32] observed also a higher coke content at the outlet than at the inlet of a fixed-bed reactor in methanol conversion over HZSM-5. At space velocities less than $82\,g(g_{cat}\bullet h)^{-1}$, the change in coke deposition with space velocity is not significant, since the oxygenate conversion is almost complete. This explains the observation [34] that the coking rate is identical regardless of space velocity in the range of 8–45 $g(g_{cat}\bullet h)^{-1}$.

The coke deposition can be described as a function of the cumulative amount of hydrocarbons formed per gram of catalyst, regardless of space velocity (shown in Fig. 15.3). From Fig. 15.3, it is straightforward to obtain the coke selectivity and the catalyst capacity at the coke content of 18%, where the catalyst was almost completely deactivated. The catalyst capacity (g hydrocarbon formed/g catalyst) and the coke selectivity (g coke/g hydrocarbon formed) are 29.3 and 0.006 at 673 K compared to 4.0 and 0.045 at 823 K, respectively. Higher temperature favoured the reactions leading to coke formation more than the reactions leading to olefins, hence a faster deactivation, in good agreement with the observation on H-ZSM-5 [59]. However, the coking rate was found to decrease with increasing temperature in the range of 613 to 693 K on SAPO-34 [33], which was explained by increased cracking of coke precursors. Most likely

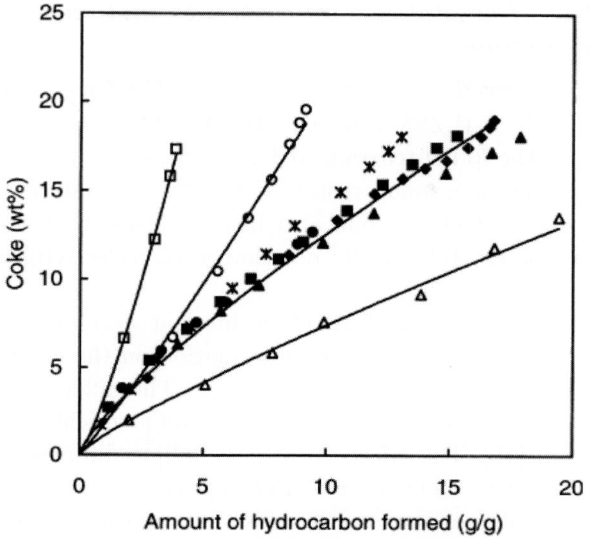

Fig. 15.3. Coke formation versus the amount of hydrocarbon formed (CAHF) at different temperatures: 823 K: □ (P_{MeOH} = 13 kPa, WHSV = 270 g/g_{cat},h); 773 K: ○ (P_{MeOH} = 13 kPa, WHSV = 270 g/g_{cat},h); 698 K and P_{MeOH} = 7.2 kPa: WHSV = *57, ■ 82, ▲ 113, • 270 and ◆ 385; 673 K: Δ (P_{MeOH} = 7.2 kPa, WHSV = 385 g/g_{cat},h) [21].

the nature of the coke molecules is very different at different temperatures. The nature of the coke molecules has been proposed to involve both olefin and polyolefin species [39], or a hydrocarbon pool consisting of a mixture of C_6–C_{12} olefins and aromatics, such as polymethylhexenes, alkylated octadienes and polymethylbenzenes [13].

It can be expected that coke formed at lower temperatures is more hydrogen-rich, while the hydrogen content of the retained molecules goes down at higher temperatures. Coke molecules can decompose with increasing temperatures, explaining lower coking rates at higher temperatures in the lower temperature range [33]. Generally, a higher operating temperature will enhance the reactions involved in coke formation such as oligomerization, cyclization, hydrogen transfer and alkylation to yield larger polyaromatic structures [60].

The rapid deactivation of the catalyst by coke formation is a challenge for the design of the MTO process, which using SAPO-34 would involve continuous catalyst regeneration. One process alternative would consist of a separate fluidized-bed regenerator with catalyst circulation to and from a fluidized-bed reactor for methanol conversion.

15.3.3. *Effect of crystal size on carbon formation and lifetime*

It has long been recognized that the crystal size of zeolites can influence the performance both of H-ZSM5 [61–65], H-SAPO-34 [22, 45, 51, 66–71] and H-STA-7 [68] in the MTO reaction. It has been shown that a long lifetime can be obtained by using small crystals. For H-ZSM5, small crystals resulted not only in a long lifetime, but also increased selectivity to aromatics [63]. However, the crystal size did not change the selectivity on SAPO-34 [22, 51, 67].

A detailed kinetic study of MTO on different sized crystals of SAPO-34 in a TEOM reactor has provided information on the crystal size effect on coke formation and deactivation [22, 66, 72]. The different sized crystals were synthesized in a controlled manner to obtain identical acid site density and distribution [45]. The effect of intracrystalline diffusion on methanol conversion to olefins (MTO) over SAPO-34 was elucidated by performing reactions on crystal sizes in the range of 0.25–2.5 μm in identical conditions [22, 66]. The conversion to olefins was found to be influenced by the crystal size, the highest reaction rate was found on 0.4–0.5 μm crystals. On the contrary, the selectivity to the different olefins was independent of the crystal size. Dimethyl ether (DME) diffusion seems to play an important role in the formation of olefins. The smallest crystals resulted in a relatively large amount of DME escaping the pores of SAPO-34 before being converted to olefins, hence giving lower olefin yields. Coking rates increased with increasing crystal size at low coke contents, levelling off at a lower level for the larger crystals than for the smaller at high coke contents. The deactivating effect of the coke formed was larger on the larger crystals.

The effect of coke deposition on DME conversion to light olefins (DTO) was also studied over SAPO-34 with crystal sizes of 0.25 and 2.5 μm using the TEOM reactor [72]. Similar to the MTO reaction, the conversion was lower and deactivation was faster on larger crystals. An induction period was very obvious on 0.25 μm crystals, but not on 2.5 μm crystals. Kinetic modelling was used to estimate the effective DME diffusivity and the first-order intrinsic rate constant at different coke contents, and the results are plotted in Fig. 15.4. The diffusivity of DME decreased exponentially with the coke content, while the intrinsic rate constant went through a maximum. The effectiveness factor of 0.25 μm crystals was close to 1 even on coked samples, indicating no diffusion limitation. The reaction rate on the 2.5 μm crystals was significantly influenced by diffusion, and coke deposition

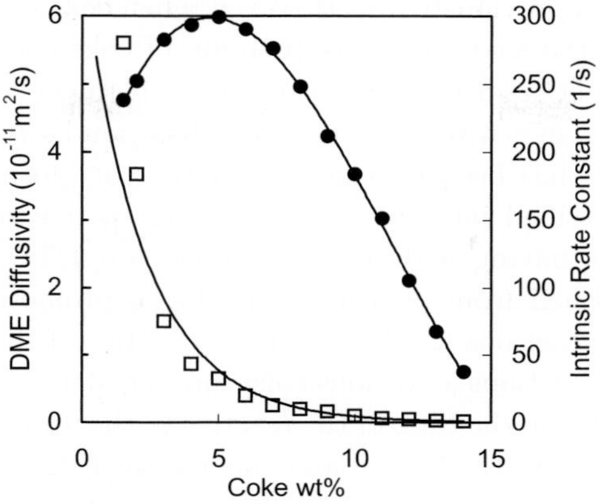

Fig. 15.4. Effective DME diffusivity and intrinsic rate constant of DTO versus coke content at 425°C [72].

reduced the diffusivity and thus the effectiveness factor, which explains the faster deactivation rate observed on large crystals.

All the results clearly point out that crystal size is a very important parameter for controlling coke formation and catalyst deactivation in the MTO process. The crystal size of SAPO-34 has recently been controlled between 1.5 and 7 μm by hydrothermal synthesis using a mixture of tetraethyl-ammonium hydroxide (TEAOH) and morpholine (Mor) as the structure-directing agents. The smaller SAPO-34 crystals showed a longer catalyst lifetime in the methanol-to-olefins reactions. The amount of coke deposited on the deactivated SAPO-34 catalyst increased with decreasing crystal size, indicating that the effectiveness of the catalyst could be improved by reducing the crystal size of SAPO-34 [67]. Crushing large crystals (5–20 μm) to 1 μm by ball milling increased the lifetime more than three times [51]. Formation of a reaction zone near the external surface caused by diffusion control in MTO has recently been directly observed by combined *in situ* UV-Vis microscopy and confocal fluorescence microscopy [73].

15.3.4. *Reaction mechanism and pathways leading to coke formation*

Figure 15.4 shows that the DME conversion rate increased slightly with coke formation up to about 5 wt% coke. Coke has obviously a promoting effect besides a deactivating effect on the formation of olefins, which has been also reported previously [74]. However, when coke was formed directly from propene as the feed, no active function of coke could be detected in the sequential MTO and DTO reactions, even if the coke by pore volume measurement was shown to be in the same location as that formed in the MTO reaction. It has been concluded that the coke formed in MTO and DTO can be classified into two categories: unreactive coke formed from adsorbed alkenes having a deactivating effect on DTO and MTO, and reactive coke formed from oxygenates having a promoting effect on the DTO and MTO reactions [22, 55, 72]. The activities of the catalyst for the MTO and DTO reactions at various coke contents depend on the nature of the coke, especially on the ratio of the reactive to the unreactive coke. Based on a comparison between the real and the apparent density of coke, it was concluded that both types of coke are located inside the cages [75], which is in good agreement with observations obtained by synchrotron powder X-ray diffraction (PXRD) study [40].

The nature of coke molecules is definitely of importance for understanding the promoting effect of coke molecules. Many different techniques have been used to characterize coke molecules. The most widely used technique is the post-extraction method [76–77]. This technique can provide ^{13}C and ^{12}C distribution in retained carbon after isotopic exchange experiments [78–80]. The most abundant molecules are aromatics with multi-methyl groups, which are typically equilibrated with isotopic C in the isotopic exchange experiments. These findings have led to the development of the hydrocarbon pool mechanism [11, 44, 78]. Since SAPO-34 has small pore openings (8-membered rings) and large cages (11 Å long and 6.5 Å wide), relatively large molecules such as branched olefins and paraffins could be trapped in the cage as coke.

The MTO reaction mechanism has been discussed in a few reviews [9, 13, 81]. At least 20 distinct mechanisms have been proposed during the last 25 years, but most of them do not account for the primary products or the kinetic induction period [11]. As noted by Chang, a large number of proposals have been made concerning the mechanism of formation of the primary hydrocarbon products. Some of the proposed mechanisms have been based on detailed experimental support, notably the carbene mechanism, the trimethyloxonium mechanism, the free radical mechanism and deprotonation of a surface bonded methyl oxonium ion, to give a surface bonded oxonium methylide [3]. Recent experimental and theoretical work has established that methanol and dimethyl ether react on cyclic organic species contained in the cages or channels of the inorganic host.

These organic reaction centres act as scaffolds for the assembly of light olefins so as to avoid the high-energy intermediates required by all "direct" mechanisms. It turns out to be useful to consider each cage (or channel) with its included organic and inorganic species as a supramolecule that can react to form various species [11]. There are a lot of theoretic [82–88] and experimental [13, 73, 79, 89–97] efforts for identifying the hydrocarbon pool or organic reactive centres in MTO catalysts. Hexamethylbenzene has been found to be the most active organic centre to undergo paring reactions to split olefins with low energy barriers [11]. The essential feature of the reaction, ring contraction followed by expansion, provides the means to extend an alkyl chain and hence eliminate an olefin from methylbenzene. Reaction of the aromatic product by methanol or dimethyl ether under MTO conditions would regenerate the original methylbenzene, completing a catalytic cycle. Haw *et al.* have recently presented a complete catalytic cycle for supramolecular methanol-to-olefins conversion in H-ZSM-5 by linking theory with experiment (Fig. 15.5) [96]. However, the authors have pointed out that HSAPO-34 has different topologies and different compositions which will most likely lead to slightly different major catalytic cycles.

The hydrocarbon pool mechanism seems to be gradually accepted for olefin formation in MTO, but there are still some experimental observations which are not easily explained: (1) SAPO-34 has a cage density of about 1.1 mmol/g as measured by ammonia temperature-programmed desorption [57]. If each cage contains a hexamethylbenzene molecule, the coke content will be 18 wt%. At this coke content, SAPO-34 was found to deactivate almost completely even in the case of no diffusion limitation on the 0.25 μm SAPO-34 crystals. (2) SAPO-34 has quite high activity at very low coke content and the activity increases with coke content from 0 to 5–6%, which cannot be fully explained by the hydrocarbon pool mechanism. (3) The coke has much lower coke density at low coke contents and reaches a similar density as hexamethylbenzene when SAPO-34 is completely deactivated [75].

The main experimental evidence of the hydrocarbon pool mechanism is from the analysis of the catalyst after conversion of the methanol in batch mode, and *in situ* NMR analysis. Considering that the aromatic ion is most likely formed from olefins following Hutchings and Hunter's oxonium methyl ylide mechanism [98], Froment reconciled the hydrocarbon pool mechanism [16]. According to the mechanism suggested by Hutchings and Hunter [98], oxonium methylylide produced from the methoxy carbenium ion is generated from methanol by dehydration. The primary carbenium ions

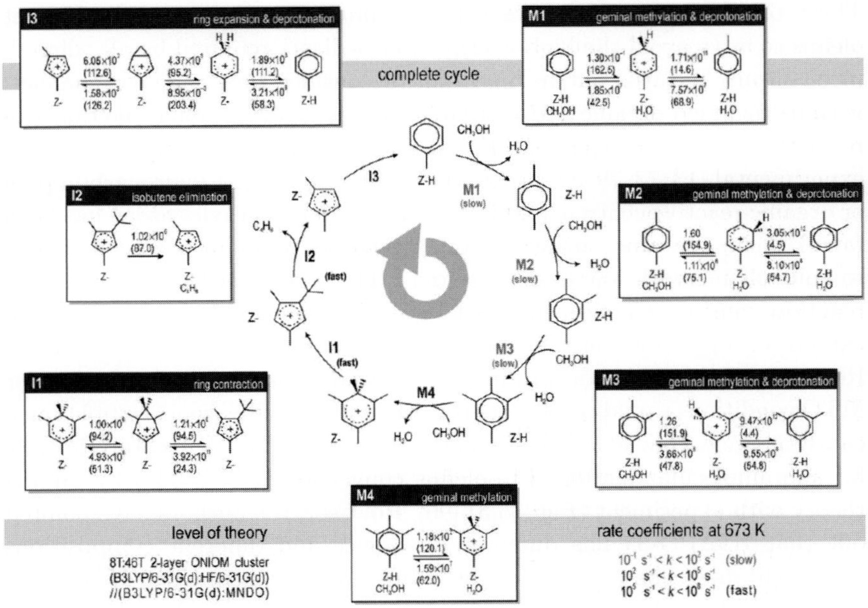

Fig. 15.5. Full catalytic cycle for carbon-atom scrambling and isobutene formation from methanol through a combined methylbenzene/cyclopentenyl cation pool in HZSM-5. Calculated rate constants at 673 K are given in s^{-1} and reaction barriers at 0 K (in brackets) are given in kJ mol^{-1}; for details see text in [96].

ethyl+, propyl+ and butyl+, whose formation involves the oxonium methyl ylide, can deprotonate into the corresponding olefins but can also undergo methylation and oligomerization into higher carbenium ions, leading to paraffinic R^+ by H-transfer and to aromatic R+ by cyclization [12, 16]. The latter evolves by alkylation and cyclization into polyring aromatics too bulky to leave the catalyst because of the structure of SAPO-34. They cover sites and block pores and thus deactivate the catalyst. Hence, hexamethylbenzene could be the product of the conversion of primary carbenium ions, but the hexamethylbenzene and primary carbenium ions could equilibrate each other, something which would explain the isotopic distribution in products and retained coke molecules. This is consistent with the observation of the olefinic nature of coke [39] and the co-existence of aromatics and carbenium ions [13]. Recent studies also revealed that the alkylation and cracking reactions of C_{3+} olefins contribute significantly to the production of C_3 olefins in H-ZSM-5, resulting in a rather low ethene selectivity [99].

Yield-conversion plots have often been used to help identify the type of coke and the reaction network. It has been found that DME is an unstable primary product and that all the alkenes are stable secondary products in the MTO reaction [31, 53]. It has also been shown that simultaneous measurements of conversion of methanol, conversion to alkenes and formation of coke are helpful in the identification of reaction pathways for coke formation. The results indicate that coke is a stable secondary product plus a stable tertiary product from olefins at high conversions of methanol. A parallel reaction pathway leading to the formation of olefins and coke might indicate similar intermediates or the hydrocarbon pool for their formation [21].

15.4. Catalyst Deactivation During the MTO Reaction

15.4.1. *Deactivation of SAPO-34 during MTO by coke formation*

As discussed above, coke forms rapidly on SAPO-34 due to its large cages and small windows. Catalyst deactivation depends not only on the coke content but also on the nature and location of the coke molecules. SAPO-34 deactivates generally as a linear function of the coke content [21, 31, 53], but it deactivates faster at higher temperatures, indicating that larger molecules are formed at higher temperatures at identical coke contents [53]. More importantly, large SAPO crystals deactivate faster than small crystals with similar coke contents, due to the fact that coke is mainly formed at a region near the external surface of the catalyst particles, gradually blocking the diffusion path of oxygenates to the inner core of the catalyst, in accordance with the observed lower coke content on larger SAPO-34 crystals. Removal of coke by combustion in air can completely recover the activity [100].

15.4.2. *Effects of pore structure on catalyst deactivation*

A large number of catalysts with different structures have been investigated in MTO aiming at increasing the olefin selectivity and lifetime. Effects of the catalyst structure on the activity and selectivity to olefins have been well-documented in several reviews [4, 15, 19, 81]. Unfortunately, most studies report only the conversion and deactivation with time-on-stream, and coke deposition and the deactivating or activating effect of coke could not be determined [43, 52, 68, 94, 101–108].

Silicoaluminophosphates (SAPOs) are a relatively new generation of crystalline microporous molecular sieves. They are synthesized by incorporating Si into the framework of the aluminophosphate (AlPO) molecular sieves. Djieugoue *et al.* studied four small-pore SAPOs: erionite-like SAPO-17, chabazite-like SAPO-34, levynite- like SAPO-35 and SAPO-18, whose structure is closely related to, but crystallographically distinct from, that of SAPO-34 [104]. The cage structures are presented in Fig. 15.6. The lifetime followed the order of SAPO-18 > SAPO-34 ≈ SAPO-17 > SAPO-35. SAPO-35, having the smallest cage size and the lowest number of 8-membered ring openings in each cage, deactivated most rapidly. The lowest olefin yield or selectivity is reported on SAPO-35, but the measurements could have been influenced by deactivation. In addition it should be mentioned that the number of acid sites is not the same, following

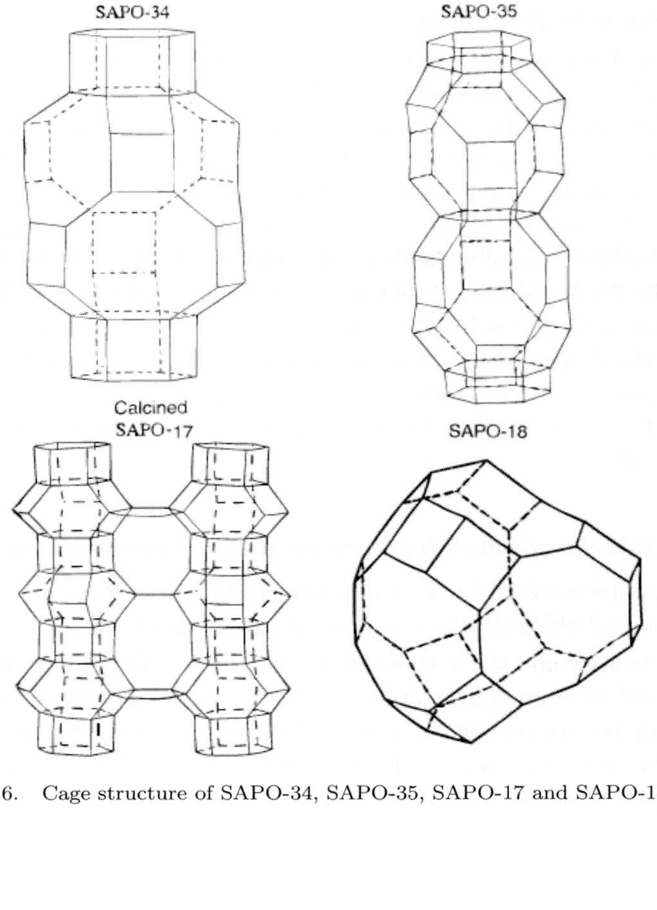

Fig. 15.6. Cage structure of SAPO-34, SAPO-35, SAPO-17 and SAPO-18 [109].

an order of SAPO-34 > SAPO-17 > SAPO-18 > SAPO-35. Therefore, the rapid deactivation of SAPO-35 cannot be explained by acid site density. It was suggested that the small cage is the main cause of the faster deactivation. Guisnet *et al.* have long recognized the profound effects of the pore structure of zeolites on coke deposition and deactivation [60, 110–112]. There are several studies comparing H-ZSM-5 and H-SAPO-34 [35, 39, 99], and the reported rate of coke deposition is less on ZSM-5 compared to SAPO-34, which can be explained by the ratio of cage size to pore size. A small cage size provides a space constraint on the growth of coke molecules [22, 66, 113]. In this regard, it is expected that the coking rate should be lower on SAPO-35. The fast deactivation of SAPO-35 would probably be related to its structure of three 8-membered ring openings while SAPO-34, -17 and -18 have six 8-membered rings in each cage. The probability of molecular diffusion in and out of the cages is related to the number of ring openings in each cage and a small number of rings could cause more pronounced deactivation by a faster reduction in diffusion caused by blockage by coke molecules. A better performance of SAPO-18 could be explained by a different shape of the cage and less acidic sites [104].

The effect of the ratio of cage to pore openings was also investigated by Park *et al.* using four kinds of 8-membered ring (8-MR) small-port molecular sieves with CHA (SAPO-34), ERI (UZM-12), LTA (UZM-9) and UFI (UZM-5) topologies [102]. The ratio of cage to pore size and deactivation are in the same order of LTA > ERI > CHA. UV-Vis spectroscopic studies revealed that larger coke molecules are formed in larger cages, causing the faster deactivation. UFI and LTA have similar cage and pore size, LTA has a three-dimensional (3-D) structure while UFI has a two-dimensional (2-D) connectivity. Clearly a 3-D structure has better stability than a 2-D structure [102], which has better stability than a 1-D structure such as SAPO-41 [106], SAPO-5 [43, 101, 114] and SAPO-11 [52].

15.4.3. *Effects of catalyst acidity and composition on deactivation*

Acid site density and acidic strength are expected to have significant effects on reactions leading to coke formation, such as oligomerization, cyclization and hydrogen transfer. These reactions require strong acidic sites. Aguayo *et al.* have used calorimetric measurements of differential adsorption of ammonia (DSC) and temperature-programmed desorption of ammonia

(TPD) to investigate the total acidity and acidic strength distribution [108]. High acidic strength resulted in high reaction rates on SAPO-18 and SAPO-34 compared to SAPO-11 and beta-zeolites. Considering the similarity of SAPO-18 and the SAPO-34 in pore structure and acidic strength, the greater density of acid sites of SAPO-34 might be responsible for the faster deactivation compared to SAPO-18. Similar observations were also reported by other groups [97, 107, 115–117]. Yuen *et al.* [117] performed comparative MTO tests on catalysts with CHA topology with varying acid strength (SAPO-34 and SSZ-13, respectively) and acid site density. They observed that an intermediate acid site density (10% Al vs. 18% Al and 3.3% Al, respectively, on H-SSZ-13) is advantageous for the stability of H-SSZ-13 catalysts. Recently, Bleken *et al.* performed a similar study with carefully synthesized zeolites (SAPO-34 and SSZ-13) with the same crystal size and acid site density to elucidate the effect of acidic strength on MTO [97]. Consistent with the results in [108, 117], acidic strength decreases the energy barrier and thus increases the reaction rates. Ethene increases in a shorter time on H-SSZ-13 than on H-SAPO-34, suggesting a faster coke formation on H-SSZ-13 due to the higher acidic strength. The acid site density has been long recognized to have profound effects on catalyst coke formation and deactivation [4, 20].

A comparison of isostructural H-SAPO-34 and H-chabazite (HCHA) in MTO showed a similar initial rate and selectivity, but also a much faster deactivation on H-chabazite with higher acid site density than the low acid-site density analogues [118]. It is expected that the coking rate is much larger on HCHA. Zhu *et al.* noticed that less Si in CHA zeolites results in a prolonged catalyst lifetime [119].

Incorporating metal into the SAPO framework has been recognized as a way of improving the olefin selectivity and prolonging the catalyst lifetime since the reports from Inui *et al.* [120–121], as part of the large research efforts on metallosilicates [122–124] in his group. Various Me-APSO-34 (Me = Ni [109, 125–141], Co, Mn, Fe [126, 130, 134–135], Cu [103], Ti, Cr, Cu, Zn, Mg, Ca, Sr and Ba [140]) have been tested in MTO. Incorporating metal into the framework of zeolites has been proven to reduce the acid density, leading to high selectivity to ethene and reduced coke formation and higher catalyst lifetime [109, 129]. Co-feeding of water increased the olefin selectivity and reduced coke formation and prolonged the catalyst lifetime. This can be explained by strong acidic sites being occupied by polar water molecules [52, 142–143].

15.5. Effects of Coke on Shape Selectivity

There are three types of shape selectivities, namely reactant, product and transition-state shape selectivity, making zeolites very attractive for many applications. Both product shape selectivity and transition-state shape selectivity have been reported in MTO. Product shape selectivity plays an obvious role in obtaining high selectivity to olefins in SAPO-34 compared to ZSM-5. The 8-membered ring openings limit aromatics diffusion out of the cages. However, there is still debate about the role of shape selectivity with respect to the selectivity of C_2–C_4 olefins on SAPO-34.

We have studied MTO and DTO on different SAPO-34 crystal sizes [22, 72], and found that the olefin selectivity was identical at a certain coke content, regardless of the crystal size, but an increase in the coke content increased the ethene selectivity and decreased the selectivity to propene and C_4–C_6. More importantly, the change in product selectivity was identical, regardless of crystal size. It led to the conclusion that transition-state selectivity governs the product selectivity [22]. The conclusion was supported by Haw *et al.*, who performed a systematic study of MTO using combined GC analysis and solid-state MAS 13C NMR, and observed an increase in ethene-to-propene ratio with increasing flush duration after stopping methanol flow, correlating to a decrease in the number of methyl groups in the polymethylbenzene intermediates. The authors concluded that the formation of olefins is governed by transition-state shape selectivity [92]. However, Dahl *et al.* reached an opposite conclusion with product shape selectivity dominating for olefin distribution [45], using ethanol and 2-propanol as probe molecules on different sized crystals (being the same as that used in [22]). Diffusion of molecules is normally slow in the case of similar molecular and channel size, and it has been defined as the configurational diffusion. Dahl *et al.* found that ethanol conversion was not limited by the diffusion of ethanol while 2-propanol conversion was controlled by 2-propanol diffusion. In fact, whether product shape selectivity or transition-state shape selectivity governs the olefin distribution is a matter of the relative rate order of the diffusion rate of products through the pore networks and the reaction rate for olefin formation. If the formation rate is lower than the diffusion rate, the transition-state shape selectivity dominates. Therefore it is difficult to directly extrapolate the results in ethanol and 2-propanol reactions to MTO. First, the diffusivity of 2-propanol can be rather different from propene due to their different structure. Second, the dehydration of

2-propanol could be a faster reaction than the MTO reaction. A diffusion limitation of 2-propanol in the dehydration reaction does not necessarily mean that the selectivity is also controlled by diffusion limitation of propene in the MTO reaction conditions, even if we assume the same diffusivity of 2-propanol and propene. Barger [144] has also proposed product shape selectivity by comparing measured ethene to propene (C_2/C_3) ratios and the thermodynamic predicted ratio in the gas phase at different temperatures. It was found that the measured ethene to propene ratio was much higher than the equilibrium ratio, but the experimental C_2/C_3 ratio increased at the same rate as the equilibrium C_2/C_3 ratio with increasing temperature. Recently, Hereijgers also reached a conclusion of product shape selectivity by analysis of ^{13}C in retained molecules inside the cages of SAPO-34 [80]. However, equilibrium effects between olefins and formation of aromatics from olefins might add complexity to the discussion.

The selectivity at the same conversion level is very different on coked SAPO-34, compared to fresh SAPO-34 [21, 53, 56]. It was explained by the influence of transition-state shape selectivity. The formation of olefins with different sizes would go through different sized intermediates and the reaction with the larger intermediates deactivates faster. Coke deposition itself is also suppressed by coke molecules in the cages through transition-state shape selectivity, and the change in coke selectivity is rather similar to that for the large olefins with increasing coke content.

15.6. Conclusion

SAPO-34 is still the most promising industrial catalyst for the MTO reaction, mainly due to its high selectivity to ethene and propene. The main challenge is the rapid deactivation due to coke formation, and much research effort has been directed towards improving the stability of the catalyst. Fundamental studies on the reaction mechanism for olefin and coke formation, as well as the effect of crystal size, cage topology, acidic strength and density on coke formation and deactivation, have gained a much better understanding of the chemistry involved in this reaction, and form a good basis for further catalyst development for the process.

The catalyst coke content can be directly described by the accumulative amount of hydrocarbons formed over the catalyst. Product selectivities are identical on SAPO-34 with different crystal sizes at a certain coke content. Coke deposition in the cages has profound effects on product selectivity by increasing the ethene selectivity and decreasing the selectivity to larger molecules including coke, due to pronounced shape selectivity effects on this catalyst.

The high heat of reaction and the rapid catalyst deactivation by coke formation would strongly influence the reactor design of the MTO process, as is the case in the FCC process.

References

[1] Vora B.V., Pujadó P.R., Miller L.W., Barger P.T., Nilsen H.R., Kvisle S., Fuglerud T., *Stud. Surf. Sci. Catal.*, 136 (2001) 537–542.

[2] Vora B.V., Marker T.L., Barger P.T., Nilsen H.R., Kvisle S., Fuglerud T., *Stud. Surf. Sci. Catal.*, 107 (1997) 87–98.

[3] Chang C.D., *Catalysis Reviews: Science and Engineering*, 25 (1983) 1–118.

[4] Chang C.D., *Catalysis Reviews: Science and Engineering*, 26 (1984) 323–345.

[5] Brown N.F., Barteau M.A., *ACS Symposium Series*, 517 (1993) 345–354.

[6] Wender I., *Fuel Processing Technology*, 48 (1996) 189–297.

[7] Ancillotti F., Fattore V., *Fuel Processing Technology*, 57 (1998) 163–194.

[8] Keil F.J., *Microporous and Mesoporous Materials*, 29 (1999) 49–66.

[9] Stöcker M., *Microporous and Mesoporous Materials*, 29 (1999) 3–48.

[10] Verkerk K.A.N., Jaeger B., Finkeldei C.H., Keim W., *Appl. Catal. A: General*, 186 (1999) 407–431.

[11] Haw J.F., Song W.G., Marcus D.M., Nicholas J.B., *Acc. Chem. Res.*, 36 (2003) 317–326.

[12] Froment G.F., *Catalysis Reviews: Science and Engineering*, 47 (2005) 83–124.

[13] Wang W., Jiang Y., Hunger M., *Catal. Today*, 113 (2006) 102–114.

[14] Chen D., Bjorgum E., Christensen K.O., Holmen A., Lodeng R., *Advances in Catalysis*, 51 (2007) 351–382.

[15] Narasimharao K., Lee A., Wilson K., *J. Biobased Mater. Bio.*, 1 (2007) 19–30.

[16] Froment G.F., *Catalysis Reviews: Science and Engineering*, 50 (2008) 1–18.

[17] Khadzhiev S.N., Kolesnichenko N.V., Ezhova N.N., *Petrol. Chem.*, 48 (2008) 325–334.

[18] Melero J.A., Iglesias J., Morales G., *Green Chemistry*, 11 (2009) 1285–1308.

[19] Mokrani T., Scurrell M., *Catal. Rev.: Science and Engineering*, 51 (2009) 1–145.

[20] Froment G.F., Dehertog W.J.H., Marchi A.J., *Catalysis*, 9 (1992) 1.

[21] Chen D., Rebo H.P., Gronvold A., Moljord K., Holmen A., *Microporous and Mesoporous Materials*, 35–36 (2000) 121–135.

[22] Chen D., Moljord K., Fuglerud T., Holmen A., *Microporous and Mesoporous Materials,* 29 (1999) 191–203.

[23] *BP Statistical Review of World Energy* (2009).

[24] Holmen A., *Catal. Today*, 142 (2009) 2–8.
[25] Fox J.M., *The Fixed-Bed Methanol-to-Gasoline Process Proposed for New Zealand*, paper presented at the Coal Gasification Conference, Australian Inst. Petro., Adelaide, Australia, March 1982.
[26] Chang C.D., Silvestri A.J., *Chemtech*, 10 (1987) 624–631.
[27] Chen J.Q., Vora B.V., Pujadó P.R., Gronvold A., Fuglerud T., Kvisle S., *Stud. Surf. Sci. Catal.*, 147 (2004) 1–6.
[28] Vora B., Chen J.Q., Bozzano A., Glover B., Barger P., *Catal. Today*, 141 (2009) 77–83.
[29] Beeckmann J.W., Froment G.F., Pismen L., *Chemie Ingenieur Technik*, 50 (1978) 960–961.
[30] Froment G.F., *Appl. Catal. A: General*, 212 (2001) 117–128.
[31] Bos A.N.R., Tromp P.J.J., Akse H.N., *Ind. Eng. Chem. Res.*, 34 (1995) 3808–3816.
[32] Aguayo A.T., Gayubo A.G., Ortega J.M., Olazar M., Bilbao J., *Catal. Today*, 37 (1997) 239–248.
[33] Aguayo A.T., del Campo A.E.S., Gayubo A.G., Tarrio A., Bilbao J., *Journal of Chemical Technology and Biotechnology*, 74 (1999) 315–321.
[34] Qi G.Z., Xie Z.K., Yang W.M., Zhong S.Q., Liu H.X., Zhang C.F., Chen Q.L., *Fuel Processing Technology*, 88 (2007) 437–441.
[35] Grønvold A., Moljord K., Dypvik T., Holmen A., *Stud. Surf. Sci. Catal.*, 81 (1994) 399–404.
[36] Chen D., Gronvold A., Rebo H.P., Moljord K., Holmen A., *Appl. Catal. A: General*, 137 (1996) L1–L8.
[37] Chen D., Rebo H.P., Moljord K., Holmen A., *Chemical Engineering Science*, 51 (1996) 2687–2692.
[38] Rebo H.P., Chen D., Brownrigg M.S.A., Moljord K., Holmen A., *Collection of Czechoslovak Chemical Communications*, 62 (1997) 1832–1842.
[39] Li J., Xiong G., Feng Z., Liu Z., Xin Q., Li C., *Microporous and Mesoporous Materials*, 39 (2000) 275–280.
[40] Wragg D.S., Johnsen R.E., Balasundaram M., Norby P., Fjellvåg H., Grønvold A., Fuglerud T., Hafizovic J., Vistad Ø.B., Akporiaye D., *J. Catal.*, 268 (2009) 290–296.
[41] Park J.W., Kim S.J., Seo M., Kim S.Y., Sugi Y., Seo G., *Appl. Catal. A: General*, 349 (2008) 76–85.
[42] Hunger M., *Microporous and Mesoporous Materials*, 82 (2005) 241–255.
[43] Campelo J.M., Lafont F., Marinas J.M., Ojeda M., *Appl. Catal. A: Gen.*, 192 (2000) 85–96.
[44] Dahl I.M., Kolboe S., *J. Catal.*, 149 (1994) 458–464.
[45] Dahl I.M., Wendelbo R., Andersen A., Akporiaye D., Mostad H., Fuglerud T., *Microporous and Mesoporous Materials*, 29 (1999) 159–171.
[46] Gayubo A.G., Aguayo A.T., del Campo A.E.S., Benito P.L., Bilbao J., *Stud. Surf. Sci. Catal.*, 126 (1999) 129–136.
[47] Gayubo A.G., Aguayo A.T., del Campo A.E.S., Tarrio A.M., Bilbao J., *Ind. Eng. Chem. Res.*, 39 (2000) 292–300.

[48] Gronvold A., Moljord K., Dypvik T., Holmen A., *Stud. Surf. Sci. Catal.*, 81 (1994) 399–404.
[49] Iglesia E., Wang T., Yu S.Y., in *Natural Gas Conversion V*, Parmaliana A., Sanfilippo D., Frusteri F., Vaccari A., Arena F. (Eds.), Elsevier, Amsterdam. (1998) 527–532.
[50] Izadbakhsh A., Farhadi F., Khorasheh F., Sahebdelfar S., Asadi M., Feng Y.Z., *Appl. Catal. A: General*, 364 (2009) 48–56.
[51] Lee Y.J., Baek S.C., Jun K.W., *Appl. Catal. A: General*, 329 (2007) 130–136.
[52] Marchi A.J., Froment G.F., *Appl. Catal.*, 71 (1991) 139–152.
[53] Chen D., Gronvold A., Moljord K., Holmen A., *Ind. Eng. Chem. Res.*, 46 (2007) 4116–4123.
[54] Chen D., Rebo H.P., Holmen A., *Chem. Eng. Sci.*, 54 (1999) 3465–3473.
[55] Chen D., Rebo H.P., Moljord K., Holmen A., *Stud. Surf. Sci. Catal.*, 111 (1997) 159–166.
[56] Chen D., Rebo H.P., Moljord K., Holmen A., *Ind. Eng. Chem. Res.*, 36 (1997) 3473–3479.
[57] Chen D., Rebo H.P., Moljord K., Holmen A., *Ind. Eng. Chem. Res.*, 38 (1999) 4241–4249.
[58] Qi G., Xie Z., Yang W., Zhong S., Liu H., Zhang C., Chen Q., *Fuel Process Technol.*, 88 (2007) 437–441.
[59] Benito P.L., Gayubo A.G., Aguayo A.T., Olazar M., Bilbao J., *Ind. Eng. Chem. Res.*, 35 (1996) 3991–3998.
[60] Guisnet M., Magnoux P., *Catal. Today*, 36 (1997) 477–483.
[61] Möller K.P., Böhringer W., Schnitzler A.E., van Steen E., O'Connor C.T., *Microporous and Mesoporous Materials*, 29 (1999) 127–144.
[62] Shiralkar V.P., Joshi P.N., Eapen M.J., Rao B.S., *Zeolites*, 11 (1991) 511–516.
[63] Sugimoto M., Katsuno H., Takatsu K., Kawata N., *Zeolites*, 7 (1987) 503–507.
[64] Suzuki K., Kiyozumi Y., Matsuzaki K., Shin S., *Appl. Catal.*, 42 (1988) 35–45.
[65] Völter J., Lietz G., Kürschner U., Löffler E., Caro J., *Catal. Today*, 3 (1988) 407–414.
[66] Chen D., Moljord K., Holmen A., *Stud. Surf. Sci. Catal.*, 130 (2000) 2651–2656.
[67] Nishiyama N., Kawaguchi M., Hirota Y., van Vu D., Egashira Y., Ueyama K., *Appl. Catal. A: General*, 362 (2009) 193–199.
[68] Castro M., Warrender S.J., Wright P.A., Apperley D.C., Belmabkhout Y., Pirngruber G., Min H.K., Park M.B., Hong S.B., *J. Phys. Chem. C*, 113 (2009) 15731–15741.
[69] Lee K.Y., Chae H.J., Jeong S.Y., Seo G., *Appl. Catal. A: General*, 369 (2009) 60–66.
[70] Abraha M.G., Wu X., Anthony R.G., *Stud. Surf. Sci. Catal.*, 133 (2001) 211–218.

[71] Wilson S., Barger, *Microporous and Mesoporous Materials*, 29 (1999) 117–126.
[72] Chen D., Rebo H.P., Moljord K., Holmen A., *Stud. Surf. Sci. Catal.*, 119 (1998) 521–526.
[73] Mores D., Stavitski E., Kox M.H.F., Kornatowski J., Olsbye U., Weckhuysen B.M., *Chemistry: A European Journal*, 14 (2008) 11320–11327.
[74] Guisnet M., *Journal of Molecular Catalysis A: Chemical*, 182 (2002) 367–382.
[75] Chen D., *Adsorption, Diffusion and Reactions in Methanol to Olefins on SAPO-34*, PhD thesis, Norwegian University of Science and Technology, Trondheim, Norway (1998).
[76] Magnoux P., Guisnet M., *Zeolites*, 9 (1989) 329–335.
[77] Magnoux P., Cartraud P., Mignard S., Guisnet M., *J. Catal.*, 106 (1987) 235–241.
[78] Dahl I.M., Kolboe S., *J. Catal.*, 161 (1996) 304–309.
[79] Bjorgen M., Olsbye U., Kolboe S., *J. Catal.*, 215 (2003) 30–44.
[80] Hereijgers B.P.C., Bleken F., Nilsen M.H., Svelle S., Lillerud K.P., Bjorgen M., Weckhuysen B.M., Olsbye U., *J. Catal.*, 264 (2009) 77–87.
[81] Kvisle S., Fuglerud T., Kolboe S., Olsbye U., Lillerud K.-P., Vora B.V., in *Handbook of Heterogeneous Catalysis*, Second Ed., Ertl G., Knözinger H., Schüth F., Weitkamp J. (Eds.), Wiley-VCH, Weinheim (2008) 2950–2965.
[82] Arstad B., Kolboe S., Swang O., *J. Phys. Chem. B*, 106 (2002) 12722–12726.
[83] Arstad B., Kolboe S., Swang O., *J. Phys. Org. Chem.*, 17 (2004) 1023–1032.
[84] Arstad B., Kolboe S., Swang O., *J. Phys. Chem. B*, 108 (2004) 2300–2308.
[85] Arstad B., Kolboe S., Swang O., *J. Phys. Chem. A*, 109 (2005) 8914–8922.
[86] Arstad B., Kolboe S., Swang O., *J. Phys. Org. Chem.*, 19 (2006) 81–92.
[87] Arstad B., Nicholas J.B., Haw J.F., *J. Amer. Chem. Soc.*, 126 (2004) 2991–3001.
[88] Svelle S., Arstad B., Kolboe S., Swang O., *J. Phys. Chem. B*, 107 (2003) 9281–9289.
[89] Haw J.F., *Topics in Catalysis*, 8 (1999) 81–86.
[90] Haw J.F., Song W.G., *Abstracts of Papers of the American Chemical Society*, 221 (2001) 37-CATL.
[91] Song W.G., Fu H., Haw J.F., *J. Phys. Chem B*, 105 (2001) 12839–12843.
[92] Song W.G., Fu H., Haw J.F., *J. Amer. Chem. Soc.*, 123 (2001) 4749–4754.
[93] Fu H., Song W.G., Marcus D.M., Haw J.F., *J. Phys. Chem. B*, 106 (2002) 5648–5652.
[94] Marcus D.M., Song W.G., Ng L.L., Haw J.F., *Langmuir*, 18 (2002) 8386–8391.
[95] Song W.G., Haw J.F., *Angewandte Chemie-Int. Edit.*, 42 (2003) 892–902.
[96] McCann D.M., Lesthaeghe D., Kletnieks P.W., Guenther D.R., Hayman M.J., Speybroeck V., Waroquier M., Haw J.F., *Angewandte Chemie-Int. Edit.*, 47 (2008) 5179–5182.

[97] Bleken F., Bjorgen M., Palumbo L., Bordiga S., Svelle S., Lillerud K.P., Olsbye U., *Topics in Catalysis*, 52 (2009) 218–228.
[98] Hutchings G.J., Hunter R., *Catal. Today*, 6 (1990) 279–306.
[99] Bjørgen M., Svelle S., Joensen F., Nerlov J., Kolboe S., Bonino F., Palumbo L., Bordiga S., Olsbye U., *J. Catal.*, 249 (2007) 195–207.
[100] Aguayo A.T., Gayubo A.G., Atutxa A., Olazar M., Bilbao J., *Journal of Chemical Technology and Biotechnology*, 74 (1999) 1082–1088.
[101] Zhu Z., Chen Q., Xie Z., Yang W., Li C., *Microporous and Mesoporous Materials*, 88 (2006) 16–21.
[102] Park J.W., Lee J.Y., Kim K.S., Hong S.B., Seo G., *App. Catal. A: General*, 339 (2008) 36–44.
[103] Kim S.J., Park J.W., Lee K.Y., Seo G., Song M.K., Jeong S.Y., *Journal of Nanoscience and Nanotechnology*, 10 (2010) 147–157.
[104] Djieugoue M.-A., Prakash A.M., Kevan L., *J. Phys. Chem. B*, 104 (2000) 6452–6461.
[105] Dejaifve P., Auroux A., Gravelle P.C., Védrine J.C., Gabelica Z., Derouane E.G., *J. Catal.*, 70 (1981) 123–136.
[106] Chen J.S., Thomas J.M., *Catal. Lett.*, 11 (1991) 199–207.
[107] Baek S.C., Lee Y.J., Jun K.W., Hong S.B., *Energy & Fuels*, 23 (2009) 593–598.
[108] Aguayo A.T., Gayubo A.G., Vivanco R., Olazar M., Bilbao J., *Appl. Catal. A: General*, 283 (2005) 197–207.
[109] Djieugoue M.A., Prakash A.M., Kevan L., *J. Phys. Chem. B*, 104 (2000) 6452–6461.
[110] Guisnet M., Magnoux P., *Appl. Catal.*, 54 (1989) 1–27.
[111] Guisnet M., Magnoux P., *Stud. Surf. Sci. Catal.*, 88 (1994) 53–68.
[112] Guisnet M., Magnoux P., Martin D., *Stud. Surf. Sci. Catal.*, 111 (1997) 1–19.
[113] Chen D., Rebo H.P., Grønvold A., Moljord K., Holmen A., *Micropor. Mesopor. Mat.*, 35–36 (2000) 121–135.
[114] Zhang D.H., Wei Y.X., Xu L., Du A.P., Chang F.X., Su B.L., Liu Z.M., *Catal. Lett.*, 109 (2006) 97–101.
[115] Mees F.D.P., van der Voort P., Cool P., Martens L.R.M., Janssen M.J.G., Verberckmoes A.A., Kennedy G.J., Hall R.B., Wang K., Vansant E.F., *J. Phys. Chem.*, 107 (2003) 3161–3167.
[116] Benito P.L., Gayubo A.G., Aguayo A.T., Olazar M., Bilbao J., *Journal of Chemical Technology and Biotechnology*, 66 (1996) 183–191.
[117] Yuen L.-T., Zones S.I., Harris T.V., Gallegos E.J., Auroux A., *Microporous Materials*, 2 (1994) 105–117.
[118] Dahl I.M., Mostad H., Akporiaye D., Wendelbo R., *Microporous and Mesoporous Materials*, 29 (1999) 185–190.
[119] Zhu Q., Kondo J.N., Ohnuma R., Kubota Y., Yamaguchi M., Tatsumi T., *Microporous and Mesoporous Materials*, 112 (2008) 153–161.
[120] Inui T., *Studies in Surface Science and Catalysis*, 67 (1991) 233–242.
[121] Inui T., Kang M., *Appl. Catal. A: General*, 164 (1997) 211–223.

[122] Inui T., *Studies in Surface Science and Catalysis*, 44 (1989) 189–201.
[123] Inui T., Ishihara Y., McKamachi K., Matsuda H., *Studies in Surf. Sci. Catal.*, 49 (1989) 1183–1192.
[124] Inui T., Matsuda H., Yamase O., Nagata H., Fukuda K., Ukawa T., Miyamoto A., *J. Catal.*, 98 (1986) 491–501.
[125] Thomas J.M., Xu Y., Catlow C.R.A., Couves J.W., *Chemistry of Materials*, 3 (1991) 667–672.
[126] Niekerk M.J., Fletcher J.C.Q., O'Connor C.T., *Appl. Catal. A: Gen.*, 138 (1996) 135–145.
[127] Inui T., *Applied Surface Science*, 121 (1997) 26–33.
[128] Inui T., *Progress in Zeolite and Microporous Materials*, Pts A–C, 105 (1997) 1441–1468.
[129] Inui T., Kang M., *Appl. Catal. A: General*, 164 (1997) 211–223.
[130] Rajic N., Gabrovsek R., Ristic A., Kaucic V., *Thermochimica Acta*, 306 (1997) 31–36.
[131] Djieugoue M.A., Prakash A.M., Kevan L., *J. Phys. Chem. B*, 102 (1998) 4386–4391.
[132] Kang M., Inui T., *Catal. Lett.*, 53 (1998) 171–176.
[133] Inoue M., Dhupatemiya P., Phatanasri S., Inui T., *Microporous and Mesoporous Materials*, 28 (1999) 19–24.
[134] Kang M., Lee C.T., Um M.H., *J. Ind. Eng. Chem.*, 5 (1999) 10–15.
[135] Kang M., *Journal of Molecular Catalysis A: Chemical*, 160 (2000) 437–444.
[136] Choo H., Hong S.B., Kevan L., *J. Phys. Chem. B*, 105 (2001) 1995–2002.
[137] Stojakovic D., Rajic N., *Journal of Porous Materials*, 8 (2001) 239–242.
[138] Dubois D.R., Obrzut D.L., Liu J., Thundimadathil J., Adekkanattu P.M., Guin J.A., Punnoose A., Seehra M.S., *Fuel Processing Technology*, 83 (2003) 203–218.
[139] Dutta P., Manivannan A., Seehra M.S., Adekkanattu P.M., Guin J.A., *Catal. Lett.*, 94 (2004) 181–185.
[140] Xu L., Liu Z.M., Du A.P., Wei Y.X., Sun Z.G., *Natural Gas Conversion VII*, 147 (2004) 445–450.
[141] Zhang X., Wang R.J., Yang X.X., Zhang F.B., *Microporous and Mesoporous Materials.*, 116 (2008) 210–215.
[142] Marchi A.J., Froment G.F., *Appl. Catal. A: General.*, 94 (1993) 91–106.
[143] Wu X.C., Anthony R.G., *Appl. Catal. A: General.*, 218 (2001) 241–250.
[144] Barger P., *Zeolites for Cleaner Technologies*, Guisnet M., Gilson J.P. (Eds.), Imperial College Press, London (2002) 239–260.

Chapter 16

AROMATIZATION OF C_6, C_7 PARAFFINS OVER PT/LTL CATALYSTS

M.F. Ribeiro and M. Guisnet

16.1. Introduction

The reforming process, developed by UOP in 1949 [1] is used to increase the octane number of naphtha by transforming the C_6–C_{10} components into aromatics (plus hydrogen) and branched alkanes over bifunctional catalysts based on Pt/chlorinated Al_2O_3. While these catalysts are efficient in transforming C_8–C_{10} alkanes into aromatics, their selectivity in hexane aromatization to benzene proceeds with a quite poor selectivity [2]. In 1980, a new catalyst, Pt supported on the potassium form of the LTL zeolite (Pt/LTL), was reported to aromatize n-hexane with high activity and selectivity [3]. On this catalyst which does not present any acidity, it is generally accepted that only Pt sites participate in the catalysis of alkane aromatization (monofunctional mechanism).

The great advantage of the corresponding commercial processes — Aromax (Chevron) [2] and RZ Platforming (UOP) [4] — was that they were operated in conditions similar to those of conventional reforming. Therefore, no significant changes in reformer designs and hardware were needed to allow their use. However, owing to the extreme sensitivity of the catalyst to sulphur poisoning, only sulphur-free feeds can be processed [5]. Moreover, the current regulations that limit the content in aromatics, especially in benzene, have significantly reduced the appeal of these processes for gasoline production. However, new challenges and opportunities can be created by the increasing market trend of BTX (benzene, toluene and xylenes) for petrochemicals [6, 7]. The first RZ Platforming unit was brought on stream in 1998 to produce benzene and toluene as feedstock for an aromatic complex [8]. More recently, Chevron Phillips Chemical Company entered into a contract to license its Aromax process and catalyst technologies for the production of benzene, toluene and hydrogen to CEPSA's refinery in Huelva [9]. Moreover, this light naphtha reforming technology is particularly suitable to refine Fischer–Tropsch products which do not contain sulphur impurities [10].

The RZ Platforming process employes adiabatic radial flow reactors that are arranged in a conventional side-by-side pattern. A furnace is used between each reactor to reheat the charge to reactor temperature. Desulphurized feed (naphtha or C6–C8) is combined with recycled hydrogen and sent to the reactor section. Typical cycle lengths are 8–12 months and the units are designed for either efficient *in situ* or *ex situ* catalyst regeneration. The RZ-100 catalyst system has met all expectations for activity, selectivity and stability for over 8 years [4].

16.2. Aromatization Mechanisms and Active Sites

On the bifunctional Pt/chlorinated Al_2O_3 reforming catalyst, the C_6^+ alkane aromatization (or dehydrocyclization) occurs through the following successive steps: dehydrogenation to alkenes, ring closure to alkyl cyclopentenes then ring expansion to cyclohexenes over the acid sites, and finally dehydrogenation to aromatics. Under the operating conditions, only traces of the olefinic intermediates appear owing to thermodynamic limitations. In the reforming process, many other reactions compete with alkane dehydrocyclization, some of them desired (naphthene dehydrogenation, alkane isomerization), the others undesirable (aromatic disproportionation and alkylation, hydrogenolysis and hydrocracking, which limits the yield in aromatics, especially in the C_6–C_7 range) [11].

Over monofunctional metal catalysts (Pt on non-acidic supports such as LTL), aromatization of alkanes with 6 or more C atoms in a straight chain involves predominantly (~90%) direct 1,6 ring closure followed by dehydrogenation with a minor path (~10%) involving 1,5 ring closure followed by ring enlargement then dehydrogenation. All these steps, including ring enlargement, were shown to be catalyzed by Pt [12]. With this monofunctional catalyst, alkane hydrogenolysis, a well-known structure-sensitive or demanding reaction which requires a site ensemble of two or more Pt atoms for its catalysis, is the main side reaction. It will be shown later that on Pt/LTL, hydrogenolysis occurs essentially in terminal positions. The various reactions which are involved in *n*-hexane conversion on a Pt/LTL catalyst are shown in Fig. 16.1.

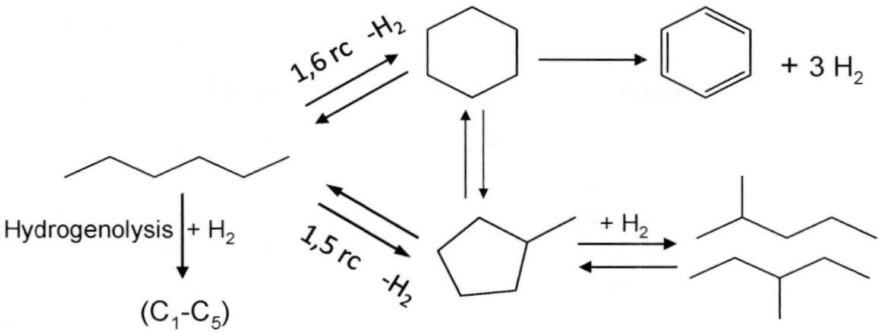

Fig. 16.1. Aromatization of *n*-hexane over Pt/LTL catalysts. Main reaction steps.

To avoid the secondary bifunctional reactions which are detrimental to the aromatization selectivity, the presence of protonic acid sites must be banished. This negative effect of protonic sites was demonstrated in the original work of Bernard [3]. Indeed, the protonic exchange of K cations caused a decrease in both the activity and the aromatization selectivity of a Pt/LTL zeolite. Other observations confirm that minimization of the support acidity is a critical parameter. Thus, protons created through Ba exchange of K cations (owing to Ba^{2+} hydrolysis) or during Pt incorporation over LTL zeolites (during transformation into Pt of the $Pt(NH_3)_4^{2+}$ cations) were detrimental to aromatization selectivity. In the first case, calcination treatment is sufficient to remove the acid sites [13]. In the second, the same occurs when incorporating Pt via incipient wetness impregnation of $Pt(NH_3)_4Cl_2$ owing to the exchange with residual potassium of the protons created during the Pt reduction [3, 14].

Not only would the total elimination of the protonic acidity be indispensable but the presence of basic sites on the Pt support could positively affect the aromatization selectivity. This was demonstrated by using a series of Pt/LTL samples with Li, Na, K, Cs in exchange position: an increase in the aromatization selectivity with the zeolite basicity was observed [15]. The same was found with a series of Pt/alkali/BEA catalysts [16, 17]. In another study, an interaction between Pt and basic sites was shown with a series of alkaline-earth exchanged Pt/LTL samples [18]. Furthermore, Derouane and Davis [19] discovered that Pt clusters supported on a non-microporous basic support (MgO) exhibited high aromatization selectivity. However, no electronic metal support interaction can be invoked to explain the high initial aromatization activity and selectivity of a well-dispersed Pt/SiO_2 catalyst [20]. This suggests that it is rather Pt cluster size than the support basicity which determines the aromatization activity and selectivity (the case of catalyst stability is discussed in Section 16.4).

16.3. What Makes the LTL Zeolite So Well-Adapted to Aromatization?

Since the original work of Bernard [3], no better performance in n-hexane aromatization was found on other Pt catalysts than those based on LTL zeolites. It is why many works have been devoted to the explanation of this remarkable adaptation of the micropore system of this zeolite to the catalysis of aromatization.

Zeolite LTL (Fig. 16.2) possesses a hexagonal framework with parallel 12-MR ring and 8-MR ring channels. The 8-ring projections along the

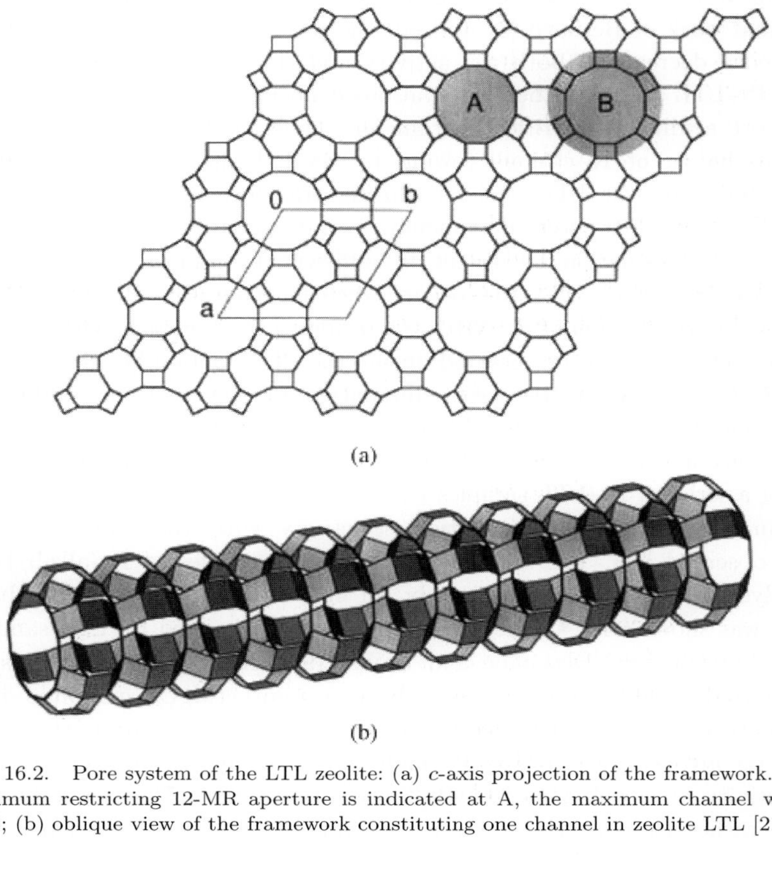

Fig. 16.2. Pore system of the LTL zeolite: (a) c-axis projection of the framework. The minimum restricting 12-MR aperture is indicated at A, the maximum channel width at B; (b) oblique view of the framework constituting one channel in zeolite LTL [21].

c-axis are deformed into elliptical windows and can contain K^+ cations. Consequently the 8-MR ring channels do not contribute significantly to the microporosity and the LTL zeolite can be considered as a monodimensional large-pore molecular sieve. The 12-MR channels, spaced about 1.84 nm, are undulated, measuring ~1.3 nm at their widest point and only 0.71 nm at the minimum restricting aperture (Fig. 16.2). Between each 12-ring, there is then a cage of ~1.3 nm width. Typical zeolite LTL syntheses have a Si/Al framework ratio of ~3. K cations are located in the main 12-MR channels and in the 8-ring side-pockets of the channel walls [21].

The main problem with the use in catalysis of monodimensional zeolites comes from the easy limitation or blockage of the diffusion of reactant and product molecules along their non-interconnected channels (Chapter 11). With the fresh Pt/LTL catalysts, these limitations or blockages can result from crystal defects, detrital material and/or too large clusters of platinum, the consequence being lower catalyst efficiency [21]. All these problems can be solved by optimizing the steps of the catalyst preparation. Particular attention has to be turned to the introduction of Pt under a highly dispersed form within the channels and without any Pt deposit on the outer surface. Additional causes appear during reaction owing to coke formation and Pt sintering.

The main hypotheses which were advanced to explain the remarkable selectivity in alkane aromatization of the Pt/LTL zeolite were discussed in detail in two review papers [22, 23]: (i) a linear orientation of the n-hexane molecules in the channels which would favour their adsorption on Pt at their terminal C atom; this was advanced to account for the relation found between preferential terminal hydrogenolysis and aromatization selectivity [24]; (ii) the structural recognition and preorganization of a cyclic transition state in the favourable environment of the LTL channels [25], i.e. the LTL zeolite would behave like an enzyme; (iii) the electronic metal support interaction already evoked above; (iv) the participation of cations (e.g. K^+) in aromatization by an electrostatic field effect [26]; (v) the positive effect of tiny cluster size. The three first hypotheses were rejected on the basis of the comparison of many Pt catalysts, while the last ones require confirmation. Another proposal was very recently advanced [27], based on the results obtained in the aromatization of an equimolar mixture of C_6H_{14} and C_6D_{14} over a Pt/LTL catalyst. Hardly any kinetic isotope effect was found, indicating that dehydrocyclization was controlled by the entry of hexane molecules into the wider lobes of the channels in which are located the Pt clusters.

16.4. Deactivation of Pt-Zeolite Reforming Catalysts

As indicated in the introduction, poisoning by sulphur is the main cause of the deactivation of the Pt/LTL catalysts. However, even in the absence of sulphur a progressive decrease in activity can also be observed, this deactivation being due to coke formation.

16.4.1. *Deactivation by coking*

Deactivation by coking was investigated during the conversion of sulphur-free alkanes, most often with the aim to compare Pt/LTL catalysts to non-microporous Pt catalysts such as Pt/SiO_2. The main conclusion was that coking and deactivation of the Pt sites were inhibited by geometric constraints imposed by the one-dimensional LTL channels. This conclusion was first advanced by Iglesia and Baumgartner [28]. These authors studied the carbon fouling of Pt/LTL during heptane reforming at 683 K over Pt/LTL and Pt/SiO_2 catalysts. Initially, Pt/SiO_2 was 2.3 and 1.7 times more active in aromatization and hydrogenolysis respectively than Pt/LTL. However, deactivation was much faster over Pt/SiO_2 (aromatization/11) than over Pt/LTL (1.1) because of a more significant coke deposition and the used Pt/LTL sample was more active (5.5 times) and much more selective in aromatization than the used Pt/SiO_2. More recently, other authors [29] observed that the amount of carbon deposits was greater in the presence of sulphur and after exposure to water vapour than in the reaction with clean feeds.

16.4.2. *Deactivation by sulphur*

Two groups of investigators [30, 31] have claimed that the very high sensitivity of Pt/LTL to sulphur poisoning is due to the agglomeration of Pt particles and the subsequent blockage of LTL zeolite channels. Vaarkamp *et al.* [30] examined the effect of poisoning with H_2S on a working Pt/Ba/LTL aromatization catalyst. The fresh sample which was highly active and selective had a Pt-Pt coordination number (determined by EXAFS) of 3.7 and a H/Pt ratio of 1.4, which were consistent with Pt clusters of 5–6 atoms. The poisoned sample was less active (3 times) and selective in aromatization and the Pt-Pt coordination number and the H/Pt ratio were respectively equal to 5.5 and 1.0, indicating Pt clusters of $\sim$13 atoms. Although most of the exposed Pt atoms could chemisorb hydrogen, their aromatization activity was greatly decreased. Therefore the authors ascribed the high sensitivity to sulphur of Pt/LTL catalysts to a combination of sulphur adsorption on the Pt surface atoms and cluster growth leading to channel blockage. A growth of the Pt clusters during the poisoning of a Pt/LTL catalyst was also observed by McVicker *et al.* [31], the high-resolution transmission electron microscopy of the samples showing moreover that Pt agglomerates were located near the micropore mouth.

The observation that sulphur affected *n*-hexane aromatization more than cyclohexane dehydrogenation led Fukunaga and Ponec [26] to reject this agglomeration mechanism, proposing as an alternative that the interaction of sulphur with K^+ reduced the stabilizing effect of this cation.

16.5. How to Prevent the Catalyst Deactivation?

Since the first works that evidenced the extremely high sensitivity to even traces of sulphur (e.g. parts per billion) [13], there have been several attempts to increase the sulphur tolerance of Pt/LTL catalysts. First, it must be highlighted that deactivation is influenced by initial Pt dispersion. So a high initial Pt dispersion is beneficial for both initial activity and stability, which means that by managing the method of preparation and pretreatment of Pt zeolites, it will be possible to prevent deactivation. The morphology of Pt clusters also greatly vary with loading and method of preparation, these variations having strong effects on the stability of the catalyst used under both sulphur-free and sulphur-containing conditions [32]. There are in the literature many works reporting the characteristics of Pt particles produced from the different procedures and the consequences they have on activity, selectivity and stability of catalysts under reaction conditions.

Another strategy that has been attempted in order to develop sulphur-tolerant Pt/LTL aromatization catalysts was to introduce promoters which might stabilize crystal growth, limiting the micropore blockage by Pt agglomerates. One approach is to use first-row transition metal (e.g. Ni) cations that act as anchoring sites for Pt particles [33]. The promoters can also act as sulphur traps, thus protecting Pt, even if temporarily, from sulphur poisoning. Alternatively, they might modify the chemical properties of Pt, making it less susceptible to poisoning. Thus, the positive effect of rare-earth elements added to Pt/LTL catalyst on aromatization stability and selectivity would result from an electronic donation to Pt particles [34]. Furthermore, Jacobs *et al.* [35] from an investigation of the effect of thulium (Tm) addition concluded that Tm has two positive effects, the first on Pt dispersion, the second on the stability. This latter effect was related to the action of Tm as a sulphur getter, delaying Pt poisoning. However, the amount and method of incorporation of Tm was critical to the catalyst performance.

As it is shown in Chapter 11, the common ways to limit the blockage of the channels of monodimensional zeolite catalysts, hence to improve their stability, are either: (i) To use samples with very small (nano-sized) crystal size, hence with shorter paths of diffusion of reactant and poison molecules. This limits their contact time with the active sites within each of the channels, hence the extent of secondary transformations leading to micropore blockage. (ii) To transform the monodimensional micropore system into a quasi-tridimensional one by creating internal mesopores generally through post-synthesis treatments (namely dealumination, desilication). Although the synthesis of LTL samples with short channel sizes and with different crystal morphologies — spherical, cylindrical, disc and clam-like – was carried out for more than 20 years by industrial investigators, only a few papers report the effect of LTL crystal size on aromatization [36, 37]. Furthermore, up to now, no work testing the possible effect of internal mesopores on the LTL aromatization has been published. In [36], LTL samples with different crystal sizes and shapes were synthesized and characterized by various techniques, in particular by scanning electron microscopy. The LTL samples with cylindrical shape were effective catalyst supports for aromatization. The crystal size was shown to affect the catalytic properties but in what concerns the stability, it was not possible to discriminate between a direct effect of the channel size and an indirect one linked to differences in Pt dispersion and the distribution (inside and outside of the channels) of the Pt clusters. However, the samples with shorter channel length exhibited improved activity, selectivity and catalyst life.

References

[1] Antos G.J., Aitani A.M. (Eds.), *Catalytic Naphtha Reforming*, Marcel Dekker, N.Y. (1995).

[2] Tamm P.W., Mohr D.H., Wilson C.R., *Stud. Surf. Sci. Catal.*, 38 (1988) 335–353.

[3] Bernard J.R., *Proc. Fifth Intern. Zeolites Conference*, Rees L.V.C. (Ed.), Heyden, London (1980) 686–695; Bernard J.R., U.S. Patent 4, 104, 320 (1978).

[4] Jensen R.H., in *Zeolites for Cleaner Technologies*, Guisnet M., Gilson J.P. (Eds.), Imperial College Press, London (2002) Chapter 4, 98.
[5] Besoukhanova C., Guidot J., Barthomeuf D., Breysse M., Bernard J.R., *J. Chem. Soc.*, Faraday Trans. I, 77 (1981) 1595–1604.
[6] Netz D., NPRA Annual Meeting, March 2007.
[7] http://www.chemsystems.com/reports/search/docs/abstracts/0607_6_abs.pdf.
[8] http://www.uop.com/objects/RZ%20Platform.pdf.
[9] http://www.cpchem.com/enu/press_releases_6993.asp.
[10] Dry M.E., *Appl. Catal. A: General*, 276 (2004) 1–3.
[11] Lepage J.F., *Catalyse de Contact*, Editions Technip, Paris (1978) Example 6, 575–622.
[12] Gates B., Katzer J.R., Schuit G.C.A., *Chemistry of Catalytic Processes*, McGraw-Hill Book Company, New York (1979) Chapter 3, 184–324.
[13] Hughes T.R., Buss W.C., Tamm P.W., Jacobson R.L., *Stud. Surf. Sci. Catal.*, 28 (1986) 725–732.
[14] Ostgard D.J., Kustov L., Poeppelmeier K.R., Sachtler W.M.H., *J. Catal.*, 133 (1992) 342– 357.
[15] Han W.J., Kooh A.B., Hicks R.F., *Catal Lett.*, 18 (1993) 193–208.
[16] Maldonado-Hodar F.J., Bécue T., Silva J.M., Ribeiro M.F., Massiani P., Kermarec M., *J. Catal.*, 195 (2000) 342–551; Bécue T., Maldonado-Hodar F.J., Antunes A.P., Silva J.M., Ribeiro M.F., Massiani P., Kermarec M., *J. Catal.*, 181 (1999) 244–255.
[17] Silva E.R., Silva J.M., Massiani P., Ramôa Ribeiro F., Ribeiro M.F., *Catal. Today*, 107–108 (2005) 792–799.
[18] Larsen G., Haller G.L, *Catal. Lett.*, 3 (1989) 103–110.
[19] Davis R.J., Derouane E.G., *Nature*, 349 (1991) 313–315; Davis R.J., Derouane E.G., *J. Catal.*, 132 (1991) 269–274.
[20] Mielczarski E., Hong S.B., Davis R.J., Davis M.E., *J. Catal.*, 134 (1992) 359–369.
[21] Treacy M.M.J., *Micropor. Mesop. Mater.*, 28 (1999) 271–292.
[22] Davis R.J., *Heterog. Chem. Rev.*, 1 (1994) 41–53.
[23] Meriaudeau P., Naccache C., *Catal. Rev. Sci. Eng.*, 39 (1997) 5–48.
[24] Tauster S.J., Steger J.J., *J. Catal.*, 125 (1990) 387–389.
[25] Derouane E.G., Vanderveken D.J., *Appl. Catal.*, 45 (1988) L15–L22.
[26] Fukunaga T., Ponec V., *J. Catal.*, 157 (1995) 550–558.
[27] Azzam K.G., Jacobs G., Shafer W.D., Davis B.H., *J. Catal.*, 270 (2010) 242–248.
[28] (a) Iglesia E., Baumgartner J.F., in *New Frontiers in Catalysis*, Guczi L. *et al.* (Eds.), *Proc. 10th Int. Congr. Catal.*, Akad Kiado, Budapest (1993) B, 993; (b) Iglesia E., Baumgartner J.F.,*Proc. 9th Int. Zeolite Conf.*, Montreal (1992) 421.
[29] Jongpatiwut S., Sackamduang P., Rirksomboon T., Osuwan S., Alvarez W.E., Resasco D.E., *Appl. Catal. A: General*, 230 (2002) 177–193.
[30] Vaarkamp M., Miller J.T., Modica F.S., Lane G.S., Koningsberger D.C., *J. Catal.*, 138 (1992) 675–685.

[31] McVicker G.B., Kao J.L., Ziemiak J.J., Gates W.E., Robbins J.L., Treacy M.M.J., Rice S.B, Vanderspurt T.H., Cross V.R., Ghosh A.K., *J. Catal*, 139 (1993) 48–61.
[32] Jacobs G., Ghadiali F., Pisanu A., Borgna A., Alvarez W.E., Resasco D.E., *Appl. Catal. A: General*, 188 (1999) 79–98.
[33] Larsen G, Resasco D.E, Durante V.A., Kim J., Haller G.L., *Stud. Surf. Sc. Catal.*, 83 (1994) 321–329.
[34] Fang X., Li F., Zhou Q., Luo L., *Appl. Catal. A: General*, 161 (1997) 227–234.
[35] Jacobs G., Ghadiali F., Pisanu A., Padro C.L., Borgna A., Alvarez W.E., Resasco D.E., *J. Catal.*, 191 (2000) 116–127.
[36] Trakarnroek S., Jongpatiwut S., Rirksomboon T., Osuwan S., Resasco D.E., *Appl. Catal. A: General*, 313 (2006) 189–199.
[37] Jentoft R.E., Tsapatsis M., Davis M.E., Gates B.C., *J. Catal.*, 179 (1998) 565–580.

Chapter 17

DEACTIVATION OF MOLECULAR SIEVES IN THE SYNTHESIS OF ORGANIC CHEMICALS

M. Guidotti and B. Lázaro

17.1. Introduction

The term *deactivation* indicates any kind of progressive loss of activity of a catalyst with increasing reaction time. Any sort of catalyst, either homogeneous or heterogeneous, is subject to deactivation and the deactivating processes take place during the intermediate time between the status of fresh catalyst (as-synthesized or as-prepared) and the one of spent catalyst (when the catalytic activity is zero). Nevertheless, deactivation phenomena are particularly relevant for heterogeneous catalysts, since the factors leading to the deactivation of solid systems are more abundant than those occurring for homogeneous ones and, for these reasons, catalyst stability (or catalyst robustness) is one of the most sought-after properties in solid materials together with good activity (in terms of reaction rate and substrate conversion) and high selectivity to desired products [1].

In petrochemistry and bulk chemicals production, the importance of deactivation phenomena has been acknowledged for a long time and a relevant section of the literature deals with methods to minimize them by careful process design [2–4] (see previous chapters in Parts IV and V). On the contrary, in the field of high added-value organic chemicals synthesis, even if the problem of catalyst deactivation plays a remarkable role as well, since the use of bulky and functionalized reactants is typically a primary cause of activity loss, the importance of catalyst robustness towards deactivation is frequently underestimated, not only at a fundamental research level, but also at small industrial scale [1, 5, 6]. Actually, in the production of fine and speciality chemicals, where the high added-value of the final compounds may account for the use of valuable and costly reactants, the solid catalyst is often considered as a consumable reagent and it is not necessarily worth being recovered, regenerated and reused. The spent catalyst is thus disposed of and substituted by a fresh one [7].

However, by following this simplicistic point of view, the peculiar advantages and benefits due to the use of a heterogeneously catalysed production process instead of a non-catalytic stoichiometric one are lost. The ever-increasing attention on the economical and environmental sustainability of organic chemical production is thus prompting the scientific community to develop solid catalysts that are less prone to deactivation.

17.2. Peculiar Factors Leading to Deactivation of Molecular Sieves in Organic Synthesis

Before considering a series of selected cases of deactivation of microporous and mesoporous catalysts active in the transformation of organic chemicals, it is necessary to recall some fundamental concepts (widely described in Chapter 1) that are particularly useful for describing and explaining the loss of activity observed in molecular sieve catalysts used in the synthesis of organic commodity and fine chemicals.

First, catalyst deactivation can be either *irreversible* or *reversible*. In the first case, the catalyst undergoes a permanent modification that alters irremediably its catalytic functionalities. In the second case, the activity of the spent catalyst can be partially or totally restored by defined chemical and/or physical treatments, such as calcination, reduction, extraction with solvents, washing, etc. These methods are usually referred to as *regeneration*, if the catalyst activity is totally (or almost totally) recovered, or as *rejuvenation*, if only a part of the activity is restored by simpler and less drastic treatments so that the final disposal of the solid is postponed. It is worth underlining however that practically in all cases only a limited number of reactivation cycles is possible and the catalyst's life always ends with a form of irreversible deactivation.

Five causes of deactivation are generally considered [8], namely: (1) poisoning, (2) fouling, (3) thermal degradation, (4) detrimental interaction with the reaction mixture and (5) mechanical degradation. Such classification is valid for any kind of heterogeneous catalyst but in the specific case of microporous and mesoporous materials applied to organic chemical synthesis, poisoning, fouling and loss of active catalytic sites by leaching or volatilization are typically the major factors leading to deactivation.

Poisoning is caused by the strong, specific chemisorption of inhibiting species onto the catalytically active sites. Such inhibiting compounds can be impurities, feed components (reactants, solvents) and/or reaction products, which possess a remarkable affinity towards the reactive site and hinder the adsorption of less strongly bound compounds. So, especially in fine organic chemicals synthesis, where the reactions are typically carried out in liquid phase and the reaction mixture may contain a wide variety of by-products originated during the previous synthetic steps, a careful purification of the reactant feed is crucial to minimize the occurrence of undesired poisoning phenomena. Nevertheless, in several cases, deactivation can also originate from poisoning of the active sites by desired reaction products (*auto-inhibition*) and this often occurs when the products possess a higher polar character than that of the starting reagents. It is the case, for instance, for the oxidation of benzene to phenol over Fe-ZSM-5, where formed phenol can adsorb tightly onto zeolite sites, thus inhibiting further access to unreacted benzene molecules.

Then, fouling is a general term to describe all the phenomena where the catalyst surface is covered by species originating from the fluid phase and that block the connection between the active sites and the reaction mixture. In the case of fine and speciality chemicals synthesis (in particular at the laboratory scale) the study of the fouling mechanism, case by case, can be cumbersome and time-consuming and the approach to minimize its effect is mostly based on non-specific removal of the deactivating species by calcination or solvent washing, rather than on targeted countermeasures.

Leaching of active species out of the catalyst is a primary reason of gradual loss of activity and it is particularly critical in liquid-phase reactions, in the presence of highly polar species, often bearing complexing and/or solvolytic moieties (such as -OH, -NH_2 or -COOH groups) [9, 10]. Therefore, it is common practice nowadays that all good papers in the literature claiming the use of a novel heterogeneous catalyst report the *hot-filtration heterogeneity test* as a basic requirement. This test was originally established to verify the stability of zeolites and zeotypes in liquid-phase oxidations [10], but it can be (and it is actually) applied to any kind of catalytic reaction carried out in liquid phase over solids.

Generally, with respect to metal-containing molecular sieves, three different scenarios are considered for heterogeneous catalysts in a liquid phase [9]: (i) the metal is stable, it does not leach out and the catalyst is truly heterogeneous; (ii) the metal leaches out from the solid, but it is not active in liquid phase as a homogeneous catalyst; (iii) the metal leaches out from the solid and it works as a highly active homogeneous catalyst. Clearly, the solid catalyst is truly heterogeneous only in the first two cases, whereas in the third it is not.

From a practical point of view, the "heterogeneity test" consists in removing the solid catalyst (by filtration or ultracentrifugation) from the reaction mixture at the reaction temperature during the course of the reaction (typically when the conversion is almost half of the maximum attainable value). Then, the liquid mixture (with no solid) is allowed to react further under identical conditions. If the reaction does not proceed further, there is a strong confirmation of the genuine heterogeneous nature of the catalyst. It is necessary to take care while removing the solid from the reaction mixture, to avoid eventual re-adsorption or change in the nature of the leached species (e.g. polymerization, degradation or oxidation/reduction to inactive secondary products). For example, in the allylic oxidation of α-pinene to verbenone over chromium-containing alumino-phosphate molecular sieves (Cr-APO-5), if the filtration is carried out at room temperature instead of the reaction temperature (80°C), the reduction of active leached Cr(VI) species into inactive Cr(III) species occurs and no further reaction is observed during the "heterogeneity test" [9]. Furthermore, other kinds of test demonstrating that the solids can be recovered and recycled without apparent loss of activity or loss of active species (measured by elemental analysis after some catalytic runs) are not enough to prove that the reaction is heterogeneous in nature [10, 11]. In fact, very low amounts (hence difficult to be detected) of highly active species in solution are sometimes enough to have a fast, homogeneously catalysed process. A remarkable example of it was reported for the Suzuki reaction (palladium-catalysed cross-coupling of aryl halides with boronic acids), where palladium contaminants down to a level of 50 ppb found in commercially available sodium carbonate were responsible for the generation of the biaryl product (Fig. 17.1) [12].

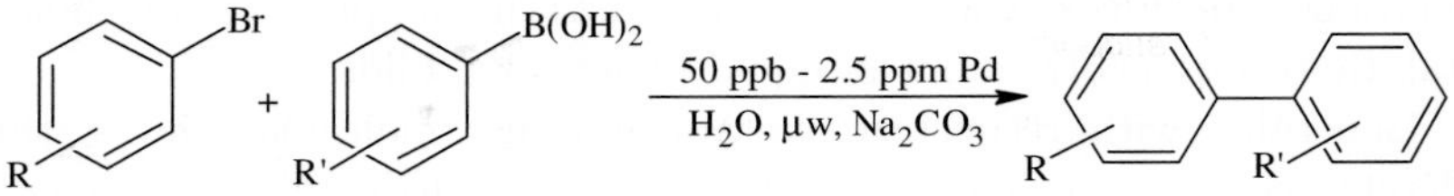

Fig. 17.1. Suzuki couplings catalysed by ultra low amounts of palladium.

Nevertheless, it must be recalled that the occurrence and the extent of metal leaching depends on the kind of reaction and on the operating conditions. For this reason a catalyst heterogeneity test has to be proven for each reaction and/or for each combination of reactants/solvents/conditions under study [13].

Nevertheless, in some very particular cases, the *hot-filtration heterogeneity test* can be non-exhaustive, since it may fail to detect active species which possess a limited lifetime in solution, but which are continuously supplied by the solid catalyst. In this case, a more stringent (but more laborious) way to detect metal leaching is the so-called *three-phase test*, according to which one of the reactants and the catalytically active sites are irreversibly bound to an insoluble solid support (e.g. covalently) (Fig. 17.2) [14].

Under such conditions, only soluble active metal species, released by the active sites, are able to reach the heterogenized reactant and therefore the formation of a heterogenized reaction product is conclusive proof for the existence of metal leaching. The test was originally conceived as a mechanistic probe based on solid-phase chemistry for the detection of reactive intermediates, but it was later applied to several domains, including the evaluation of minimal traces of Pd in the Heck and Suzuki coupling of aryl halides over a Pd-containing mercaptopropyl-modified mesoporous silica SBA-15 [15, 16].

Finally, the interaction with the reaction mixture can lead to deactivation due to detrimental modification of the chemical nature of the site, rather than to leaching. For instance, the aggregation processes

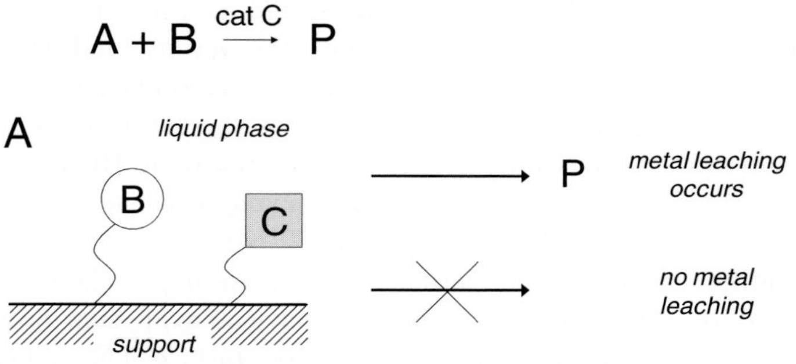

Fig. 17.2. Sketch of a generic three-phase test over a molecular sieve. A and B, reactants; C, solid catalyst; P, product. B and C are immobilized over a solid support and are in contact with A in liquid phase under reaction conditions. If P is observed, metal leaching does occur.

taking place in titanium(IV) isolated centres grafted onto mesoporous silica catalysts, active in the batch liquid-phase epoxidation of alkenes with hydrogen peroxide, give rise to a gradual diminution of catalytic activity after a few catalytic cycles, due to the aggregation of Ti(IV) single sites into TiO_2 nano-sized domains located on the silica support, which in turn are inactive for epoxidation and cause hydrogen peroxide decomposition [17, 18].

A thorough understanding of the processes taking place at the catalyst surface or, in the case of molecular sieves, in the catalyst pores relies on analytical investigation techniques. Spectroscopic, thermoanalytical, microscopic, kinetic and modelling studies can all contribute to a deeper insight into the onset of deactivation mechanisms (see Part II of this book). However, some specific tests or methods have been designed and set up to study the nature and the influence of deactivating processes in organic synthesis on molecular sieves. As an example, the nature of fouling species can be systematically studied by the method of mineralization and analysis, originally developed for the study of heavy by-products of hydrocarbon cracking over acid zeolites (thoroughly described in Chapter 4), that yet can be applied to other organic reactions in gas/vapour or liquid phase [19]. Such an approach has been also applied successfully to the acylation of aromatics with acetic anhydride [20–23], in benzene hydroxylation to phenol with nitrous oxide [24] or in cyclohexanone ammoximation over TS-1 [11]. In addition, the same method can also be used to perform competitive adsorption tests with solutions that contain one or more specific reactant(s), in order to measure the affinity of the porous material towards each species and evaluate their tendency to inhibit the reaction (*auto-inhibition*) [20].

Analogously, the selective poisoning is used as a diagnostic technique to investigate the location, the amount, the nature and the role of catalytically active sites. However, it can be exploited not only as a means of investigation, but also as a tool to improve the catalytic performance of the materials. As an example, Ti- and Zr-containing BEA zeolites are active catalysts for the alkylation of 2-methoxynaphthalene with propylene oxide in liquid phase in the autoclave at 10 bar and 150°C [25]. Under these conditions, a major competing reaction, leading to the deactivation of the catalyst, is the oligomerization of propylene oxide that provides for the deposition of these oligomers on the catalyst surface. The main reaction products are one O-alkylated product (I) and four C-alkylated products, (II) through (V), deriving from 2,6- and 1,2-alkylation at the naphthalenic ring (Fig. 17.3).

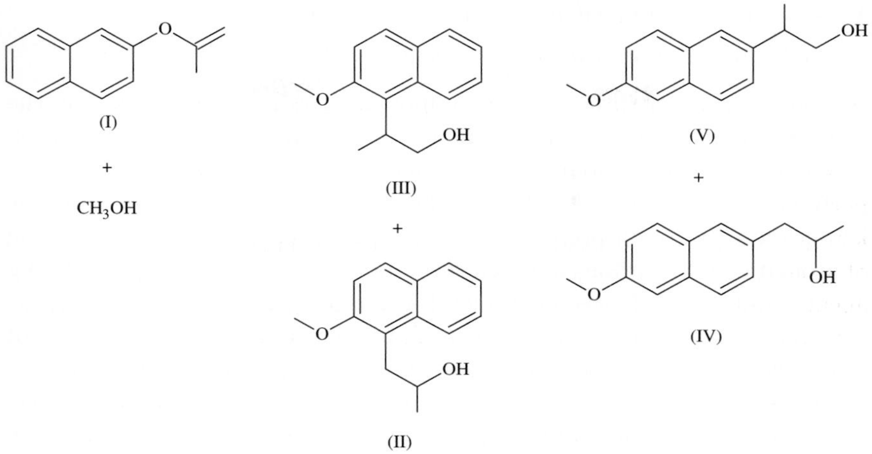

Fig. 17.3. Products in the alkylation of 2-methoxynaphthalene with propylene oxide over Ti- and Zr-containing BEA zeolites.

The conversion of propylene oxide (acting as limiting agent) can be as high as 50% with a selectivity towards the desired product (V) ranging from 12% to 20%. The stereoselectivity, in terms of ratio of 2,6- to 1,2-product is 1.6 and 3.0 for Ti-BEA and Zr-BEA, respectively. Nevertheless, the shape-selective feature of the zeolites can be enhanced by selectively poisoning the acid sites on the outer surface of the catalyst, by treating them with various compounds, among which the most effective was tris[2-(diphenylphosphino)ethyl]-phosphine (TetraPhos-II).

This kind of targeted "passivation" blocks the activity on the external surface where there are no space restrictions for the reactants in the alkylation reaction and therefore no shape selective effects can take place. Thanks to this technique, the 2,6- to 1,2-product ratio can increase from 1.6, over "non-passivated" Ti-BEA, to a value of 4.0, over the "passivated" catalyst.

17.3. Exemplar Cases

Seven widely known organic chemical synthesis reactions have been chosen, among the hundreds of cases reported in open and patent literature, as representative examples of the above-mentioned general concepts. They will be briefly commented on, with particular attention paid to the studies regarding the catalyst deactivation and the methods to minimize its effects.

17.3.1. *Acid-catalysed processes*

17.3.1.1. *Acylation of aromatics on acid molecular sieves*

Aromatic acylation processes are important synthetic pathways for the synthesis of aromatic ketones that are intermediates for the production of a wide range of compounds used as pharmaceuticals, pesticides, flavours, perfumes, UV adsorbers and dyes [26–30]. At industrial scale, aromatic ketones are still often prepared via homogeneous Friedel–Crafts acylation of aromatic hydrocarbons in batch reactors with acid chlorides as acylating agents and using Lewis acid metal chlorides ($AlCl_3$, $ZnCl_2$, etc.) as catalysts. Thus, major environmental problems arise from the use of acid chlorides and especially from the production of huge quantities of corrosive hydrochloric acid to be neutralized, treated and properly disposed of. In the last 20 years, a large number of research groups have investigated the liquid-phase acylation of aromatics and heteroaromatics over acid zeolites, using often anhydrides as acylating agents, in the presence of a solvent and operating the reaction in batch reactors. The sustainable acetylation process based on zeolites was then successfully applied by Rhodia to anisole, veratrole and thioether substrates. Most of the reports deal with aromatic acetylation over large-pore (HFAU and HBEA) and medium-pore zeolites (HMFI). As could be expected from the classical mechanism of electrophilic aromatic substitution, the acylation rate strongly depends on the degree of activation of the aromatic ring by the chemical nature of the substituents and the formation of polyacylated aromatics is mainly inhibited by the deactivating effect of the acyl group. Transacylation reactions can also occur, as demonstrated, for instance, in 2-methoxynaphthalene acetylation by using deuterium-labelled substrates.

Nevertheless, the formation of small (and even minimal) amounts of polyacylated species and the strong competition between reactant, solvent and product molecules for adsorption within the zeolite micropores (for their occupancy) and at the protonic acid sites are the main factors causing catalyst deactivation in this class of reactions [22, 31, 32]. This is an example of *auto-inhibition* due to the strong adsorption of acetylated products (which are bulkier and more polar than the substrates) and also of the co-produced polar acetic acid onto the catalytically active acid sites. Such inhibition is responsible for a rapid decrease in reaction rate with, as a consequence, a plateau in the yield in acetylated products after short reaction times under batch conditions.

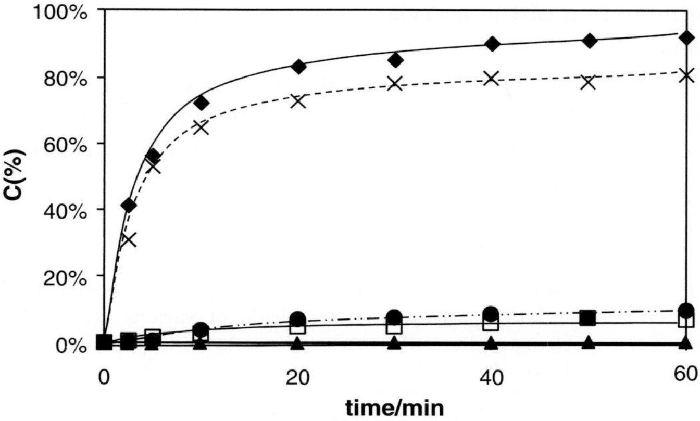

Fig. 17.4. Acetylation in batch reactor of anisole (♦), 2-methoxynaphthalene (×), toluene (•), 2-methylnaphthalene (□) and fluorobenzene (▲) over a HFAU zeolite. Conversion (%) vs. reaction time. Reaction conditions: 500 mg zeolite; substrate: acetic anhydride ratio = 5; $PhNO_2$; 373 K; 1 bar; 60 min (adapted from Ref. [33]).

The effect is much more pronounced with hydrophobic substrate molecules such as methyl- and fluoro-substituted aromatics (Fig. 17.4), because of the larger difference in polarity between substrate and product molecules [23, 33]. This constitutes an additional limitation for the substitution of Friedel–Crafts catalysts by acidic zeolites in the acylation of scarcely polar, weakly activated and hence poorly reactive substrates.

Furthermore, owing to their long residence time within the zeolite micropores, the polar product molecules undergo secondary condensation reactions, giving rise to bulkier and generally more polar products which are consequently more strongly retained and sometimes sterically blocked within the pores. For instance, in the micropores of a HBEA zeolite used for long reaction times for anisole acetylation, di- and tri-acetylated anisoles (resulting from secondary acetylation of the side chain of the desired *p*-methoxyacetophenone) and various other heavy products are found. Such bulky and polar trapped compounds block the access of the reactants to the micropores and they are thus the main cause for catalyst deactivation during the acetylation reaction [22, 23, 31–34]. Furthermore, the detrimental effect of heavy compounds derived from self-condensation of the acylating agents over the protonic zeolite can be confirmed by deliberate deactivation studies. For instance, the addition of dehydroacetic acid (3-acetyl-4-hydroxy-6-methyl-2-pyrone, derived by multiple condensation of acetic anhydride catalysed by an acid zeolite), leads to an almost complete inhibition of the acetylation of poorly activated aromatics, such as *m*-xylene [23].

Tables 17.1 and 17.2 report a summary (and a semi-quantitative distribution) of the heavy polyfunctionalized species found in the micropore network of an acid HBEA zeolite used as a catalyst for the acetylation of benzenic and naphthalenic rings.

Table 17.1. Composition of organic deposits strongly retained inside HBEA zeolite pores after 1 h of acetylation with acetic anhydride of various aromatic substrates. Reaction conditions: 500 mg zeolite; substrate: acetic anhydride ratio = 5; $PhNO_2$; 373 K; 1 bar; 60 min (adapted from Ref. [23]).

Substrates	Carbon content[a] (wt%)	Organic compound extracted after HF mineralization of spent catalyst[b]	Composition (wt%)
Anisole	1.45	monoacetylated anisole	93
		diacetylated anisole	6
		others	1
2-Methoxynaphthalene	2.01	unreacted methoxynaphthalene	6
		monoacetylated methoxynaphthalene	43
		diacetylated methoxynaphthalene	32
		others	19
m-Xylene	1.90	monoacetylated xylene	37
		diacetylated xylene	35
		AA derivatives[c]	14
		others	14
Toluene	1.80	AA derivatives	6
		monoacetylated toluene	4
		diacetylated toluene	81
		triacetylated toluene	7
		others	2
2-Methylnaphthalene	2.64	AA derivatives	21
		monoacetylated methylnaphthalene	43
		diacetylated methylnaphthalene	17
		others	19
Fluorobenzene	1.74	AA derivatives	80
		Others	20

[a]From C, H, N, S elemental analysis on the used catalyst after Soxhlet extraction and before HF mineralization.
[b]Mineralization and analysis method, as described in Chapter 4 of this book.
[c]Compounds derived from condensation and/or oligomerization of acetic anhydride.

Table 17.2. Main families of heavy organic compounds strongly retained inside HBEA zeolite pores during acetylation of anisole with acetic anhydride (adapted from Ref. [31]).

Family	Chemical structure	Molecular weight (M)
Methoxyacetophenone and acetylated derivatives	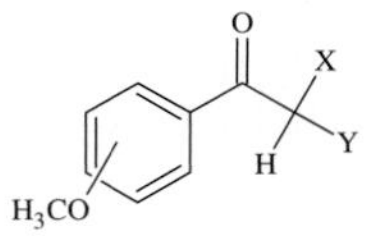	$M = 150$, X = Y = H $M = 192$, X = Ac, Y = H $M = 234$, X = Y = Ac
2-Acetyl, 3-methyl, methoxyindene-1-one	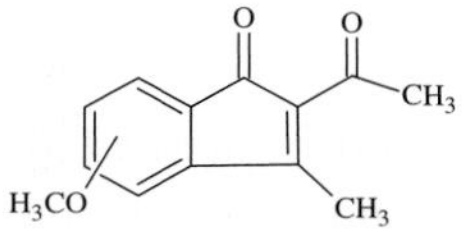	$M = 216$
Dimethoxydypnone	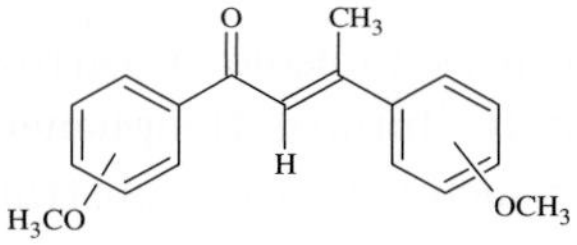	$M = 282$
Methoxy alpha-methyl styrene and acetylated derivative	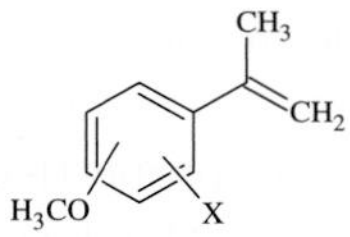	$M = 148$, X = H $M = 190$, X = Ac
Dimethoxybenzophenone	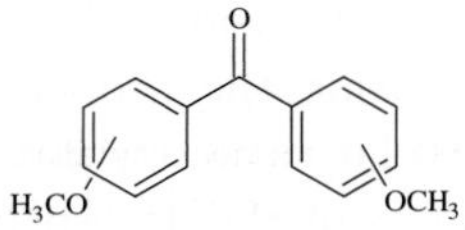	$M = 242$
2-Methylchromone and acetylated derivative	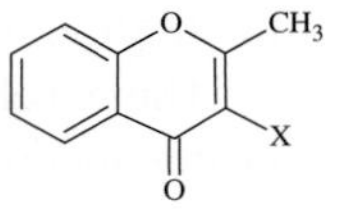	$M = 160$, X = H $M = 202$, X = Ac
2,4,6-Heptanone	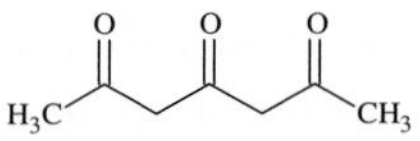	$M = 142$
2,6-Dimethylpyran-4-one and acetylated derivative	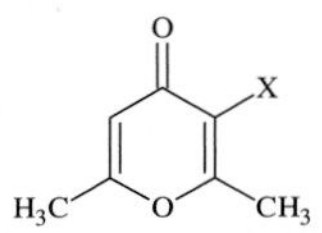	$M = 124$, X = H $M = 166$, X = Ac
2,6-Dimethylpyran-5,6-dihydro-4-one and acetylated derivative	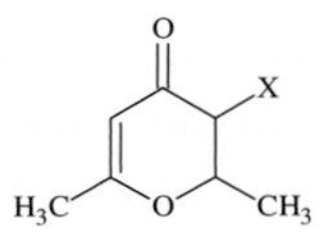	$M = 126$, X = H $M = 168$, X = Ac

(*Continued*)

Table 17.2. (*Continued*)

Family	Chemical structure	Molecular weight (M)
Heavy products	$C_{18}H_{15}O_3X$	$M = 280$, X = H $M = 322$, X = Ac
	$C_{20}H_{16}O_3$	$M = 304$
	$C_{20}H_{17}O_3X$	$M = 306$, X = H $M = 348$, X = Ac
	$C_{22}H_{20}O_4$	$M = 324$

As the acetylation of poorly activated substrates is limited by both their low reactivity and the strong inhibiting effect of the polar product molecules, most of the successful studies deal with the acetylation of aromatic ethers (anisole, veratrole, methoxynaphthalene) and heterocyclic compounds (furans, thiophenes, pyrroles). Indeed, the liquid-phase acetylation of such activated substrates can be efficiently and selectively carried out over zeolite catalysts under mild conditions. The key point is to favour the desorption of the bulky and polar desired product. This can be done by: 1) choosing the adequate zeolite type and tuning the physico-chemical properties (namely, the use of large-pore three-dimensional zeolites in the form of nano-sized crystals and with hydrophobic properties); 2) using a flow fixed-bed reactor rather than a batch reactor; 3) choosing the proper operating conditions: high substrate/acetic anhydride molar ratio, eventual use of a suitable solvent, temperature high enough to favour the desorption but low enough to avoid extensive secondary reactions, and high space velocity values to enhance the sweeping of the products out of the catalyst. These rules are also valid for the acylation with bulkier acylating agents (benzoylation, etc.) and their strict application can allow the acetylation of poorly activated substrates such as toluene [35, 36]. So far, these adsorbed inhibiting compounds are removed by calcination under air of the spent zeolite, even if the use of high temperatures can be a drawback (for safety reasons) at industrial scale in fine chemicals plants.

17.3.1.2. *Beckmann rearrangement*

The Beckmann rearrangement consists of an acid-catalysed stereospecific rearrangement of ketoximes to amides and it is a topic of great research interest, since not only ketoximes but also some aldoximes and esters and ethers of oximes can be starting substrates.

The Beckmann rearrangement of the cyclohexanone oxime is a noteworthy transformation in industrial chemistry, because it represents a key step in the production of ε-caprolactam, the monomer of nylon-6 and resins. Today, the ε-caprolactam capacities reach around 4 millions tons per year and the classical route of its production involves two successive steps: (i) cyclohexanone ammoximation with hydroxylamine sulphate in the presence of ammonia, forming ammonium sulphate as a by-product; and (ii) Beckmann rearrangement of cyclohexanone oxime in concentrated sulphuric acid or oleum; ammonia treatment is used to recover caprolactam from the produced caprolactam sulphate with further formation of ammonium sulphate. Such a process carries serious drawbacks, such as the formation of ammonium sulphate as a by-product and a good solution consists in the use of heterogeneous catalysis over solid acid materials.

Because of these problems, a great deal of papers and patents have been published for more than 20 years on the Beckmann rearrangement over zeolite catalysts [37]. In the late 1980s, it was observed that extremely highly siliceous HMFI samples, virtually with no acidity, are active in the vapor-phase Beckmann rearrangement with a high selectivity to ε-caprolactam (75–80%) at total conversion of cyclohexanone oxime (Fig. 17.5) [38].

From this pioneering investigation, the industrial-scale process, developed by Sumitomo Chemical Co. in 2003, is a combination of the EniChem ammoximation process based on TS-1 (see Section 17.3.2.1)

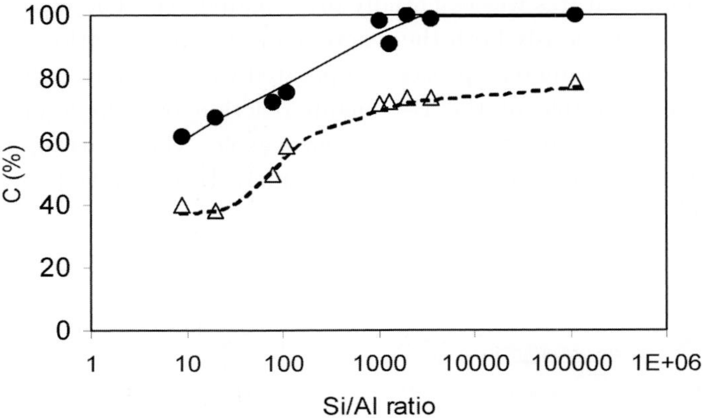

Fig. 17.5. Influence of the Si/Al ratio of HMFI zeolites on the conversion of cyclohexanone oxime (•) and on the selectivity to ε-caprolactam (Δ) (adapted from Ref. [38]).

and of the vapor-phase Beckmann rearrangement process developed by Sumitomo and based on a highly siliceous HMFI [39, 40]. The long delay required for this industrial development was likely due not only to the indispensable improvements to maximize the Beckmann rearrangement, but also to the difficulty of controlling the catalyst stability [41]. In fact, catalyst deactivation by carbonaceous compounds is always observed as a major problem in the Beckmann rearrangement over all kinds of zeolites. Experiments carried out over weakly acidic borosilicate MFI zeolites, using depth profiling XPS and SIMS techniques, led to the conclusion that catalyst deactivation is mainly due not to excessive amounts of carbon species deposited on the surface or to a pronounced and irreversible deboronation of the zeolite, but rather to pore blocking by nitrogen-containing species formed via condensation reactions of unsaturated nitrile by-products [42]. Probably, carbonaceous compounds can result from various condensation reactions involving molecules of cyclohexanone oxime, caprolactam and other side products. The hypothesis of caprolactam condensation was also supported by the observation that the yield in desorbed caprolactam oligomers (in particular in tetramers) is directly correlated to a deactivation constant b, determined under various conditions [43]. The higher the yield, the higher the constant b value, hence the faster the deactivation. Even if several patents claim the possibility of restoring the catalyst activity by a simple treatment with polar compounds (e.g. with aliquots of water, ammonia, amine or a carboxylic acid in the inlet feed), the classical mode of coke removal by combustion of the trapped species at elevated temperatures was essentially investigated [44]. Therefore, in order to conduct continuously both the reaction and the regeneration step at the same time, the commercial process is operated with a fluidized-bed system.

Another example of the Beckmann reaction over high-added value substrate is the production of benzoxazole via rearrangement of salicylaldoxime over protonic zeolites, namely H-Y (HFAU) and HBEA (Fig. 17.6) [45].

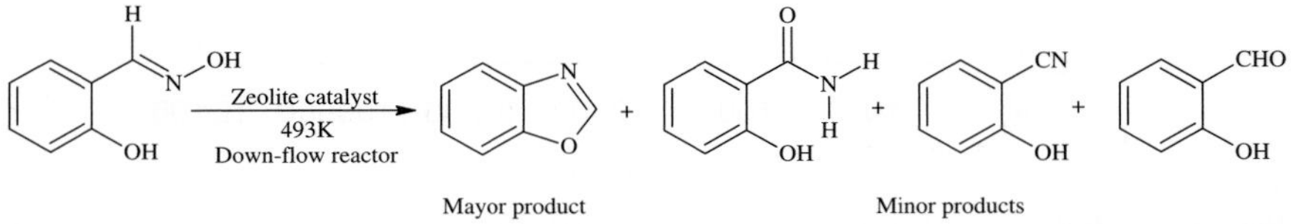

Fig. 17.6. Beckmann rearrangement of salicylaldoxime producing benzoxazole over protonic FAU and BEA zeolites.

Table 17.3. Catalytic performances of fresh H-Y and HBEA zeolites (values in parenthesis) and after 1 to 3 regeneration cycles. Reaction conditions: 700 mg zeolite, 5% solution of oxime in 1 : 1 benzene : acetonitrile (v/v) mixture, 4 mL h^{-1} (WHSV 0.29 h), 3 h, 498 K (adapted from Ref. [47]).

	Fresh	I cycle	II cycle	III cycle
Salicylaldoxime conversion (%)	100 (100)	98.8 (99)	97.3 (97.9)	98.1 (98.3)
Benzoxazole selectivity (%)	71.9 (79.6)	71.7 (79)	70.9 (78.1)	71.7 (77)

Nevertheless, these catalysts are susceptible to deactivation and it is likely due to the deposition on the active sites of reactants and/or products as well as of water, formed during dehydration of the oxime. By increasing the reaction temperature, the catalyst performance improves and the product desorption from the catalyst surface is easier. Alternatively, the deactivated materials can be regenerated by solvent extraction, such as dichloromethane, to remove most of the trapped by-products, followed by oxidative treatment at 500°C for 5 h. The results obtained demonstrate that the spent zeolites can be regenerated without framework damage and loss of catalytic activity or selectivity even after three cycles (Table 17.3). Currently, studies addressed at the extension of this protocol to other types of substituted salicylaldoximes are underway.

17.3.2. *Oxidation processes*

17.3.2.1. *Ammoximation of ketones*

The ammoximation process is an important synthetic route for the production of ketoximes, such as cyclohexanone oxime, *para*-hydroxyacetophenone oxime or methyl ethyl ketone oxime. Cyclohexanone oxime is a key intermediate in the manufacture of ε-caprolactam, the monomer of nylon-6, *para*-hydroxyacetophenone oxime is an intermediate of interest in the industrial synthesis of the analgesic paracetamol (Fig. 17.7), and methyl ethyl ketone oxime is widely used as an additive in paints and

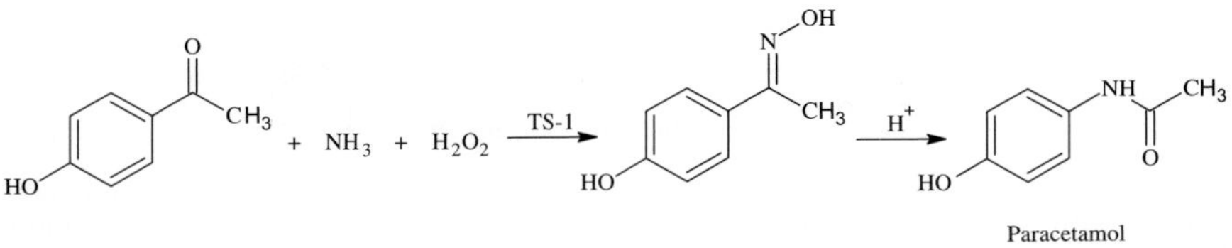

Fig. 17.7. Paracetamol synthesis via ammoximation of *para*-hydroxyacetophenone.

lacquers as well as an antioxidant for drying materials, a corrosion inhibitor and a functional chemical building block for synthesizing agrochemicals and pharmaceuticals [11, 46, 47].

Many conventional processes for oxime synthesis present environmental disadvantages, such as the use of poisonous oximation agents and the production of large amounts of ammonium sulphate, a valueless by-product with problems and costs in disposal and purification [11, 47].

Great efforts have been made to develop alternative environmentally benign process over heterogeneous catalysts and now cyclohexanone ammoximation, with ammonia and hydrogen peroxide, over titanium silicalite-1 (TS-1; also known as Ti-MFI) is a cleaner process exploited at industrial level [48]. In particular, the recent commercial plant built in Japan combines ammoximation plus the Beckmann rearrangement and produces ε-caprolactam (*ca.* 60,000 tons year^{-1}) without co-producing ammonium sulphate. Such a major achievement prompted the investigation of other titanium-containing molecular sieves too.

Generally, when the ammoximation reaction is performed over Ti-containing zeolites, the catalyst activity decreases with the time-on-stream (typically, over 100–150 hours; Fig. 17.8) and this is observed on TS-1 too, although it is typically a stable and robust catalyst in other oxidation

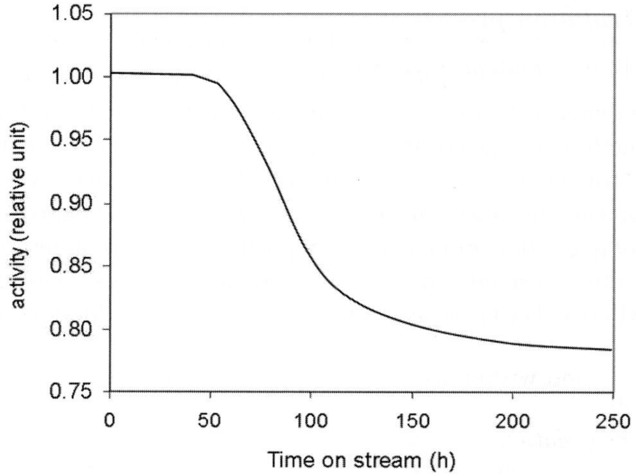

Fig. 17.8. Activity loss of TS-1 vs. time-on-stream during the cyclohexanone ammoximation with ammonia and hydrogen peroxide. Reaction conditions: cyclohexanone/H_2O_2 = 1 : 1.2 (mol), cyclohexanone/NH_3 = 1 : 1.5 (mol), cyclohexanone/t-butanol = 1 : 3 (w/w), WHSV = 6 h, 70°C (adapted from Ref. [11]).

processes. This deactivation is due to several reasons: (a) detrimental interaction with the reaction medium and slow dissolution of the zeolitic framework, (b) "coke" formation on the internal surface (within the micropores of zeolite), (c) fouling and pore blocking with heavy products and (d) leaching of active sites [11, 46, 47]. Carbonaceous deposits ("coke") produces a decrease in catalytic activity of zeolites by two mechanisms: active-site suppression and pore blocking. The two pathways can occur simultaneously, although one of the two mechanisms is usually predominant. The "coke" deposits are mainly located on the micropores of the zeolite and consist of "soft coke" and "refractory coke". Most of the "soft coke" is deposited on the weak acid sites, near the framework Ti centres and it can be removed by oxidation under O_2 at 350°C. The "refractory coke", which is deposited on the strong acid centres as well as on the silanol groups at crystal defect sites, can be removed by oxidation under O_2 until 700°C [11]. In particular, in the case of ammoximation performed over Ti-MWW, a choice of several regeneration methods has been suggested [47]. The method to prolong the zeolite lifetime can be: calcination under air at 550°C for 6 h; treatment with an aqueous solution of the structure-directing agent piperidine at 170°C for 24 h, in order to force a zeolite framework rearrangement; and addition of a colloidal silica gel, which is helpful in repairing the crystalline structure, by re-inserting Si into the zeolite lattice, and also in suppressing Si leaching.

The content and the nature of the carbonaceous deposits adsorbed on spent TS-1 can be evaluated by a mineralization and analysis method (Chapter 4). The soluble coke was extracted by dimethyl ether and analysed by GC/MS, and the insoluble coke was separated by centrifugal sedimentation and analysed by NMR and FTIR. In the cyclohexanone ammoximation over TS-1, the former is a complex mixture, among which the dimer of cyclohexanone and substituted cyclohexane are most likely responsible for the formation of polycyclic or substituted aromatic products, and the latter is N-containing polycyclic aromatic and/or polyalkene species that may come from the polymerization of soluble coke. Table 17.4 shows the species found within spent TS-1 [11].

17.3.2.2. *Epoxidation of alkenes*

Epoxidation is one of the most important C=C bond functionalization methods and epoxides are important synthetic intermediates for the synthesis of several fine and specialty chemicals both at laboratory and industrial scale. Commercially, only the epoxidation of propene relies

Table 17.4. Ammoximation of cyclohexanone. Major components of soluble coke extracted from spent TS-1 catalyst (adapted from Ref. [11]).

Compound	Formula	Molecular weight
Cyclohexanol	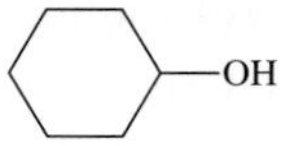	100
Phenol	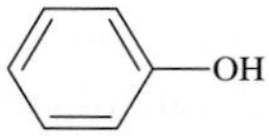	94
2-Cyclohexenecyclohexanone	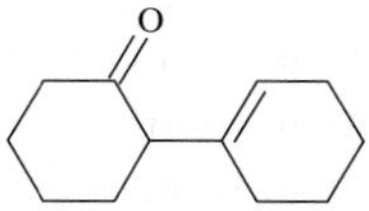	178
2-Cyclohexlidenecyclohexane	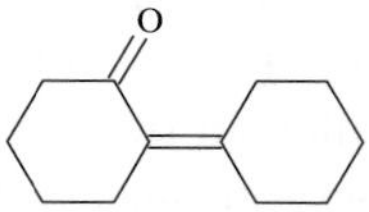	178
1,1′-Bicyclohexyl-1,1′-diol	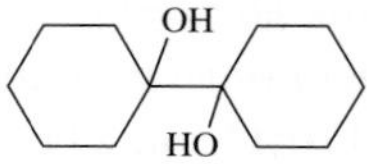	198
Nitrocyclohexane	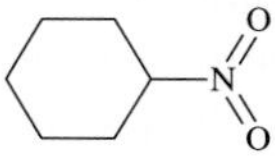	129
ε-Caprolactam	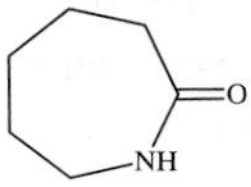	113
2-Tert-butylcyclohexanone	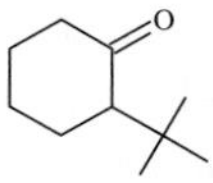	154
3-Tert-butylcyclohexanone	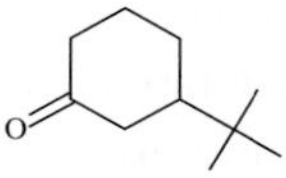	154
4-Tert-butylcyclohexanone	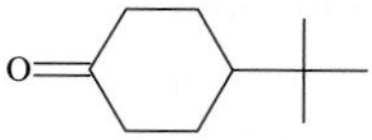	154
2-(Tert-butylperoxy)-cyclohexanamine	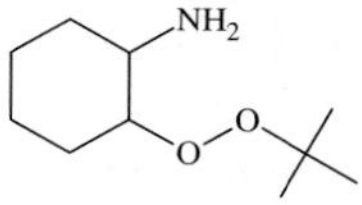	187

on zeolite-based process technology [49, 50]. In fact, although alkene epoxidation is generally considered a well-established methodology, a truly versatile and eco-friendly heterogeneous system to be applied in organic and fine chemicals synthesis is still vague and conventional processes based on hazardous and aggressive oxidants are thus widely used. Most of the research team focused their attention on titanium-containing (either micro- or mesoporous) molecular sieves, keeping in mind the outstanding performance of TS-1 in alkene epoxidation [51]. In fact, bulky unsaturated substrates, such as cyclic terpenes, fatty acid derivatives or polycyclic aliphatic compounds can be epoxidized only over large-pore titanosilicate zeolites or mesoporous molecular sieves in liquid phase with oxidants, such as hydrogen peroxide or organic hydroperoxides [17, 52–54]. For this kind of catalysts, the major reason of loss of activity is their double redox-active and acid nature. In fact, a Ti(IV) site in a silica matrix possesses, together with its redox-active property, an intrinsic Lewis acid character too, which leads to the formation of oxirane ring-opening products (diols or hydroxyethers) or rearrangement derivatives (aldehydes, ketones). Such "bifunctional" activity of titanosilicates materials can be exploited to form a large variety of oxidized compounds via bifunctional cascade reaction [55, 56], but, since alcohols, polyols, ethers, etc. all show a high affinity for Ti(IV) centres, it can also be a cause of auto-inhibition by the oxidized by-products themselves via poisoning and fouling. In particular, when H_2O_2 is used as an oxidant, the predominant oxirane-ring opening secondary products are diols, whereas when alcohols are present (as solvents or as side products of oxidation with organic hydroperoxides) hydroxyethers can be formed too and they all can adhere at the catalyst surface (thanks to the affinity of (poly)hydroxylated molecules with the silanol-rich surface of titanosilicates) and slow down or, in the worst case, block the reaction.

When Ti-containing silica molecular sieves (either ordered or non-ordered) are used in the batch liquid-phase epoxidation with *tert*-butylhydroperoxide (TBHP) of a series of unsaturated terpenes (i.e. limonene, isopulegol, α-terpineol, terpinen-4-ol, carvotanacetol), in all cases, high terpene epoxidation rates are observed in the first hours of reaction. Then, a remarkable decrease in reaction rate is recorded and, with some less reactive substrates such as limonene, the epoxidation does not proceed further (Fig. 17.9) [57, 58].

Several authors, studying the batch epoxidation of cyclohexene or unsaturated terpenes, agree on the detrimental role of polyhydroxylated secondary products which lead to deactivation [57–60] and recently the

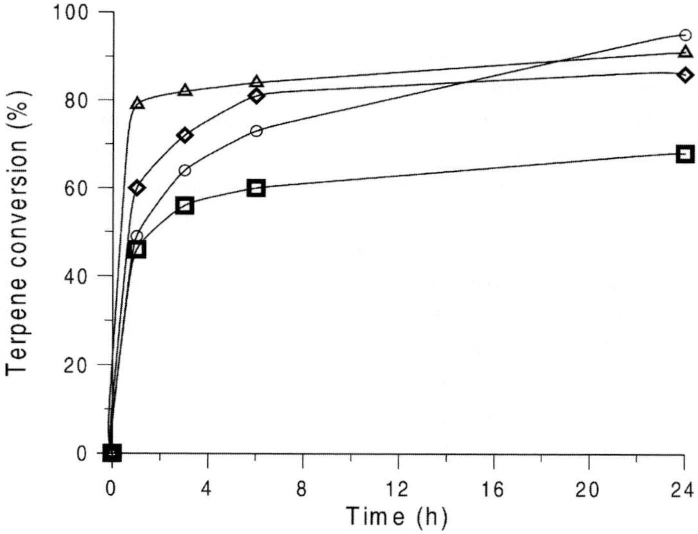

Fig. 17.9. Conversion profiles for the batch liquid-phase epoxidation of unsaturated terpenes over Ti-MCM-41. Carvotanacetol (△), terpinen-4-ol (◇), α-terpineol (○) and limonene (□). Reaction conditions: CH_3CN solvent, 30 wt% catalyst, TBHP : terpene = 1.1 (mol), 85°C (adapted from Ref. [57]).

presence of adsorbed diols has also been confirmed by TGA measurements and *in situ* ^{13}C-CP-MAS-NMR spectra of a used Ti-MCM-48 catalyst [18].

A method to reduce the occurrence of fouling by polyhydroxylated by-products can be the silylation of free silanol groups (with alkylchlorosilanes or alkylsilazanes), to reduce the hydrophilic character of the catalyst [61–63]. For instance, in the epoxidation of cyclohexene with TBHP over Ti-MCM-41, the catalyst can be recycled up to five times with a slight decrease in activity of ca. 10%, whereas the activity of an analogous non-silylated sample decreases to ca. 85% of the initial value after a single recycling [59].

Such major loss in catalytic activity of the non-silylated catalyst can be attributed to the poisoning of the active Ti sites by adsorbed glycols, rather than to structural damages, since their XRD analysis patterns did not change significantly [64]. Nevertheless, polyhydroxylated molecules can be responsible for Ti leaching from titanosilicate matrixes as well and can cause deactivation via gradual loss of active metal. Detailed studies, for instance, on the formation of triols during the epoxidation of crotyl alcohol, have revealed that Ti leaching is caused by the reaction of 1,2,3-butanetriol with $Ti(OSi{\equiv})_4$ centres in the presence of hydrogen peroxide [65]. The triol is able to chelate the Ti, breaking the Ti–O–Si framework bonds, and such mechanism gives rise to chelated Ti species in solution. This can occur not only on mesoporous titanosilicates (such as Ti-MCM-41) that are intrinsically less robust to hydrolysis by protic solvents [51], but also on the highly stable TS-1. Furthermore, the Ti species formed by triol chelation can be themselves active oxidation catalysts and could contribute to by-product formation or hydrogen peroxide decomposition. Hence, once Ti-leaching is initiated by triols, the effect could become a parallel catalytic pathway leading to a loss in selectivity towards desired products.

In most cases, finally, the suggested method to recover the initial activity of liquid-phase titanosilicate epoxidation catalysts is based on high-temperature calcination (under air or pure oxygen at temperatures as high as 500°C) in order to remove the species strongly adsorbed onto the catalyst surface by combustion and to re-obtain clean and accessible Ti(IV) sites [18, 58].

17.3.2.3. *Oxidation of phenols to quinones*

Dihydroxybenzenes are valuable intermediates for organic chemical synthesis. Phenol oxidation to hydroquinone was the first example of large-scale industrial exploitation of TS-1 and since 1986 (when the process was introduced into the market by EniChem) ca. 10,000 tons of hydroquinone and pyrocatechol are produced yearly by this pathway.

When the reaction conditions are optimized, 94% of selectivity, based on phenol, and 84% selectivity, based on H_2O_2, at 30% phenol conversion are obtained, with an almost equimolar distribution of catechol and hydroquinone [51, 66]. This process was also exploited in fine chemicals synthesis, in the above-mentioned Rhodia process for the manufacture of vanillin [67]. In this case, as in other oxidation processes, a critical factor is the careful control of the reaction conditions to avoid over-oxidation, leading to the formation of heavy tars and fouling species. This is normally achieved by operating with an excess of phenol over hydrogen peroxide. Moreover, a fast diffusion of the diphenols out of the pores is required in order to avoid over-oxidation and H_2O_2 decomposition. For this purpose, crystal sizes as small as possible are preferred, and, if it is feasible, with a deactivated external surface [68]. However, in some studies based on the time course of the reaction and the solubility of tar deposits, it was concluded that catechol and hydroquinone are produced on different sites: the external and internal ones, respectively [69]. In any case, the thermal and structural stability of this solid makes the regeneration feasible, by burning off the organic deposits under oxygen or air atmosphere.

Alternatively, the hydroxylation of phenol can be carried out via the gas-phase oxidation with nitrous oxide over Fe-MFI (Fe-ZSM-5) [70]. The addition of benzene to the feed mixture improves the selectivity and catalytic stability of the zeolite, which presents an intrinsic good stability [71]. The main cause of catalyst deactivation is "coke" formation during the reaction and the heavy organic deposits can be burnt out by flowing helium with nitrous oxide, with gradual temperature elevation from 450°C to 550°C in order to regenerate the solid and provide a complete restoration of its activity.

17.3.2.4. *Oxidation of benzene to phenol*

The oxidation reaction of benzene to phenol is a reaction of pivotal relevance due to the ever-increasing use of phenolic resins in several areas, such as the plywood, construction, automotive and appliance industries, and of ε-caprolactam (via hydrogenation of phenol to cyclohexene, its hydration to cyclohexanol/cyclohexanone and then ammoximation; cf. Section 17.3.2.1) and bisphenol A (intermediate in the manufacture of epoxy resins) [50, 70]. From the point of view of the fine chemicals industry, vanillin, a major flavour ingredient, may also be obtained from phenol [67].

Phenol is primarily produced by the cumene hydroperoxide process, which suffers from some drawbacks: in particular, use of a hazardous intermediate (the hydroperoxide); poor eco-sustainability; and the simultaneous co-production of acetone (with 1 : 1 stoichiometry), a commodity with a lower demand than that for phenol. For these reasons, it is desirable to use new processes, such as direct oxidation of benzene to phenol, independent of acetone co-production. Currently, the two most attractive alternatives are the direct hydroxylation of benzene by nitrous oxide over iron-containing MFI zeolites [70, 72–77] and the benzene hydroxylation by hydrogen peroxide over TS-1B (a zeolitic solid derived from TS-1) [78, 79].

In the first process (also known as the Solutia AlphOx benzene oxidation process), based on iron-containing ZSM-5 zeolites, the reaction is driven at ca. 400°C and excellent selectivity to phenol (up to 99%) at high benzene conversions (higher than 30%) are obtained [50, 72]. The major disadvantage of the AlphOx route is due to the limited availability of nitrous oxide. In addition, the deactivation of these catalysts is relatively rapid and frequent regeneration contributes to worsening the economic sustainability of the process. The catalysts, in fact, suffer from deactivation due to "coke" formation in the zeolite pores. The species composing the "coke" can be classed into two kinds: (1) compounds formed by phenol over-oxidation, such as dihydroxybenzene, benzoquinone, etc. and (2) compounds formed by coupling benzenic molecules. The strong retention of the polar phenol molecules within the micropore network is likely to be responsible for the formation of a large fraction of these deactivating by-products. The "coke" content and composition can be determined according to the mineralization and analysis method described in Chapter 4. The "soluble coke" is a mixture of phenols, simple or condensed polyphenols and products derived from further condensation (Table 17.5). The "insoluble coke" consists of high-molecular-weight unextractable substances [24].

The analysis of the by-products, the effect of iron on the rate of deactivation and formation of "coke" give some useful indications about the reaction network and the nature of active sites, as summarized in Fig. 17.10 [80]. Regeneration of deactivated catalyst can be performed by burning out the coke under air flow at 650°C, but the addition of water steam to the feed improves activity, selectivity and durability of the catalyst, thanks likely to an efficient cleaning up of carbonaceous deposits from the surface [24]. Otherwise, the catalysts can be regenerated in oxygen diluted with helium and in nitrous oxide diluted with helium, nitrous oxide being more efficient. Analogously, it was observed that deactivation is faster over Fe-ZSM-5 with a large crystal size than with a relatively smaller one. Anyway, as a main factor, solids with a lower acid character suffer from less deactivation [77]. Another possibility in order to improve the catalyst lifetime consists in the

Table 17.5. Main components of "soluble coke" in benzene hydroxylation over Fe-MFI (Fe-ZSM-5) with N_2O (adapted from Ref. [24]).

General formula	Typical molecule of the family
$C_nH_{2n-6}O$	Phenol
$C_nH_{2n-6}O_2$	1,2-Benzenediol
$C_nH_{2n-12}O$	1-Naphtol
$C_nH_{2n-14}O$	1,1′-Biphenyl-2-ol
	1,1′-Biphenyl-3-ol
	1,1′-Biphenyl-4-ol
$C_nH_{2n-16}O$	Dibenzofuran
$C_nH_{2n-14}O_2$	1,1′-Biphenyl-2,2′-diol
$C_nH_{2n-16}O_2$	Oxanthrene

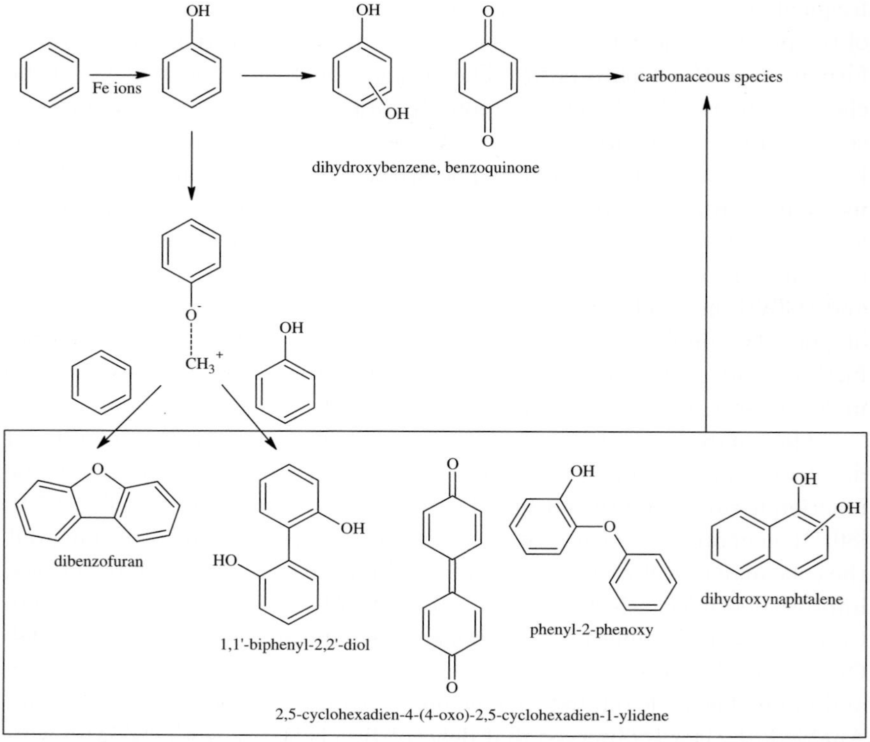

Fig. 17.10. Diagram of reactions occurring during benzene hydroxylation over Fe-MFI.

use of a highly active hierarchical Fe-ZSM-5. The integration of very small microporous domains into a highly accessible mesoporous matrix alleviates the detrimental effects of pore blocking by carbonaceous deposits [81].

Catalytic hydroxylation with N_2O over Fe-containing zeolites has been later applied to a wide series of substituted aromatics as well (e.g. toluene, halobenzenes, naphthalene, biphenyl) and, in these cases, the need for high reaction temperatures may lead to degradation of the products (lowering selectivity) and to catalyst deactivation due to "coking" [70].

As an alternative to the use of N_2O, Polimeri Europa developed a process for the hydroxylation of benzene by using hydrogen peroxide and a modified TS-1 zeolite, named TS-1B [78, 79]. A method for the direct hydroxylation of benzene has been long investigated since the discovery of TS-1, but since phenol is more reactive than benzene, the formed phenol is rapidly converted into over-oxidized species (cathecol,

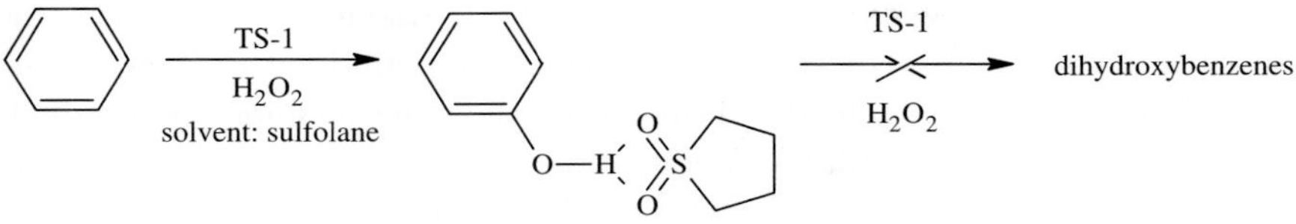

Fig. 17.11. The adduct formation between phenol and sulfolane avoids over-oxidation during benzene hydroxylation.

hydroquinone, benzoquinones, etc.). Moreover, the quick polymerization of these compounds gives rise to the formation of fouling deposits within the zeolite pores. However, the Polimeri Europa process is based on a couple of key innovations. From the point of view of the reaction mixture, the substitution of the classical solvents (i.e. acetone or methanol) for sulfolane leads to a dramatic improvement of selectivity thanks to the formation of stable and bulky complexes with phenol (Fig. 17.11) that cannot enter the TS-1 micropore network. The further oxidation of phenol into dihydroxyarenes is therefore more difficult thanks to such steric exclusion. In addition, sulfolane is able to dissolve effectively polymeric species and heavy fouling products that can deactivate the catalyst by plugging the pores and blocking the active sites. Analogously, the modification of the TS-1 catalyst is essential to have a material displaying a maximized selectivity to phenol. A post-synthesis treatment with NH_4HF_2 and H_2O_2 of TS-1 removes the more exposed, and hence less selective, Ti sites which are likely to be responsible for the undesired over-oxidation reactions to catechol, quinones and heavy polyphenols. The resulting lower defectivity can therefore contribute to an enhanced hydrophobic character of the zeolite and hence to a material less prone to fouling. The resulting catalyst, called TS-1B, attains a selectivity computed on benzene of up to 84.6% and on hydrogen peroxide of up to 61.4% [79].

17.3.2.5. *Baeyer–Villiger oxidation*

The Baeyer–Villiger oxidation is used to convert ketones into esters or lactones, which are versatile intermediates in the synthesis of a variety of chemicals. A heterogeneous catalyst which showed high activity and selectivity in this reaction is Sn-BEA zeolite and the results are particularly promising in the Baeyer–Villiger oxidation of cyclic ketones with hydrogen peroxide, in which selectivities to the desired lactones higher than 98% are obtained [82–85]. The excellent selectivity is due to the fact that the carbonyl group is specifically activated by the Lewis acid character of the Sn sites in the zeolite. For this reason, every compound with a Lewis base character in the reaction medium (including the reactants and the product as well) are potential inhibitors for the reaction. Detailed mechanistic studies on the Baeyer–Villiger oxidation of cyclohexanone show a negative effect of water on the reaction rate by competitive adsorption on Sn sites [86]. In any case, a competitive adsorption effect is more important in the case of the lactone product. When the adsorption occurs, the catalyst does not suffer from a permanent deactivation and its previous activity can be restored by filtering the solid, washing it and desorbing the product [84]. Together with product inhibition, Sn-BEA may also undergo deactivation due to the presence of inhibiting organic deposits. For instance, in the oxidation of *p*-anisaldehyde it is observed that the loss of activity is likely due to the blocking of the active sites and/or channels by organic material [87]. The simple catalyst recovery method, according to which the solid is filtrated and re-activated at 200°C under vacuum, is not completely efficient and, after four cycles, some activity is lost (Table 17.6).

Thermogravimetric and spectroscopic investigations show that up to 4 wt% of organic compounds are retained by the catalyst after conventional re-activation. On the contrary, only when the catalyst is regenerated by calcination at 500°C for 3 h in air is the initial activity fully recovered [87, 88].

Sn-BEA finds application in the synthesis of flavours and fragrances compounds such as δ-decalactone and Melonal [82, 85, 87, 89] (Fig. 17.12).

Table 17.6. Recycling of Sn-BEA in the Baeyer–Villiger oxidation of *p*-anisaldehyde with hydrogen peroxide. Reaction conditions: 0.5 g aldehyde, 3.0 g CH_3CN, 1.2 eq. H_2O_2 (50%) (1.2 eq), 50 mg zeolite (adapted from Ref. [87]).

Cycle	Catalyst recovery	TON	Conversion (%)
1	filtration[a]	282	51
2	filtration	199	36
3	filtration	237	43
4	filtration	222	38
4 + 1	filtration + calcination[b]	282	51
1	filtration	277	50
1 + 1	filtration + extraction[c]	207	37

[a]Filtration and re-activation at 200°C under vacuum.
[b]Calcination at 500°C in air for 3 h.
[c]Soxhlet extraction with ethanol or acetone.

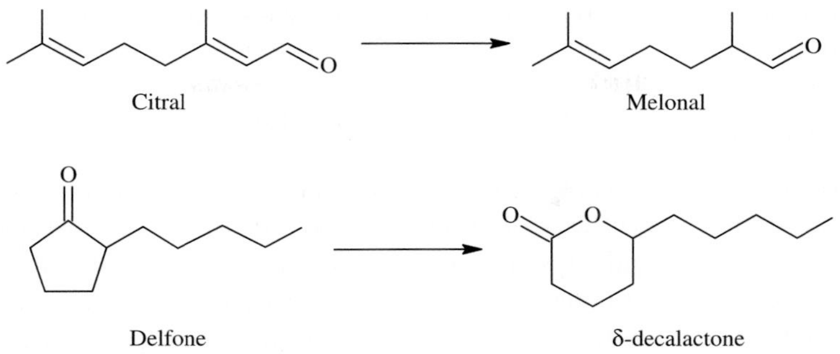

Fig. 17.12. Synthesis of Melonal by Baeyer–Villiger of citral and δ-decalactone by Baeyer–Villiger of delfone.

Table 17.7. Recycling of Sn-BEA in the Baeyer–Villiger oxidation of citral with hydrogen peroxide. Reaction conditions: 3.7 mmol citral, 1.5 eq. H_2O_2 (50%), 50 mg Sn-BEA, tert-amylalcohol, 100°C, 1 h (adapted from Ref. [89]).

		Selectivity (%)			
Cycle	Citral conversion (%)	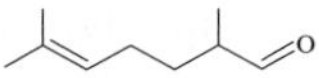	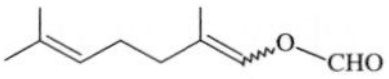	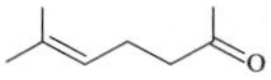	Other products
1	53	10	85	0	5
2	58	9	87	0	4
3	56	11	85	0	4
4	56	11	84	0	5
5	48	9	85	0	6
6[a]	62	12	77	0	11

[a]After activation by calcination.

In the case of δ-decalactone, the catalyst remains active for long reaction times, and no detectable metal leaching is observed [82]. In the synthesis of Melonal, the catalyst can be reused after a simple activation at 200°C and 2 Torr for 2 h. A final further calcination step provides a more active catalyst at the cost of some selectivity [89] (Table 17.7). It is possible that a competitive adsorption of products and secondary products with Lewis basic character may deactivate or inhibit the catalyst [87].

For the same class of reactions, the use of mesoporous molecular sieves, such as Sn-MCM-41, has also been described. Some loss in activity is observed when the catalyst is recycled three times, especially after the first recycling. The decrease in catalytic activity cannot be attributed to the leaching of Sn species, but to catalyst poisoning by organic deposits [90].

In this case too, the activity can be restored by a simple calcination in air at 500°C.

17.4. Conclusion

Deactivation is a widespread phenomenon when molecular sieves are used as catalysts in the synthesis of organic compounds. It is often overlooked, especially in fine and specialty chemicals synthesis carried out under batch conditions, and heterogeneous catalysts are frequently considered as consumable reactants that have to be changed and disposed of when they are inactive. Such an approach can be the reason for the poor attention paid to a deep understanding of the deactivation mechanisms in many research works at the laboratory scale, and of the countermeasures to improve the lifetime of catalysts. Actually, recycling and catalyst recovery tests are often performed only to confirm the genuine heterogeneous character of the solid system, rather than to study the possibility to use it for long reaction times or to restore its initial catalytic activity. The lack of this kind of information is often a major limitation hindering or, at least, slowing down the development of processes at the industrial level based on zeolites or zeotypes and induces many small-scale producers of fine chemicals to go on using conventional non-catalysed (or homogeneously catalysed) synthetic pathways, even if they show scarce environmental sustainability. In addition, factors leading to zeolite deactivation are so many and various that most of the theoretical studies so far are based on a "learning by doing" *a posteriori* approach which tries to rationalize the data directly collected from experimental observations and which is valid only for specific combinations of catalyst, reagents, reaction medium and conditions. For these reasons, a multidisciplinary interaction among experts in organic chemical synthesis, zeolite synthesis, analytical chemistry, chemical modelling and chemical process scale-up can play a crucial role when sustainable heterogeneously catalysed organic chemical processes have to be studied and developed.

Acknowledgements

The authors gratefully acknowledge the financial support of the EU IDECAT Network of Excellence and EU NANO-HOST Marie Curie Initial Training Network (grant no. 215193). The authors also thank Prof. Michel Guisnet (Technical University of Lisbon) and Dr. Michael Renz (Polytechnical University of Valencia) for fruitful discussion.

References

[1] Moulijn J.A., van Diepen E., Kapteijn F., *Appl. Catal. A: Gen.*, 212 (2001) 3–16.

[2] Farrauto R.J., Bartholomew C.H., *Fundamentals of Industrial Catalytic Processes*, Kluwer Academic Publishers, London (1997).

[3] Sie S.T., *Appl. Catal. A: Gen.*, 212 (2001) 129–151.

[4] Chorkendorff I., Niemantsverdriet J.W., *Concepts of Modern Catalysis and Kinetics*, Wiley-VCH, Weinheim (2007).

[5] Murzin D.Y., Salmi T., *Trends Chem. Eng.*, 8 (2003) 137–148.

[6] Guisnet M., Guidotti M., in *Catalysts for Fine Chemical Synthesis,* Derouane E. (Ed.), Wiley-VCH Verlag GmbH & Co., Weinheim (2006) 39–67.

[7] Trimm D.L., *Appl. Catal. A: Gen.*, 212 (2001) 153–160.

[8] Bartholomew C.H., *Appl. Catal. A: Gen.*, 212 (2001) 17–60.

[9] Arends I.W.C.E., Sheldon R.A., *Appl. Catal. A: Gen.*, 212 (2001) 175–187.

[10] Sheldon R.A., Wallau M., Arends I.W.C.E., Schuchardt V., *Acc. Chem. Res.*, 31 (1998) 485–493.

[11] Zhang X., Wang Y., Xin F., *Appl. Catal. A: Gen.*, 307 (2006) 222–230.

[12] Arvela R.K., Leadbeater N.E., Sangi M.S., Williams V.A., Granados P., Singer R.D., *J. Org. Chem.*, 70 (2005) 161–168.

[13] Clerici M.G., in *Fine Chemicals through Heterogeneous Catalysis*, Sheldon R.A., van Bekkum H. (Eds.), Wiley-VCH, Weinheim (2001) 538–549.

[14] Rebek J., Gavina F., *J. Am. Chem. Soc.*, 96 (1974) 7112–7114.

[15] Richardson J.M., Jones C.W., *J. Catal.*, 251 (2007) 80–93.

[16] Crudden C.M., McEleney K., MacQuarrie S.L., Blanc A., Sateesh M., Webb J.D., *Pure Appl. Chem.*, 79(2) (2007) 247–260.

[17] Kholdeeva O.A., Trukhan N.N., *Russian Chem. Rev.*, 75(5) (2006) 411–432.

[18] Guidotti M., Pirovano C., Ravasio N., Lázaro B., Fraile J.M., Mayoral J.A., Coq B., Galarneau A., *Green Chem.*, 11 (2009) 1421–1427.

[19] Guisnet M., Magnoux P., *Appl. Catal.*, 54 (1989) 1–27.

[20] Rohan D., Canaff C., Magnoux P., Guisnet M., *J. Mol. Catal. A: Chem.*, 129 (1998) 69–78.

[21] Fromentin E., Coustard J.M., Guisnet M., *J. Mol. Catal. A: Chem.*, 159 (2000) 377–388.

[22] Moreau V., Fromentin E., Magnoux P., Guisnet M., *Stud. Surf. Sci. Catal.*, 135 (2001) 4113–4120.

[23] Guidotti M., Canaff C., Coustard J.M., Magnoux P., Guisnet M., *J. Catal.*, 230 (2005) 375–383.

[24] Meloni D., Monaci R., Solinas V., Berlier G., Bordiga S., Rossetti I., Oliva C., Forni L., *J. Catal.*, 214 (2003) 169–178.

[25] Brait A., Davis M.E., *Appl. Catal. A: Gen.*, 204 (2000) 117–127.

[26] Kouwenhoven K.W., van Bekkum H., in *Handbook of Heterogeneous Catalysis*, Vol. 7, 2nd ed., Ertl G. (Ed.), Wiley-VCH., Weinheim (2008) 3578–3591.

[27] Marion P., Jacquot R., Ratton S., Guisnet M., in *Zeolites for Cleaner Technologies*, Guisnet M., Gilson J.P. (Eds.), Imperial College Press, London (2002) 281–300.
[28] de Vos D.E., Jacobs P.A., *Microp. Mesop. Mater.*, 82 (2005) 293–304.
[29] Sartori G., Maggi R., *Chem. Rev.*, 106 (2006) 1077–1104.
[30] Beiblova M., Zilkova N., Cejka J., *Res. Chem. Intermed.*, 34(5–7) (2008) 439–454.
[31] Rohan D., Canaff C., Fromentin E., Guisnet M., *J. Catal.*, 177 (1998) 296–305.
[32] Derouane E.G., Dillon C.J., Bethell D., Derouane-Abd Hamid S.B., *J. Catal.*, 187 (1999) 209–218.
[33] Guidotti M., Coustard J.M., Magnoux P., Guisnet M., *Pure Appl. Chem.*, 79 (2007) 1833–1836.
[34] Guisnet M., Guidotti M., in *Catalysts for Fine Chemical Synthesis*, Derouane E. (Ed.), Wiley-VCH Verlag GmbH & Co., Weinheim, (2006) 69–94.
[35] Botella P., Corma A., Lopez-Nieto J.M., Valencia S., Jacquot R., *J. Catal.*, 195 (2000) 161–168.
[36] Sheemol N., Tyagi B., Jasra R.V., *J. Mol. Catal. A: Chem.*, 215 (2004) 201–208.
[37] Dahlhoff G., Niederer J.P., Hoelderich W.F., *Catal. Rev.*, 43 (2001) 381–441.
[38] Sato H., Ishii N., Hirose K., Nakamura S., in *Proc. 7th Intern. Zeolite Conf.* (1986) 755–762.
[39] Ichihashi H., Kitamura M., *Catal. Today*, 73 (2002) 23–28.
[40] Ichihashi H., Ishida M., Shiga A., Kitamura M., Suzuki T., Suenobu K., Sugita K., *Catal. Surveys Asia*, 7 (2003) 261–270.
[41] Tatsumi T., in *Fine Chemicals through Heterogeneous Catalysis*, Sheldon R.A., van Bekkum H. (Eds.), Wiley-VCH, Weinheim (2001) 185–204.
[42] Albers P., Sibold K., Haas T., Prescher G., Hoelderich W., *J. Catal.*, 176 (1998) 561–568.
[43] Takahashi T., Kai T., Nakao E., *Appl. Catal. A: Gen.*, 262 (2004) 137–142.
[44] Sugita K., Kitamura M., Hoshino M., Patent *EP1593433* (2005): to Sumitomo Co.
[45] Thomas B., George J., Sugunan S., *Ind. Eng. Chem. Res.*, 48 (2009) 660–670.
[46] Perego C., Carati A., Ingallina P., Mantegazza M.A., Bellussi G., *Appl. Catal. A: Gen.*, 221 (2001) 63–72.
[47] Song F., Liu Y., Wang L., Zhang H., He M., Wu P., *Appl. Catal. A: Gen.*, 327 (2007) 22–31.
[48] Masami F., Oikawa M., Patent *EP 1717222* (2006): to Sumitomo Co..
[49] Tatsumi T., in *Modern Heterogeneous Oxidation Catalysis*, Mizuno N. (Ed.), Wiley-VCH, Weinheim (2008) 125–156.
[50] Cavani F., Teles J.H., *Chem. Sus. Chem.*, 2 (2009) 508–534.
[51] Clerici M.G., in *Metal Oxide Catalysis*, Jackson S.D., Hargreaves J.S.J. (Eds.), Wiley-VCH, Weinheim (2009) 705–754.

[52] Ravasio N., Zaccheria F., Guidotti M., Psaro R., *Topics Catal.*, 27 (2004) 157–168.
[53] Guidotti M., Psaro R., Sgobba M., Ravasio N., in *Catalysis for Renewables: From Feedstock to Energy Production*, Centi G., van Santen R.A. (Eds.), Wiley-VCH, Weinheim (2007) 257–272.
[54] de Vos D.E., Sels B.F., Jacobs P.A., *Adv. Synth. Catal.*, 345(4) (2003) 457–473.
[55] Bisio C., Gatti G., Marchese L., Guidotti M., Psaro R., in *Nanomaterials for Energy and Environment*, Pan Stanford Press, Singapore (2010) in press.
[56] Guidotti M., Moretti G., Psaro R., Ravasio N., *Chem. Commun.* (2000) 1789–1790.
[57] Berlini C., Guidotti M., Moretti G., Psaro R., Ravasio N., *Catal. Today*, 60 (2000) 219–225.
[58] Guidotti M., Ravasio N., Psaro R., Ferraris G., Moretti G., *J. Catal.*, 214(2) (2003) 247–255.
[59] Pena M.L., Dellarocca V., Rey F., Corma A., Coluccia S., Marchese L., *Microp. Mesop. Mater.*, 44–45 (2001) 345–356.
[60] Cativiela C., Fraile J.M., Garcia J.I., Mayoral J.A., *J. Mol. Catal. A: Chem.*, 112 (1996) 259–267.
[61] Guidotti M., Batonneau-Gener I., Gianotti E., Marchese L., Mignard S., Psaro R., Sgobba M., Ravasio N., *Microp. Mesop. Mater.*, 111 (2008) 39–47.
[62] Fraile J.M., Garcia J.I., Mayoral J.A., Vispe E., *Appl. Catal. A: Gen.*, 245 (2003) 363–376.
[63] Igarashi N., Hashimoto K., Tatsumi T., *Microp. Mesop. Mater.*, 104 (2007) 269–280.
[64] Corma A., Domine M., Gaona J.A., Jordà J.L., Navarro M.T., Rey F., Pérez-Pariente J., Tsuji F., McCulloch B., Nemeth L.T., *Chem. Commun.*, (1998) 2211–2212.
[65] Davies L.J., McMorn P., Bethell D., Bulman Page P.C., King F., Hancock F.E., Hutchings G.J., *J. Catal.*, 198 (2001) 319–327.
[66] Clerici M.G., Bellussi G., Patent *US 4,410,501* (1983): assigned to Snamprogetti.
[67] Ratton S., *Chem. Today*, 16 (1998) 33–37.
[68] Corma A., *J. Catal.*, 216 (2003) 298–312.
[69] Tuel A., Moussa-Khouzami S., Ben Taarit Y., Naccache C., *J. Mol. Catal.*, 68 (1991) 45–52.
[70] Panov G., Dubkov K.A., Kharitonov A.S., in *Modern Heterogeneous Oxidation Catalysis*, Mizuno N. (Ed.), Wiley-VCH, Weinheim (2008) 217–252.
[71] Ivanov D.P., Sobolev V.I., Pirutko L.V., Panov G.I., *Adv. Synth. Catal.*, 344 (2002) 986–995.
[72] Ivanov D.P., Rodkin M.A., Dubkov K.A., Kharitonov A.S., Panov G.I., *Kinet. and Catal.*, 41 (2000) 771–775.
[73] Yuranov I., Bulushev D.A., Renken A., Kiwi-Minsker L., *J. Catal.*, 227 (2004) 138–147.

[74] Waclaw A., Nowinska K., Schwieger W., *Appl. Catal. A: Gen.*, 270 (2004) 151–156.
[75] Selli E., Rossetti I., Meloni D., Sini F., Forni L., *Appl. Catal. A: Gen.*, 262 (2004) 131–136.
[76] Zhai P., Wang L., Liu C., Zhang S., *Chem. Eng. J.*, 111 (2005) 1–4.
[77] Gopalakrishnan S., Yada S., Muench J., Selvam T., Schwieger W., Sommer M., Peukert W., *Appl. Catal. A: Gen.*, 327 (2007) 132–138.
[78] Balducci L., Bianchi D., Bortolo R., D'Aloisio R., Ricci M., Tassinari R., Ungarelli R., *Angew. Chem. Int. Ed.*, 42 (2003) 4937–4940.
[79] Bianchi D., Balducci L., Bortolo R., D'Aloisio R., Ricci M., Spanò G., Tassinari R., Tonini C., Ungarelli R., *Adv. Synth. Catal.*, 349 (2007) 979–986.
[80] Perathoner S., Pino F., Centi G., Giordano G., Katovic A., Nagy J.B., *Topics in Catal.*, 23 (2003) 125–136.
[81] Xin H., Koekkoek A., Yang Q., van Santen R., Li C., Hensen E.J.M., *Chem. Commun.*, (2009) 7590–7592.
[82] Boronat M., Concepción P., Corma A., Renz M., *Catal. Today*, 121 (2007) 39–44.
[83] Bare S.R., Kelly S.D., Sinkler W., Low J.J., Modica F.S., Valencia S., Corma A., Nemeth L., *J. Am. Chem. Soc.*, 127 (2005) 12924–12932.
[84] Corma A., Nemeth L.T., Renz M., Valencia S., *Nature*, 42 (2001) 423–425.
[85] Boronat M., Concepción P., Corma A., Navarro M.T., Renz M., Valencia S., *Phys. Chem. Chem. Phys.*, 11 (2009) 2876–2884.
[86] Boronat M., Corma A., Renz M., Sastre G., Viruela P.M., *Chem. Eur. J.*, 11 (2005) 6905–6915.
[87] Mifsud M., Ph.D. thesis, Instituto de Tecnologia Quimica, Universidad Politecnica de Valencia (2007). Accessed on the Internet in March 2010.
[88] Corma A., Fornés V., Iborra S., Mifsud M., Renz M., *J. Catal.*, 221 (2004) 67–76.
[89] Corma A., Iborra S., Mifsud M., Renz M., *J. Catal.*, 234 (2005) 96–100.
[90] Corma A., Navarro M.T., Nemeth L., Renz M., *Chem. Commun.*, 21 (2001) 2190–2191.

INDEX